# Network Security and Data Privacy in 6G Communication

This book proposes robust solutions for securing a network against intrusions for data privacy and safety. It includes theoretical models, commercialization of validated models, and case studies.

Features:

- Explains the integration of technologies, such as artificial intelligence, the Internet of Things, and blockchain for network security in a 6G communication system.
- Highlights the challenges, such as spectrum allocation and management, network architecture and heterogeneity, energy efficiency and sustainability, antenna, and radio frequency.
- Discusses theories like quantum-safe cryptography, zero-trust networking, and blockchain-based trust management.
- Covers emerging technologies including homomorphic encryption, secure multi-party computation, post-quantum cryptography, and distributed ledger technology for security and privacy in 6G communication systems.
- Presents light and deep secure algorithms to detect fake incidents in wireless communication.

The text is primarily written for senior undergraduates, graduate students, and academic researchers in fields including electrical engineering, electronics and communications engineering, and computer science.

**Wireless Communications and Networking Technologies: Classifications, Advancement and Applications**
*Series Editor: D.K. Lobiyal, R.S. Rao and Vishal Jain*

The series addresses different algorithms, architecture, standards and protocols, tools and methodologies, which could be beneficial in implementing a next-generation mobile network for communication. Aimed at senior undergraduate students, graduate students, academic researchers, and professionals, the proposed series will focus on the fundamentals and advances of wireless communication and networking, such as mobile ad hoc networks (MANET), wireless sensor networks (WSN), wireless mess networks (WMN), vehicular ad hoc networks (VANET), vehicular cloud networks (VCN), vehicular sensor networks (VSN), reliable cooperative networks (RCN), mobile opportunistic networks (MON), delay-tolerant networks (DTN), flying ad hoc networks (FANET), and wireless body-sensor networks (WBSN).

*Wireless Ad-hoc and Sensor Networks*
*Architecture, Protocols, and Applications*
Pawan Singh, Sudesh Kumar, Sachin Kumar Gupta, Abhay Kumar Rai, Abdu Saif

*Robotics and Smart Autonomous Systems*
*Technology and Applications*
Rashmi Priyadarshini, Ram Mohan Mehra, Amit Sehgal, Prabhu Singh

*Intelligent Networks*
*Techniques, and Applications*
Vivek Kumar Singh, Anil Kumar Sagar, Parma Nand, Rani Astya, Omprakash Kaiwartya

*Network Security and Data Privacy in 6G Communication*
*Trends, Challenges, and Applications*
Rajendra Kumar, Vishal Jain, Danish Ather, Vinay Kukreja, and Manoj Singhal

For more information about this series, please visit: www.routledge.com/Wireless%20Communications%20and%20Networking%20Technologies/book-series/WCANT

# Network Security and Data Privacy in 6G Communication

## Trends, Challenges, and Applications

Edited by
Rajendra Kumar, Vishal Jain, Danish Ather,
Vinay Kukreja, and Manoj Singhal

CRC Press
Taylor & Francis Group
Boca Raton London New York

CRC Press is an imprint of the
Taylor & Francis Group, an **informa** business

Cover image: peampath2812/Shutterstock.com



First edition published 2025
by CRC Press
2385 NW Executive Center Drive, Suite 320, Boca Raton FL 33431

and by CRC Press
4 Park Square, Milton Park, Abingdon, Oxon, OX14 4RN

*CRC Press is an imprint of Taylor & Francis Group, LLC*



ISBN: 978-1-032-75817-6 (hbk)
ISBN: 978-1-032-95083-9 (pbk)
ISBN: 978-1-003-58312-7 (ebk)

DOI: 10.1201/9781003583127

Typeset in Sabon
by Newgen Publishing UK

# Contents

# Preface

The book entitled "Network Security and Data Privacy in 6G Communication: Trends, Challenges, and Applications" is about proposing robust solutions for securing a network against intrusions for data privacy and safety. The book includes theoretical models, commercialization of validated models, case studies, a review of existing practices, etc. The various contributions in the book also present challenges related to detection and prevention of cyber-attacks in the 6G environment.

System implementations in the 6G environment are being dignified to be more critical and challenging than ever before because of network security and data privacy threats. The 6G networks are growing to offer extraordinary speeds, ultra-low latency, and massive connectivity, therefore they will be enabling a wide range of emerging technologies, including blockchain, augmented and virtual reality, IoT sensors, autonomous vehicles, and many more. This increased connectivity and data exchange bring forth new security and privacy concerns that need to be addressed proactively. Some of the key aspects within the scope of network security and data privacy in the 6G environment include threat landscape expansion, secure communication, edge security, AI-driven security, privacy-preserving technologies, zero-trust architecture, regulatory compliance, supply chain security, flexibility and reliability, user awareness and education, etc.

The expansion and interconnection of devices in 6G networks will expand the attack surface for the intruders. As a challenge, the new types of threats will be targeting 6G infrastructure, applications, and services. There will be the requirement for advanced security measures to counter the potential risks. The study of such challenges and threats will be required to have a proper plan of action. The research and development divisions at public and private levels will need to ensure secure communication, edge security, AI-driven security using privacy-preserving technologies at all the levels. The integration of edge computing in 6G networks brings computing capabilities closer to end-users and devices. The adoption of artificial intelligence and machine learning for network management will also lead to the emergence of AI-driven security solutions. More personal data will be generated and transmitted in 6G networks, so the incorporated learning will become essential to protect user privacy and comply with data protection regulations. The 6G-based future systems will be based on zero-trust architecture to ensure secure access controls and prevent lateral movement of threats. The data privacy regulations will become stricter, therefore organizations operating in the 6G environment will need to comply with evolving privacy laws and demonstrate transparency in data handling and processing. The focus on promoting user awareness and education about security and privacy best practices will be essential to empower individuals to safeguard their data and privacy in the 6G environment.

This book comprises 13 chapters from contributors of different countries. The summary of each chapter is presented below:

**Chapter 1:** The progeny apprehension: new security challenges and emerging threats in the 6G network

The advent of wireless communication networks serves as a proof of human invention and technological advancement, this chapter presents the new security challenges and emerging threats in the 6G network. The transition from the 5G to 6G network offers to revolutionize connections in ways and to extents that seemed unthinkable as we step into the next generation. The opportunity to completely change the data-transfer speed, internet access, and the framework of technological interactions enables 6G to stand at the start of the next generation wireless communication.

**Chapter 2:** Japan's 2030 vision for 6G – a case study

This chapter presents the development and review of progress in the 6G market in Japan. The study also presents the 2030 vision of 6G technology in Japan. This technology is expected to be available in the 2030s and the 6G research journey is on its way. Many developing and developed countries like Japan, the USA, Canada, Europe, UK, Australia, Singapore, China, India, and Malaysia have started experiments working on the major challenges in the development of the proposed system. Japan is working on a comprehensive strategy for 6G technology and they are expecting that 6G technology will be 10 times faster than 5G.

**Chapter 3:** Embedded communication system design, implementation of the MB-OFDM transceiver and its optimization for short-range IoT in 5G/6G applications

In this chapter, MB-OFDM and modulation techniques are implemented, analyzed, and optimized for different ARM-based embedded architectures. A software reference model for the MB-OFDM system is developed in MATLAB. The bit error rate (BER) performance of the MB-OFDM technique combined with different modulation schemes over AWGN channels is studied. The software reference model is used to implement and benchmark the system in C using the ARM development suite and Xilinx tools. Higher modulation techniques provided very high data rates compared to QPSK and DCM. Image data is serialized and used as input for evaluating the performance of the system. Performance metrics such as mean square error (MSE) and peak signal-to-noise ratio (PSNR) are computed for the MB-OFDM technique for different images considered and compared with other modulation techniques.

**Chapter 4:** Blockchain and IoT for data privacy in the 6G network: challenges and solutions

This chapter comprehensively explores the various dimensions of the blockchain and IoT for data privacy in the 6G network. The rise of the Internet of Things (IoT) and blockchain have made ubiquitous computing a reality by connecting a vast number of intelligent, autonomous gadgets for a variety of uses. IoT technology with 6G support offers a framework for quick and latency-free data processing and gathering, while blockchain provides a distributed laser for transparent transactions.

**Chapter 5:** Securing the future: navigating network security and data privacy challenges in the 6G era

This chapter dives into the effects and difficulties related with defending touchy data in the advancing scene of 6G conditions. We investigate the many-sided transaction between state-of-the-art advancements and the need to maintain strong safety efforts, underlining the basic significance of adjusting methodologies to address arising dangers.

**Chapter 6:** 6G network security and artificial intelligence

ML algorithms provide notable benefits for 6G networks, the scientific community has mostly overlooked security problems regarding artificial intelligence (AI) models. Nevertheless, ensuring security is an essential component of AI algorithms as malicious individuals might corrupt an entire AI model. This chapter discusses several topics, including how emerging technologies can be used to secure 6G networks. Also, this chapter provides a survey of the many techniques for securing 6G networks against different types of attacks.

**Chapter 7:** The future of 6G in marketing

This chapter studies data to assist marketers assess 6G technology's impact on marketing. In conclusion, 6G technology gives marketers several possibilities. 6G speeds will let advertisers target customers, engage them, and track their activity. This helps marketers create more effective advertising. Overall, the arrival of 6G technology is expected to influence how businesses advertise and market to their consumer base.

**Chapter 8:** Emerging technologies for security and privacy in 6G wireless communication networks

The 6G wireless transmissions can be made more secure by the utilization of modern beamforming and signal processing techniques, which are part of the physical layer security. Traditional trust assumptions are called into question by the paradigm of zero-trust security models, which advocates for continuous authentication and authorization. This chapter provides a glimpse into the disruptive technologies that are going to determine the landscape of privacy and security in 6G wireless communication networks. These developments are critically important in laying a foundation that is secure, trustworthy, and privacy-focused for the hyper-connected future of wireless communication.

**Chapter 9:** Applications of 6G in the healthcare sector and the importance of network security and data privacy-encryption of medical images using the SHA 256 blockchain algorithm

Ensuring network security and data privacy in the healthcare sector within the context of 6G technology is of paramount importance. As 6G promises advanced connectivity and capabilities, this chapter discusses challenges for safeguarding patient information and healthcare systems. 6G technology is poised to bring about groundbreaking changes in the healthcare sector with its lightning-fast speeds, minimal latency, and advanced capabilities.

**Chapter 10:** Unveiling the future: exploring 6G network security and data collection

The purpose of the research is to enhance network security. Network susceptibilities, attacks, and interruptions cause economic loss and security threats. In a network system, data collected can be helpful in the detection of security threats. This kind of data is referred to as cybersecurity data in this context. By analyzing and applying this data, network systems may be able to identify disturbances and cyberattacks and attain 6G levels of security.

**Chapter 11:** Network security and data privacy in the 6G environment: impacts and challenges

This chapter presents the impacts and considers the challenges emerging within the sixth-generation (6G) environment concerning safeguarding network security and preserving data privacy. The evolution towards 6G networks brings forth unparalleled advancements in connectivity and data communication, yet it also introduces a host of challenges regarding network security and data privacy at all levels.

**Chapter 12:** 6G networks: pioneering advanced communication techniques for call centers and beyond

This chapter examines the environment of 6G wireless networks evolving recently and their capacity to do the necessary make-over for call centers and other systems associated with communications. It deals with a detailed scrutiny of the 6G evolutional process of mobile networks and the advanced communication techniques that the 6G may possibly offer. Additionally, it takes a deeper look at how the 6G technology can improve user experience in the call centers by giving rise to greater data rates, enhanced reliability, and user-centric network designs; it shall also provide a segment on the integration of modern technologies, such as AI and edge computing into the 6G framework.

**Chapter 13:** Enhancing distributed denial of service cloud security threats using artificial neural networks in 6G networks

This chapter explores the issues connected with future cloud computing services in light of the unprecedented technological advances of the sixth generation (6G). As a result of the invention of new technology, some major challenges have emerged, including security issues from a wide range of attacks, computational costs, and power consumption, which have all become serious concerns in cloud computing, distributed denial of service attacks (DDoS) are among the most common types of security threats that are encountered.

*Editors*
Rajendra Kumar
Vishal Jain
Danish Ather
Vinay Kukreja
Manoj Singhal

# Contributors

**Shikha Aggarwal**
Amity University
India

**Nazia Ahmed**
Imam Abdulrahman Bin Faisal University
Dammam

**Seidu Alhassan**
Tamale Technical University
Ghana

**Sajid Ali**
School of Information Science
University of Education, Lahore (Multan Campus)
Pakistan

**Muhammad Muzamil Aslam**
School of Digital Science
Universiti Brunei darussalam
Gadong, Muara, Brunei Darussalam, Pakistan

**Danish Ather**
Amity University
Tashkent, Uzbekistan

**G. Prakash Babu**
Acharya Institute of Technology
India

**Eduard Babulak**
National Science Foundation
USA

**Dina Darwish**
Ahram Canadian University
Egypt

**Sanket N Dessai**
Oriental University Indore
India

**Renu Devi**
Maharshi Dayanand University Rohtak
India

**Liping Du**
School of Computer and Communication Engineering
University of Science and Technology
Beijing, China

**Satyanarayan Dubey**
Oriental University Indore
India

**Flávio de São Pedro Filho**
Federal University Brazil
Brazil

**Masanori Fukui**
Faculty of Software and Information Science,
Iwate Prefectural University,
Takizawa, Iwate, Japan

**Anchal Garg**
University of Bolton
Greater Manchester, United Kingdom

**Sumeet Gill**
Maharshi Dayanand University Rohtak
India

**Sulemana Ibrahim**
Tamale Technical University
Ghana

**Anupriya Jain**
Professor
School of Computer Applications
MRIIRS Faridabad
India

**Vishal Jain**
Department of Computer Science and Engineering
School of Engineering and Technology
Sharda University
India

**Alan James**
Tata Elxsi, Kerala
India

**I. Christina Jane**
Mar Ephraem College of Engineering and Technology
Tamil Nadu, India

**Siju John**
Amal Jyothi College of Engineering
Kerala, India

**Jomin Joy**
Amal Jyothi College of Engineering
Kerala, India

**Balaram Yadav Kasula**
University of The Cumberlands
Williamsburg, KY, USA

**Christian Kaunert**
Professor
Dublin City University
Ireland
And
University of South Wales
United Kingdom

**Shanu Khare**
Chandigarh University
India

**Rajneesh Kler**
Amity University in Tashkent
India

**Utku Köse**
Suleyman Demirel University
Turkey

**Rajendra Kumar**
Sharda School of Engineering & Technology
Sharda University
Greater Noida, India

**S.N. Kumar**
Amal Jyothi College of Engineering
Kerala, India

**Naresh Kumari**
Department of Computer Engineering
Government Polytechnic
Sonipat, India

**Mohammed Majeed**
Tamale Technical University
Ghana

**Rama Maliya**
Karnavati University
Gandhinagar, India

**Deepti Mehrotra**
Amity University
India

**Ankita Nayak**
Trident Academy of Technology
India

**Jeel Patel**
Karnavati University
Gandhinagar, India

**Hemant Patidar**
Oriental University Indore
India

**B.C.M. Patnaik**
KIIT School of Management
KIIT DU
India

**Muralidharan R**
NUST, Sultanate of Oman
Oman

**Ajay Rastogi**
Teerthanker Mahaveer University
India

**Ipseeta Satpathy**
KIIT School of Management, KIIT DU
India

**Vaidehi Shah**
Karnavati University
Gandhinagar, India

**Nimiti Sharma**
Karnavati University
Gandhinagar, India

**Seema Sharma**
Associate Professor
School of Computer Applications MRIIRS Faridabad
India

**Bhupinder Singh**
Sharda University
India

**Prasanna Kumar Singh**
NIET, Greater Noida
India

**Navjot Singh Talwandi**
Chandigarh University
India

**Payal Thakur**
Chandigarh University
India

**Ali Tufail**
School of digital Science
Universiti Brunei Darussalam
Gadong, Muara, Brunei Darussalam, Pakistan

**Prashant Upadhyay**
Vignan's Foundation for Science Technology and Research (VFSTR)
India

**Pawan Whig**
Vivekananda Institute of Professional Studies-TC
India

**Nikhitha Yathiraju**
University of The Cumberlands
Williamsburg, KY, USA

**Jonas Yomboi**
Valley View University
Ghana

**Tanveer Baig Z**
Amity University in Tashkent
Uzbekistan

# About the editors

**Rajendra Kumar** is presently working as Associate Professor in Computer Science and Engineering Department at Sharda University, Greater Noida. He holds Ph.D., M.Tech. and B.E. degrees (all in Computer Science). He has 26 years of teaching, research, and administrative experience at various accredited institutes and universities, such as Chandigarh University (NAAC A+). His field of interest includes human computer interaction, vein-pattern recognition, IoT, blockchain, and theoretical computer science. He has published/presented more than 45 papers in peer-reviewed journals of repute and conferences held in India and abroad. Three patents, four edited books, and two monographs are also in his credit. He is the author of five textbooks and has co-edited nine conference proceedings. He has chaired many sessions at international conferences and has been a member of the organizing committee of eight international conferences held in Malaysia, Thailand, Indonesia, Singapore, and online. He is Editor-in-Chief of *ADI Journal of Recent Innovations* (AJRI), Indonesia. He has been member of the Board of Studies of UPTU (now AKTU), Lucknow. He is a senior member of IEEE, IACSIT, IAENG, UACEE, SCIEI, and CSTA.

**Vishal Jain** is presently working as an Associate Professor at the Department of Computer Science and Engineering, Sharda School of Engineering and Technology, Sharda University, Greater Noida, U. P., India. Before that, he worked for several years as an Associate Professor at Bharati Vidyapeeth's Institute of Computer Applications and Management (BVICAM), New Delhi. He has more than 16 years of experience in academia. He holds Ph.D. (CSE), M.Tech (CSE), MBA (HR), MCA, MCP, and CCNA qualifications. He has authored more than 100 research papers in reputed conferences and journals. He has authored and edited more than 50 books with various reputed publishers, and is the series editor of ten book series. He is a life member of CSI, ISTE and senior member of IEEE. His research areas include information retrieval, semantic web, ontology engineering, data mining, ad hoc networks, and sensor networks. He received a Young Active Member

Award for the year 2012–13 from the Computer Society of India, Best Faculty Award for the year 2017, and Best Researcher Award for the year 2019 from BVICAM, New Delhi.

**Danish Ather** is an accomplished book editor with a robust academic and professional background. Currently an Associate Professor at Amity University in Tashkent, Uzbekistan, Dr. Ather brings over 18 years of experience in teaching, research, and administration. He holds dual Ph.D. degrees in Computer Science and Computer Science & Engineering. His editorial expertise is highlighted through his role as Editor of the proceedings submitted yearly to IEEE USA for the SMART Series conferences. Dr. Ather has authored a book titled "*Level Up Your Programming With Core Dragon*" and published 54 research papers in international journals and conferences, with 15 indexed in Scopus. He is a Senior Member of the IEEE Society and has received numerous awards, including a Certificate of Appreciation from the Ministry of Digital Technologies, Republic of Uzbekistan. His technical proficiency and dedication to advancing knowledge in fields such as IoT, AI, and programming make him a valuable asset in the realm of academic publishing.

**Vinay Kukreja** is currently occupying the position of Professor and Director (Research) at Chitkara University, Punjab. His illustrious pedagogical career spans 18 years, during which he has expertly mentored a substantial number of Ph.D. and Master of Engineering students. In tandem with his teaching and advisory roles, he has an impressive scholastic portfolio consisting of 500 Scopus articles and over 50 patents, marking his profound contributions to his field. He has added to the academic canon by authoring three books and editing three. In the arena of competition, he secured prestigious first place at the 2018 SIH Hackathon, an event patronized by the Ministry of Housing & Urban Affairs in India. His academic pursuits and intellectual curiosities gravitate towards a diverse array of research topics, including but not limited to, machine learning, deep learning, agile software development, image processing, data analysis, and structural equation modeling.

**Manoj Singhal** obtained his Ph.D degree in Computer Science from Kurukshetra University, Kurukshetra, India in 2012. He has held a variety of positions at King Khalid University (ABET accredited), Abha (Kingdom of Saudi Arabia), Higher College of Technology, Muscat (Oman), Academy of Business and Engineering Sciences, Accurate Institute of Management & Technology, Greater Noida Institute of Technology, Uttar Pradesh, India. He is currently working as a Professor in the Department of Information

Technology at GL Bajaj Institute of Technology & Management, Greater Noida, Uttar Pradesh, India. He has published two books (*Discrete Mathematics* and *Computer Graphics*) and more than 25 research papers in international and national journals. He has guided several projects to final year students. His fields of interest are in theoretical computer science, graph theory, and neural-network-based image processing. He is a member of the International Association of Engineers.

Chapter 1

# The progeny apprehension

## New security challenges and emerging threats in the 6G network

*Ankita Nayak, Ipseeta Satpathy, Vishal Jain, B.C.M. Patnaik, and Flávio de São Pedro Filho*

## 1.1 INTRODUCTION

The advent of wireless communication networks serves as a proof of human invention and technological advancement. The transition from the 5G to 6G network offers the opportunity to revolutionize connections in ways and to extents that seemed unthinkable as we step into the next generation. The chance to completely change data-transfer speed, internet access, and the framework of technological interactions enables 6G to stand at the start of the next generation wireless communication. We will further learn the intricate details of the 6G network and its anticipated features and applications. 6G refers to the sixth-generation wireless communication network, which is the next generation network that will replace the extensively used 5G network. As 5G has already culminated substantial advancements in communication, lower latency, and data speed, the 6G expansion hopes to continue providing an even better service [1]. The major goals of 6G are to accomplish unattainable data-transfer speed, realize instantaneous connectivity, and create the latest innovations, such as quantum computing, holography, and machine learning, and weave them into our daily virtual lives. The ability to transfer data at terabits' speed is one of the features of 6G that would render multiple orders hundred times faster than the prior generation. This increase in speed could open the possibilities to previously unachievable goals like effortless virtual and augmented reality circumstances, fully immersive holographic communication, and immediate possession of enormous quantities of data. It might also promote immediate interaction on a scale that is unprecedented [2]. Apart from the gigabits-per-second speed of the 5G network, achieving ultra-high data-transfer speed is one of the primary goals of the 6G network. This could make it possible for files to be downloaded and published quickly, creating opportunities for fully exciting interactive media experiences, quick file transfers, and high-quality video streaming. One of the other major goals of the 6G network would be to reduce the waiting period or delays. The 6G network promises to allow virtually immediate communication through decreased data-transfer delays.

DOI: 10.1201/9781003583127-1

This is vital for application with constraints of time where decisions need to made in a matter of seconds, which may later determine the distinction between life and death, like in self-driving cars and surgical procedures, which are performed remotely. It is projected that 6G is going to a substantial higher number of connected devices for each unit area. The inclusion of the Internet of Things (IoT) and advanced technologies into parts of our daily lives depends on the capability of growing further. Another important aspect of 6G networks, apart from traditional communication features, is introducing advanced technologies. The potential benefits of the 6G network are anticipated to be heavily influenced by artificial intelligence (AI), machine learning, and quantum data processing, which will enable savvy decision making, customized offerings, and formerly unprecedented computing capacity [3]. The prospect of the 6G network is huge but the challenges of achieving these unusual hurdles demands creative thinking, and collaboration across varied disciplines. Understanding and fixing such challenges is vital over the successful introduction and permanent functioning of the 6G network as a whole. Discovering and assigning acceptable frequency spectrum bands is one the major challenges of the 6G network. New frequency ranges need to be identified, including millimeter wave and terahertz frequency. With these higher frequencies, there is also the fear of signal transmission and environment factors being vulnerable at higher frequencies. Advancement in connectivity will require higher efficient-energy use because of increasing environmental concerns. To ensure long-term sustainability and reduce energy-consumption challenges. The incorporation the 6G network's modern technologies provides additionally data security and privacy issues. Safeguarding of user information to prevent attacks via the internet and ensure the accuracy of communications is vital as computer systems become smarter and more sophisticated. In order for the 6G network to be widely embraced and relied upon, robust safety precautions and confidentiality procedures should be developed. Worldwide uniformity and seamless integration are two important components for the success of the new generation communication networks [4]. The various technological contexts and regulations makes it difficult to devise a single 6G standard. To ensure collaboration throughout the broad network structures, fixing the international problems, and integrating technical requirements are all required for the smooth deployment of the 6G network. The legal and social problems arising from including advanced technologies within 6G networks need to be evaluated and carefully considered. Issues pertaining to algorithm bias, and the impact of omnipresent connectivity on people's privacy requires aggressive actions to mitigate negative effects. Community inflammation, guidelines, and moral structures will be essential in settling the legal and moral landscape of 6G networks. The creation and roll out of the 6G network will require anticipating and proactively combating new hazards to ensure the continued existence of new generation wireless communication.

Everyone around the globe is awaiting the official introduction of the 6G network, however it is important to remember that the innovation has its own challenges that will need to be tackled considerably. Cybersecurity is a risk because of the anticipated speed and scale at which the 6G network is expected to work. This network has the potential for large attacks because of the higher connectivity, extremely low delays, and substantial machine density, which at first is abandoned for various cyber-attacks [5]. Advanced hacking, data theft, and distractions for essential services are a few instances of these risks. Strong cybersecurity measures, encryption standards, and detection of system intrusions must be developed to minimize the risk and protect the confidentiality of data [6]. Novel architectures and technological advancements, such as terahertz communication, holographic communication, and AI-driven network oversight will be introduced in the deployment of the 6G network. This advancement brings additional capabilities that have never been seen before, and hence needs to be carefully considered. Likewise, questions about privacy and biased decision making arise by using AI in the 6G network. It is essential to have a standard governed framework within the environment of the 6G network to ensure safe and reasonable usage. The future viability of this network depends upon the adverse environmental impact of the 6G network. Usage of numerous small cells, innovative antennas, and additional elements will have a substantial impact on the environment. The need to develop technologies that are energy efficient, the integration of alternative power resources, and the implementation of environmental consciousness throughout the lifecycle of the 6G network is necessary for forecasting and mitigating these environmental risks. The global reach of this technology offers geopolitical hazards, which have to be cautiously handled. The competition pertaining to the dominance in the development and implementation of 6G nations, means there might be challenges involving the protection of intellectual property rights, global norms, and political instability. Promoting free standards, creating frameworks, and promoting global collaborations are required for the deployment and establishment of the 6G network [7].

## 1.2 OVERCOMING DISPARITIES: BRINGING TOGETHER SECURITY TECHNIQUES FOR 6G NETWORKS BASED ON PAST EXPERIENCES

Transitioning wireless communication to carry formerly unprecedented levels of speed and connectivity from 3G to 5G has brought and resolved a number of privacy concerns. Sluggish encryption protocols seemed to be one of the primary weaknesses of the 3G era. The significant feature of 3G was the adoption of more sophisticated and efficient broadcast technologies. WCDMA (Wideband Code Division Multiple Access) and CDMA 2000 allowed for faster transmission of data and increased capacity. With the increase in speed and capabilities, the concerns pertaining to security also increased. Hackers found it easy to steal and employ private information, which was shared in the airways when older standards of encryption were in use. Incidents of spying, whereby malicious individuals might spy communications and threaten the privacy and threaten the confidentiality of data, attracted attention to the fragility of the 3G network [8]. The attention then shifted to enhancing communication speed via the advent of the 4G/LTE network, however this shift in network also had it is challenges. The potential for attacks developed, which resulted in devices becoming more interconnected particularly since the development of the IoT. With the increase in 4G network communication, a greater number of gadgets ended up being suspectable targets for attacks via the internet. The rise of 4G technology smartphone apps additionally presented security risks, and reports of spyware infecting phones were also made, which resulted in privacy and data theft issues [9]. The shift to an all-IP design was a new security concern and made the potential for communication attacks an important issue. With the upgradation to the 5G network, cloud computing and software-defined networking or SD-N notal routes of attack were additionally brought by the change. A vital aspect of 5G is network slicing, which allows the creation of multiple digital networks on one physical network. While it allows for modification alongside maximizing the efficiency for a vast array of applications, there are nonetheless challenges for safeguarding these slices from the cross-slice attacks. A security breach within a slice might affect others, leading to a resilient seclusion technique. Furthermore, doubts pertaining to cloud computing security also arised in the 5G network, increasing the resilience on the infrastructure hosted in the cloud. Data breaches in the system that utilized the cloud-based systems or illegal access to servers in the cloud can have a significant impact on 5G networks and their users. Due to the constantly changing and dispersed nature of cloud-based services, security factors became more complicated [10]. Keeping the level of equilibrium becomes more challenging as the amount of data and quantity increases. Moreover, the privacy challenges pertain to legitimate and regulatory issues along with technical difficulties. It is important to have privacy laws, such as General Data Protection Regulations (GDPRs). Ensuring that the data is managed carefully and openly has grown into a moral and legal requirement. Through the 6G network, novel security challenges would arise. There are benefits and challenges of the 6G network's inclusion of AI, computing edge technologies, and quantum computing. As an example, implementing AI to anticipate and detect threats could enhance privacy, yet it raises questions of how AI could possibly be mistreated [11]. A revolutionary change in the wireless communication, 6G guarantees never-seen-before speed, exceptionally low delays, and ground-breaking uses. These developments brought with them new challenges, demanding and grasping the dynamic changing security. The integration of sophisticated technologies, including AI, data learning, and quantum computer technologies in 6G, is a notable feature that presents new risks. These advanced technologies offer distinctive security challenges besides their multiple benefits in the areas of system optimization, statistical analysis, and increased computational capacity [12]. The integration of AI in the 6G network, for instance, raises the possibility of AI-driven online hackers, where the algorithms can harm the vulnerabilities at the formerly unheard-of speed and scales. Protecting the AI models, preventing hostile assaults, and ensuring the moral utilization of AI has become crucial. A further crucial aspect of 6G is quantum computing that offers protection and a number of benefits, as well as drawbacks. The creation of quantum Qu-resistant techniques of encryption is additionally vital, considering that the use of technology has the ability to completely transform the decryption via subverting the current algorithms of cryptography. In order to safeguard the privacy and reliability of communications, quantum-safety cryptography methods have to be implemented promptly in response to the imminent threat of the quantum-enabled attacks, such as Shor's approach, which focuses on traditional encryption. Another source for concern in relation to 6G networks is their widened attack surface [13]. The variety of entry points for possible cyber-attacks has increased greatly because of the growing number of gadgets that are wearable, connected automobiles, the world of internet devices, and advanced infrastructure. Every connected device is a suspectable weak point that hackers can take advantage of. Solid measures must be taken to protect the diverse as well as interconnected environment, such as encrypted communication standards, productive authentication system, and device-level security. Another important aspect of the 6G network is network slicing, a technique allowing the construction of numerous virtual networks for various scenarios. It boosts the allocation of resources, and adaptability, but instead it additionally poses security hazards. It is essential to guarantee the security and solitude of each network slice with the goal to stop the cross-slice attacks. As network slicing is flexible in nature, it can be created, changed, and retired as required as a result an adaptive security protocol is required, which can respond quickly to a changing network configuration [14]. 6G utilizes terahertz (THz) band frequencies to communicate, which creates new and increased security challenges. THz frequencies makes the ultra-high data rates possible, although there are certain difficulties regarding signal penetration and their propagation. Hackers may take advantage of the vulnerabilities brought by the THz waves. To secure the 6G network, it is essential to alleviate these challenges. The change in privacy models can be brought by the transition towards completely decentralized and dispersed network designs in 6G. Traditional centrally controlled security models will be less effective in an environment where nodes and peripheral devices are more actively engaged. The primary goal of the 6G network safety strategy is including edge computing assets for ensuring communication between peers' accuracy, as well as averting malicious access. The consequence for privacy has become more apparent as the 6G network aims to offer a fully immersive experience via augmented reality (AR), virtual reality (VR), and holographic communication along with traditional communication. The privacy of users is of great importance when substantial quantities of user data are gathered and analyzed over personalized aware services. Maintaining user privacy as you deliver customized services requires fragile testing, which requires a data management process and dependable secure privacy. Brain computer interfaces (BCIs) and biological interfaces in 6G takes the privacy consideration into the new dimensions. The risk of illicit use of neural network data as well as moral questions arise due to the interaction between the mind of the person and network [15].

## 1.3 QUANTUM STORM CRYPTOGRAPHY: CREATING ROBUST 6G SECURITY

Data processing is set to undergo an upheaval as a result of quantum computing, which provides both unprecedented possibilities along with significant challenges with regard to 6G security. Quantum computers are those that use quantum bits and qubits, as compared to traditional computers, which utilize bits for analyzing information either as 0s or 1s. As compared to classical computers, quantum computers are capable of performing intricate calculations at a dramatically faster rate because of the built-in analogy and properties of qubits. Even though it shows enhancements through an array of fields, it additionally it presents questions concerning the security of the used encryption in the 6G network. Quantum computing has the ability to effectively carry out techniques that can break the prevalent cryptographic method, which is the main factor affecting the 6G network. The classical algorithm poses one of the most primary dangers of quantum computers. Commonly utilized cryptosystems with public-keys such as RSA and the ECC whose integrity is dependent on the challenge of incorporating huge numbers, are directly jeopardized through this capability. The weakness of the traditional encryption technique is the major concern in the context of 6G security, where massive quantities of confidential information are diagonally circulated by the wireless networks. Data shared over the 6G network may lose its confidentiality, integrity, and reliability if quantum competitors take advantage of holes in the security in the standard encryption. The growing popularity of quantum computing increases the possibility of malicious partners capturing and deciphering the data; hence strengthening the systems beforehand becomes vital [16]. Cybersecurity specialists are working hard to safeguard which would be resistant to quantum technology as a way to solve these challenges. Creating such encryption algorithms that would be safe from quantum attacks is of prime importance, hence the field of post quantum cryptography is essential. Post-cryptographic standards are hash-based, code-based, and lattice-based cryptography, which provides a strong base for 6G network security by applying computational challenges, which are thought to be traditional and quantum machines. The shift from classical to post-quantum safety precautions offers a major challenge. There needs to be seamless application of quantum-resistant algorithms and seamless integration. In accordance with the basic principles of the theory of quantum mechanics, the QKD (quantum key distribution) offers an encrypted communication approach in a quantum era. Operational problems such as range limitations and network design incorporation needs to be resolved for the 6G network to be used widely. Global organizations, academic institutions, and other businesses are collaborating together to establish standards and guidelines for quantum-resistant security on an international scale. It needs cybersecurity experts to handle and analyze the forthcoming challenges and develop a technical infrastructure and gadgets that would be impermeable to quantum shifts like Quantum Safe Hardware Modules [17]. Conventional cryptographic techniques have threats from quantum computing; hence it is essential to establish safety protocols which can withstand the threat. In the rapidly changing quantum era, resilience and security have been studied as potential approaches. Looking into and implementing quantum-resistant cryptographic techniques like hash-code, and lattice-based cryptography have been the important components of quantum-resistant security measures. Due to mathematical complexity, lattice-based cryptography, which is hinged upon lattice structures, permits data encryption and key-swapping procedures, which stay secure in the presence of active quantum computers [18]. The techniques for achieving quantum-resistant security are hash-based cryptography, code-based cryptography, QKD. The major function of hash-based cryptography is to employ a hash function to generate fixed-size output, ensuring the accuracy of data and verification. Code-based cryptography uses error-correcting methods, which are resistant to quantum attacks. QKD uses quantum mechanical principles to safeguard the exchange of keys, encoding information in photons, and finding attempts to spy. The National Institute of Standards and Technology (NIST) is focusing on unifying cryptographic algorithms, which are immune to quantum attacks. Identifying and standardizing encryption algorithms, which are impermeable to both classical and quantum threats is the major goal of NIST. In the post-quantum world, safeguarding data and interactions necessitate quantum-resistant security features like encryption algorithms, standardization, and education [19].

## 1.4 A TWO-PRONGED APPROACH TO AI-ENHANCED SECURITY: INVENTIVENESS AND CAUTION

AI in the 6G network broadens novel avenues along with the security risks. Countermeasures may harm AI algorithms, hindering decision-making, and tampering with communication services. Resource allocation driven by AI can be manipulated by these attacks, which would result in inefficient usage of networks or delays in communication amenities. Another security risk associated with AI is the instructional data that is introduced by the 6G network's dependency on machine-learning algorithms. Discriminating erroneous choices could result from skewed information that is used to make the AI models. A skewed algorithm may inadeptly benefit some users or tasks in 6G services wherein the AI algorithm might regulate a major function like directing traffic. It could result in an uneven distribution of resources and potential degraded services for certain populations. To retain the transparency and interpretation of AI in the 6G network is challenging because of their severe complexity [20]. AI algorithms are turning into complex black-box systems, which may be challenging to fully understand via traditional methods that they develop. This absence of transparency might render it more challenging to identify and fix vulnerabilities in security, which could render difficulties in foreseeing and counteracting the potential hazards. The possibility of cross-slice attacks grows in the light of 6G network slicer capabilities in which AI may be deployed to optimize and control a variety virtual network concurrently. The integrity and reliability of additional network portions might be in danger if an intruder regulates to gain access to one slice. Given that the network slicing in the 6G network is dynamic and operated by AI it is necessary to take robust precautions to ensure every slice's security and seclusion as a way to prevent access and adjacent network travel. User data is incorporated into the 6G network through AI. Massive datasets are used by the AI programs to learn and arrive at defensible judgement [21]. As massive volumes of data are processed for customized services and optimization of the network there are higher chances of security breaches. Unapproved access to systems powered by AI could jeopardize the confidential information and lead to violations in privacy along with the improper use of personal data and information. AI algorithms must assign priority to powerful guidelines, which include encoding and impermeable training data sets if they are going to mitigate the security hazards in the 6G era. AI algorithms ought to be transparent and comprehensible, in addition, inspection and auditing is necessary to detect the flaws and potential security breaches. A secure protocol that includes confidentiality and invulnerable datasets for training are important for security risks in AI. To identify the irregularities and the probable security breaches, regular tracking and auditing is critical [22]. Although AI has the potential to improve 6G network security, its application must be cautious to prevent data leaks. It is imperative that we exercise caution and be aware of the possible threats to our health and safety. It is crucial for AI systems to be empathetic and transparent, which calls for precise AI. Techniques for the explainability of AI (XAI) are becoming more sophisticated and assist in identifying biases, addressing vulnerabilities, and promoting confidence in AI-driven security systems. Because of these difficulties, experts and professionals in the field are now able to examine cutting-edge strategies like federated learning and the XAI to help optimize the security of AI-based systems in terms of accountability, resilience, and understandability [23, 24]. It is vital to possess a comprehensive comprehension of artificial intelligence (AI), including its prospective advantages for 6G networks and security, as well as the principal concerns it may bring about for users' privacy and confidence. This suggests that the innovative method must be approached with prudence and risk management [25].

## 1.5 SAFEGUARDING THE FUTURE: COMPREHENSIVE ANALYSIS OF 6G SECURITY LIMITS

Federated learning is a method that makes it possible for AI networks to be trained independently across various devices without the need for a central repository to hold data that could improve the integration of AI in 6G security. Sensitive information is more likely to be displayed when using a decentralized system, yet individual security is maintained. AI can be applied for behavior evaluation and identifying anomalies in addition to enabling swift actions during security incidents. Furthermore, control and visibility challenges stem from the flexibility of edge circumstances. The dispersed nature of computing in edge renders conventional security safeguards more challenging to carry out. Traditional safeguarding often depends upon control and monitoring. The design may become difficult for security measures to gain an exhaustive understanding of the outermost ecosystem, which might end up resulting in blind spots, which might be exhausted with people of malicious intentions. In addition, it can be challenging to set up consistent safety procedures throughout the various devices in the edge environment because of the lacking centralized control [26]. The 6G network, however, has restricted capacity for devices in edge to add another level of security risk. A significant number of gadgets in edge have limitations on the memory, interpreting, and power capacity. Subsequently it can be challenging to set robust safety procedures implementing these sorts of appliances without compromising the major purpose. The edge program devices' efficiency could be hindered by highly resource-intensive safety procedures, which require a careful compromise between security and operational efficiency. The wireless interaction listening as well as illicit entry offers security challenges over computing-intensive edge in the environment at the 6G network. It is important to ensure the confidentiality and privacy of the data. Exchange of data, its place of origin, the destination, suitable digital encoding, and confidentiality measures are vital, resulting in robust encryption and authentication methods. Due to the computing resource in edge surrounding being dispersed, there exists an important danger with security since illicit access to gadgets occurs [27]. A broadened strategy needs to be implemented for dealing with these challenges, which involves intrusion prevention and a detection system aimed at remote settings in conjunction with robust authentication as well as access to management systems. In situations including mobile devices, encoding is crucial to earn safeguarding of information in computing in edge. Highly sensitive data may be protected via homomorphic security and confidentiality procedures. Putting into effect a lack of confidence safety strategy decreases the power source target surface. The device endurance is improved through updating software frequently and safe startup procedures [28]. Developing comprehensive safety protocols for 6G computing in edge necessitate interaction among business and organizations and experts from cybersecurity. Solid security precautions must be taken to the computing edge environment in the 6G network to be able to ward away attacks on computers and retain the reliability of a collaborative computing environment. Authentication and validation of every communication, and a no-trust safety protection framework is vital. Authorized users and devices are able to utilize the materials through the authentication process like protected boot and multi-factor authentication [29]. The cryptography protocols resolve connectivity problems and ensure data secrecy through securing the data during transit as well as idle times. To regulate authorizations in an edge environment, authorization strategies such as fragmented access of control is important. Unregulated altercations might be evaded via the habitual inspections and assessments regarding access regulates. A protected quantum computing system necessitates management of patches and periodic updates of software. Firm safety protocols in 6G computing settings need interaction between vendor player guidelines and experts in cybersecurity [30].

## 1.6 BOLSTERING THE PERIPHERY AND THE CYBERSPACE COMBAT: MANAGING SITUATIONS IN THE 6G ENVIRONMENT

As the 6G wireless network is made to accommodate an extensive array of connected gadgets, they would hence produce a huge amount of data. This enormous amount of data is collected by different gadgets and the IoT contributes, posing questions regarding the extent to which user usage, actions, and choices can be tracked and captured. The potential risk in this is the possibility of such a huge accumulation of personal data being used improperly or obtained despite the authorization. The location-based services in 6G networks are accurate, there is an increased risk of coarse administering, confidentiality of users, as well as intrusive practices that would compromise solitude and privacy. The 6G network predominantly relies upon biometric authentication and confidentiality hazards related to their data processing and storage, which may result in theft identification and intrusion. The IoT in the 6G environment gives intriguing hazards of privacy due to insecure gadgets, which could be misused by malicious individuals to gain private data and change real-world circumstances [31]. The technique of privacy by the design is involved in controlling access, confidentiality, use of encryption, and additional confidentiality protections to mitigate violations of privacy. In order to regulate, user-centered control of information is vital, which can be made feasible through apparent terms of conditions as well as authorized systems for management. Interpreting private data nearer to the source averting illicit access which allows processing of data underneath the predetermined boundaries, using edge computing in the 6G network enhances data privacy. Highly secure encryption techniques, which promote 6G data privacy, safeguard user communication along with stored files and thereby render it more difficult for malicious individuals to get hold of user confidential information [32]. The execution of building trust, allowing decision making, promoting responsibility, and accountability in the 6G environment all depends upon transparent communication between network operators and customers. Adherence to the industry protocols and privacy laws, such as GDPR international protocols, must be ensured by 6G technology. Consumers must be aware about the potential threats of data privacy to make the decision accordingly. Strong cryptography is important and safe storage techniques and biological authentication is necessary. To maintain user trust and to ensure user data privacy, ethical AI practices must be used. In order to reduce the effect on user privacy, identifying and alleviating threats as well as reacting swiftly to incidents involving security breaches and checking for security along with surveillance is crucial. Although 6G shifts the interpersonal interaction, security of users is of prime importance. An ethical and transparent mindset is necessary while using the advanced technology [33]. There is a prime need of creating and establishing efficient situation management designs critical for ensuring the security of these innovative technologies within the changing highly technical world of 6G. An extensive and pre-emptive approach is one which admits the challenges caused by the capabilities as well as limitations of the system in question required in the 6G environment. Having a meticulous strategy is an essential component of the 6G network. Along with identifying the threats, detailed plans for overcoming the threats must be outlined. As the vast array of gadgets are connected and are essential 6G programs, recovery plans and strategies created for the 6G environment must take into account all possible ways cyber-attacks can cause technical breakdown [34]. Traversing the digital battlefield demands understanding and recognizing the specific aspects that consist of a productive response to an incident. It involves quick data analysis for quick asymmetry recognition, threat detection which utilizes the AI and machine learning, and connections that assure collaboration among 6G elements of the ecosystem. To reduce the consequences of instances as well as maintain system credibility, it is important to shift swiftly from identification of threat detection to making the decision. A plan of action will be required for developing and executing incident resource plans for 6G safety. To ensure that action plans remain successful dealing with the shifting hazards, incident managers should take on the position of proactive guards by constantly assessing and evaluating it. Having a proactive approach is feasible to identify shortcomings of response strategies for incidents, implementing the necessary information prior to a real-life event taking place [35]. Further, the development of a robust incident handling structure for the 6G network focuses on the importance of responders to incidents as system defensive forces. Making the frequently altering attack environment, the foundation ought to remain versatile and fluid. In order to facilitate smooth and integrated reactions, occurrence sorting, extent assessment, and planned reaction behaviors ought to be involved. Efficient closure of incidents depends on an unambiguous and collaboration among various involved parties. Cyber resilience will be crucial for coping with fresh dangers. This involves foreseeing instances, operating across them, and recuperating from it. Innovative safeguards, ongoing surveillance, and threat information insertion are a few of these strategies. Ongoing tracking offers entities instant insight through safety incidents, and intelligence helps in mitigating their individual risks [36]. There needs to be a prevalent safety strategy to increase cyber defenses in the 6G environment. Ongoing security instruction, red teaming, and threat tracking are also a part of it. Detecting threats involves a forward-thinking process for pursuing feasible threats and discovering the risks prior to them becoming real. Use of red teams allows businesses to evaluate their protections and locate subjects for betterment through replicating genuine situations involving attacks. Constant education about safety fosters an anticipatory security culture by ensuring the fact that everyone involved knows everything about the standards of cybersecurity [37]. The resilience of cyberspace requires swift recovery strategies and continual improvement in 6G networks in the wake of an invasion. Setting up occurrence plans for recovery via an emphasis on accelerating the reinstatement of services whilst lowering interruptions is essential for rapid recovery. Businesses must carry out deep post-event analysis as a means of continual growth, highlighting lessons acquired and implementing alterations to enhance overall resilience. With each episode, the business becomes stronger by an aggressive interpretation, enabling the business to gain insight from past errors and alter their tactics accordingly. Maintaining the reliability and safety of these advanced new networks calls for productive plans to be established for incident response and cyber preparedness to be created in the 6G environment. Applying modern technologies, a strategy that is proactive along with the evaluation and review of methods, business is able to successfully negotiate the murky cyber landscape, protecting the authenticity of the 6G network [38]. To improve the periphery and mitigate potential threats, it is imperative to address a number of critical components in order to navigate the complexities of the 6G environment. One of these projects is data alignment and organization, which calls for robust data governance frameworks to ensure that the right data is collected, processed, stored, and anonymized. Additional security measures, like data encryption, are also put into place to protect users' privacy. By incorporating security by design concepts at every stage of the 6G technical development, threats will be effectively detected, strengthening the network infrastructure's resilience [39]. Offering better control choices with transparent terms of service and basic privacy settings can improve user safety and foster confidence. To guarantee compliance, organizations must strictly abide by laws like the GDPR. For enterprises and countries to maintain decision-making openness and accountability and minimize possible negative effects from algorithmic bias, ethical issues around AI are essential [40].

## 1.7 THE MORAL PRINCIPLES GUIDING PRIVACY: ENSURING LONG-TERM 6G ENVIRONMENT MANAGEMENT

As a result of this change in the 6G network, it is now even more important to establish efficient emergency response plans, keeping in mind that these highly developed wireless networks are typically more vulnerable and complex. Because of this, 6G networks are gradually exposed to a variety of new risks due to their interconnectedness, which makes the incorporation of proactive and flexible cybersecurity measures necessary. These barriers provide futile customary safety measures, demanding for an inventive approach that can be flexible to an active changing threat environment. The 6G wireless technology are linked together, they are likely prone to a varied number of threats, leading to the application of strategic and adaptable safety measures. Such challenges provide futile customary security safety protocols, advocating for an inventive approach, which can adjust to the evolving environment. With the widespread use of 6G technology, the social repercussions of safety standards have become more prominent. Moral scrutiny discusses the challenges like anonymity, governance data management, and the ethical use of cutting-edge technologies. It is important to match the safety standards with the expectations and standards of the society [41]. There is a need for moral scrutiny in order to honor the rights of individual's privacy and comply with the societal norms. An appropriate balance among the necessity of reliable safety and reverence in regard to the societal conventions serves as one of the challenges. Assessing the ethical consequences of safety practices is an important aspect of the 6G generation that needs careful consideration. The moral implications of the safety of security decisions become incorporated into technological innovations, which are widely implemented. The code of ethics encompasses managing of data, concerns about privacy, and responsibility for the deployment of the latest technologies. It is essential to examine and resolve the ethical concerns to ensure that privacy techniques not only adhere to the legal regulations but are also in accordance with the social standards and values [42]. A responsive plan needs to incorporate diligent and proactive approaches, which utilizes the computational intelligence and neural networks. The ability to predict and mitigate the potential risks is vital, along with the strategies to respond to these situations needs to be considered with the evolving and changing times to keep up with the trend and emerging threats. Another factor that is included, apart from technological fixes, is the human-centered approach that needs to be imbibed in the 6G environment that needs to strengthen the cyber resilience. Creation of structures that have the ability to cope, and bouncing back from and dealing with adverse circumstances can be referred to as cyber resilience. Extensive education, public relation efforts, and the development of robust cultural environment alongside extensive technological protections are required [43]. The 6G protocols require taking into consideration an array of social factors considering accountability, Candor, and ethical utilization of technology. Developing an appropriate state of equilibrium among moral norms, and inventiveness promotes between customers and clients. Secure hacking approaches with a confidentiality science and technology, complying to worldwide techniques contribute an increasingly culpable and durable protection system. Having safety demands in line with the standards of society is a challenging endeavor that calls for interaction from a wide range of interested parties. The technological professionals and administrations require a regular interaction to create the ethical rules, which dictate safety processes. With a diversity of opinions, safety methods are not solely effective but are also courteous of cultural variances and individual freedom [44]. Understanding the need for international collaboration is crucial for resolving privacy and security issues in the context of 6G. In light of the global reach of telecommunications networks, it is critical to develop ethical standards and guidelines that are consistently applied. In order to protect people's right to privacy and to effectively handle new dangers, international coordination is required. This cooperative endeavor may encompass exchanging optimal methodologies, harmonizing regulatory approaches, and enabling the interchange of data. The need for interdisciplinary cooperation among engineers, ethicists, legislators, and legal specialists is growing as the 6G ecosystem develops. This alliance has the potential to ensure that regulatory frameworks are flexible enough to keep up with the rapid changes in the digital ecosystem by promoting a thorough grasp of the ethical implications of technical breakthroughs [45]. By incorporating various viewpoints, stakeholders can create sophisticated privacy management plans that balance technology advancement with moral considerations [46]. Establishing openness and accountability in 6G network governance is essential. Holding stakeholders accountable for security lapses or improper handling of personal data is essential to ensuring transparency in data management procedures [47]. Putting in place unbiased oversight and auditing processes will boost confidence in the 6G ecosystem's privacy protections. The application of AI and machine-learning algorithms in 6G networks poses particular difficulties for upholding moral standards. Although these technologies can improve network performance and security procedures, they also give rise to worries about algorithmic bias and prejudice. Impact analyses, algorithmic transparency, and ongoing research are required to solve these issues and provide equitable results for all users. Setting ethical issues in privacy management as a top priority is critical as 6G technology develops. This means abiding by the laws that are in place and actively tackling newly emergent ethical conundrums through multidisciplinary cooperation, global cooperation, accountability, and transparency. Through the incorporation of ethical concepts into the governance and construction of 6G networks, interested parties can create a digital infrastructure that is both dependable and long-lasting [48].

## 1.8 CONCLUSION

A new period of automated dialog, creativity, and safety risks will be marked in the 6G network. Technology has evolved, its descendants have additionally created hazards and challenging complexities, and it is also youngsters who constitute an imminent danger to modern safety standards. A common thread as we dive through the number of such issues, the complex layout of sixth generation. The use of ethernet networks in combination with the use of edge computing and robotics develop an intricate network, having the potential for creating flaws. A safeguarding network using 6G demands an integrated and multitiered approach that protects the network instead of only bolstering safeguarding boundaries of 6G periphery. In addition, it goes over the technical barriers we deal with the multifaceted concerns. The constantly changing and ubiquitous character of cyber threat calls for an abrupt change in safety tactics from acute to pre-emptive. According to the 6G network needs to be centered on forward-thinking, and prescient indicates the AI and machine-learning abilities. Plans for incidents ought to use versatile structures that require to change and anticipate the subsequent generations of threats linked to within networks of 6G instead of reactive response rules and regulations. Collaboration additionally becomes necessary for bolstering the safeguards' defense towards the novel dangers. The interactions under progeny apprehension is vital for strengthening the barriers with the emerging dangers. The domain of 6G implies that efforts to collaborate within the government agencies, corporate partners, as well as digital age professionals is vital. The exchange of data eventually becomes a lot more than merely the implication, it likewise turns into an essential requirement to earn collaborative studies aimed at development of uniformed safety protocols with the goal to address the resulting generations of dangers, which surpasses business and national boundaries. This paper emphasizes the fact that everyone needs to get involved in digital security. There needs to be a culture that is mindful of security measures and requires comprehensive education initiatives, a wide range of awareness, initiatives, and development of general understanding regarding important safeguards among the everyday consumers.

## REFERENCES


[1] M. Wang, T. Zhu, T. Zhang, J. Zhang, S. Yu, and W. Zhou, "Security and privacy in 6G networks: New areas and new challenges," *Digit. Commun. Netw.*, vol. 6, no. 3, pp. 281–291, 2020.

[2] V. Ziegler, P. Schneider, H. Viswanathan, M. Montag, S. Kanugovi, and A. Rezaki, "Security and trust in the 6G era," *IEEE Access*, vol. 9, pp. 142314–142327, 2021.

[3] M. Ylianttila, R. Kantola, A. Gurtov, L. Mucchi, I. Oppermann, Z. Yan, T. H. Nguyen, F. Liu, T. Hewa, M. Liyanage, and A. Ijaz,"6G white paper: Research challenges for trust, security and privacy," In arXiv [cs. CR], 2020.

[4] D. Kumar and S. Chavhan, "Shift to 6G: Exploration on trends, vision, requirements, technologies, research, and standardization efforts," *Sustain. Energy Technol. Assessments*, vol. 54, no. 102666, p. 102666, 2022.

[5] P. Porambage, G. Gur, D. P. M. Osorio, M. Liyanage, A. Gurtov, and M. Ylianttila, "The roadmap to 6G security and privacy," *IEEE Open J. Commun. Soc.*, vol. 2, pp. 1094–1122, 2021

[6] Z. Wang, Y. Du, K. Wei, K. Han, X. Xu, G. Wei, W. Tong, P. Zhu, J. Ma, J. Wang, G. Wang, X. Yan, J. Xiang, H. Huang, R. Li, X. Wang, Y. Wang, S. Sun, S. Suo, ... X. Su, "Vision, application scenarios, and key technology trends for 6G mobile communications," Sci. China Inf. Sci., vol. 65, no. 5, pp. 151301:1–151301:27. 2022. https://doi.org/10.1007/s11432-021-3351-5

[7] S. Aggarwal, N. Kumar, and S. Tanwar, "Blockchain-envisioned UAV communication using 6G networks: Open issues, use cases, and future directions," *IEEE Internet Things J.*, vol. 8, no. 7, pp. 5416–5441, 2021.

[8] R. Rana and R. Kumar, "Performance analysis of AODV in presence of malicious node", *Acta Electron. Malays.*, vol. 3, no. 1, pp. 1–5, 2019.

[9] W. Hurst, S. Evmorfos, A. Petropulu, and Y. Mostofi, "Unmanned vehicles in 6G networks: A unifying treatment of problems, formulations, and tools," In arXiv [eess.SY], 2024. http://arxiv.org/abs/2404.14738

[10] S. A. H. Mohsan and Y. Li, "A contemporary survey on 6G wireless networks: Potentials, recent advances, technical challenges and future trends," In arXiv [cs.NI], 2023. http://arxiv.org/abs/2306.08265

[11] A. T. Jawad, R. Maaloul, and L. Chaari, "A comprehensive survey on 6G and beyond: Enabling technologies, opportunities of machine learning and challenges," *Comput. Netw.*, vol. 237, no. 110085, p. 110085, 2023.

[12] A. Bandi and S. Yalamarthi, "Towards artificial intelligence empowered security and privacy issues in 6G communications," in *2022 International*

*Conference on Sustainable Computing and Data Communication Systems (ICSCDS)*, 2022. IEEE.

[13] Y. Ghildiyal et al., "An imperative role of 6G communication with perspective of industry 4.0: Challenges and research directions," *Sustain. Energy Technol. Assessments*, vol. 56, no. 103047, p. 103047, 2023.

[14] R. Sedar, C. Kalalas, F. Vazquez-Gallego, L. Alonso, and J. Alonso-Zarate, "A comprehensive survey of V2X cybersecurity mechanisms and future research paths," *IEEE Open J. Commun. Soc.*, vol. 4, pp. 325–391, 2023.

[15] C. De Alwis et al., "Survey on 6G frontiers: Trends, applications, requirements, technologies and future research," *IEEE Open J. Commun. Soc.*, vol. 2, pp. 836–886, 2021.

[16] C. Mangla, S. Rani, N. M. Faseeh Qureshi, and A. Singh, "Mitigating 5G security challenges for next-gen industry using quantum computing," J. *King Saud Univ. Comput. Inf. Sci.*, vol. 35, no. 6, p. 101334, 2023.

[17] T. Q. Duong, J. A. Ansere, B. Narottama, V. Sharma, O. A. Dobre, and H. Shin, "Quantum-inspired machine learning for 6G: Fundamentals, security, resource allocations, challenges, and future research directions," *IEEE Open J. Veh. Technol.*, vol. 3, pp. 375–387, 2022.

[18] S. Sharif, S. Zeadally, and W. Ejaz, "Space-aerial-ground-sea integrated networks: Resource optimization and challenges in 6G," *J. Netw. Comput. Appl.*, vol. 215, no. 103647, p. 103647, 2023.

[19] T. Garg, G. Kaur, and P. S. Rana, "Protocol security in 6th generation (6G) networks," in *International Conference on Computation Intelligence and Network Systems*, Springer, 2023, pp. 47–62.

[20] S. Zhang and D. Zhu, "Towards artificial intelligence enabled 6G: State of the art, challenges, and opportunities," *Comput. Netw.*, vol. 183, no. 107556, p. 107556, 2020.

[21] N. Kato, B. Mao, F. Tang, Y. Kawamoto, and J. Liu, "Ten challenges in advancing machine learning technologies toward 6G," *IEEE Wirel. Commun.*, vol. 27, no. 3, pp. 96–103, 2020.

[22] F. Irram, M. Ali, M. Naeem, and S. Mumtaz, "Physical layer security for beyond 5G/6G networks: Emerging technologies and future directions," *J. Netw. Comput. Appl.*, vol. 206, no. 103431, p. 103431, 2022.

[23] A. Kumari, R. Gupta, and S. Tanwar, "Amalgamation of blockchain and IoT for smart cities underlying 6G communication: A comprehensive review," *Comput. Commun.*, vol. 172, pp. 102–118, 2021.

[24] R. C. Reyes, R. V. Sevilla, G. S. Zapanta, J. V. Merin, R. R. Maaliw, and A. Ferrer Santiago, "Safety gear compliance detection using data augmentation-assisted transfer learning in construction work environment," in *2022 IEEE International Conference on Electronics, Computing and Communication Technologies (CONECCT)*, IEEE 2022.

[25] A. Habbal, M. K. Ali, and M. A. Abuzaraida, "Artificial Intelligence Trust, Risk and Security Management (AI TRiSM): Frameworks, applications, challenges and future research directions," *Expert Syst. Appl.*, vol. 240, no. 122442, p. 122442, 2024.

[26] P. Janay and A. Sarkara, "Secure data transmission in the era of 6G: Challenges and solutions," *Algorithm Asynchronous*, vol. 1, no. 1, pp. 8–15, 2023.

[27] N. Lal, S. M. Tiwari, D. Khare, and M. Saxena, "Prospects for handling 5G network security: Challenges, recommendations and future directions," *J. Phys. Conf. Ser.*, vol. 1714, no. 1, p. 012052, 2021.
[28] Y. Siriwardhana, P. Porambage, M. Liyanage, and M. Ylianttila, "AI and 6G security: Opportunities and challenges," in *2021 Joint European Conference on Networks and Communications & 6G Summit (EuCNC/6G Summit)*, IEEE, 2021.
[29] R. Khan, P. Kumar, D. N. K. Jayakody, and M. Liyanage, "A survey on security and privacy of 5G technologies: Potential solutions, recent advancements, and future directions," *IEEE Commun. Surv. Tutor.*, vol. 22, no. 1, pp. 196–248, 2020.
[30] P. Scalise, M. Boeding, M. Hempel, H. Sharif, J. Delloiacovo, and J. Reed, "A systematic survey on 5G and 6G security considerations, challenges, trends, and research areas," *Future Internet*, vol. 16, no. 3, pp. 1–38, 2024.
[31] J. Park, S. Samarakoon, M. Bennis, and M. Debbah, "Wireless network intelligence at the edge," *Proc. IEEE Inst. Electr. Electron. Eng.*, vol. 107, no. 11, pp. 2204–2239, 2019.
[32] M. Boban, A. Kousaridas, K. Manolakis, J. Eichinger, and W. Xu, "Connected roads of the future: Use cases, requirements, and design considerations for vehicle-to-everything communications," *IEEE Veh. Technol. Mag.*, vol. 13, no. 3, pp. 110–123, 2018.
[33] M. Mitev, A. Chorti, H. V. Poor, and G. Fettweis, "What physical layer security can do for 6G security," In arXiv [cs.CR], 2022. http://arxiv.org/abs/2212.00427
[34] Y. Chen, W. Liu, Z. Niu, Z. Feng, Q. Hu, and T. Jiang, "Pervasive intelligent endogenous 6G wireless systems: Prospects, theories and key technologies," *Digit. Commun. Netw.*, vol. 6, no. 3, pp. 312–320, 2020.
[35] S. H. A. Kazmi, F. Qamar, R. Hassan, K. Nisar, and M. A. Betar, "Security of federated learning in 6G Era: A review on conceptual techniques and software platforms used for research and analysis," *Comput. Netw.*, vol. 110358, pp. 1–32, 2024.
[36] M. A. Lopez, G. N. N. Barbosa, and D. M. F. Mattos, "New barriers on 6G networking: An exploratory study on the security, privacy and opportunities for aerial networks," in *2022 1st International Conference on 6G Networking (6GNet)*, IEEE, 2022.
[37] R. M. Sundhari and K. Jaikumar, "IoT assisted Hierarchical Computation Strategic Making (HCSM) and Dynamic Stochastic Optimization Technique (DSOT) for energy optimization in wireless sensor networks for smart city monitoring," *Comput. Commun.*, vol. 150, pp. 226–234, 2020.
[38] I. Kanno, "Trends and use cases of 5G," In *Policies and Challenges of the Broadband Ecosystem in Japan*, Springer Nature Singapore, 2022, pp. 105–122.
[39] W. Guo, "Explainable artificial intelligence for 6G: Improving trust between human and machine," *IEEE Commun. Mag.*, vol. 58, no. 6, pp. 39–45, 2020.
[40] H. Yue, Q. Jiang, C. Yin, and J. Wilson, "Retracted article: Research on data aggregation and transmission planning with Internet of Things

technology in WSN multi-channel aware network," *J. Supercomput.*, vol. 76, no. 5, pp. 3298–3307, 2020.

[41] A. Adel, "Future of industry 5.0 in society: Human-centric solutions, challenges and prospective research areas," *J. Cloud Comput. Adv. Syst. Appl.*, vol. 11, no. 1, p. 40, 2022. https://doi.org/10.1186/s13677-022-00314-5

[42] A. Bentajer, M. Hedabou, F. Ennaama, and S. Elfezazi, "Development of design for enhancing trust in cloud's SPI stack," *Int. J. Comput. Exp. Sci. Eng.*, vol. 6, no. 1, pp. 13–18, 2020. https://doi.org/10.22399/ijcesen.370873

[43] M. U. Hassan, M. H. Rehmani, and J. Chen, "DEAL: Differentially private auction for blockchain-based microgrids energy trading," *IEEE Trans. Serv. Comput.*, vol. 13, no. 2, pp. 263–275, 2019.

[44] J. Wan, J. Li, M. Imran, D. Li, and Fazal-e-Amin, "A blockchain-based solution for enhancing security and privacy in smart factory," *IEEE Trans. Industr. Inform.*, vol. 15, no. 6, pp. 3652–3660, 2019.

[45] K. Hameed, M. Barika, S. Garg, M. B. Amin, and B. Kang, "A taxonomy study on securing blockchain-based industrial applications: An overview, application perspectives, requirements, attacks, countermeasures, and open issues," *J. Ind. Inf. Integr.*, vol. 26, no. 100312, p. 100312, 2022.

[46] G. Rathee, F. Ahmad, R. Sandhu, C. A. Kerrache, and M. A. Azad, "On the design and implementation of a secure blockchain-based hybrid framework for industrial internet-of-things," *Inf. Process. Manag.*, vol. 58, no. 3, p. 102526, 2021.

[47] R. Kumar, *Information and Communication Technologies*, Laxmi Publications, 2009.

[48] A. Sasikumar, S. Vairavasundaram, K. Kotecha, V. Indragandhi, L. Ravi, G. Selvachandran, and A. Abraham, "Blockchain-based trust mechanism for digital twin empowered industrial internet of things," *Future Gener. Comput. Syst.*, vol. 141, pp. 16–27, 2023.

Chapter 2

# Japan's 2030 vision for 6G

## A case study

*Nimiti Sharma, Rajendra Kumar, Rama Maliya, Vaidehi Shah, Jeel Patel, and Masanori Fukui*

### 2.1 INTRODUCTION

The research and development are in the primary phase, and the first sixth-generation (6G) network will not be projected to be used for commercial operations until 2030. The groundwork is already being laid, with advancements in 5G-Advanced, as well as crucial stages on the way to 6G. The current era of communication focuses on the transition from 5G to 6G mobile systems, emphasizing the rapid growth of 5G subscribers worldwide. It also discusses the increasing focus on 6G research and development in academia and industry, with various countries, institutions, and companies initiating 6G research programs. The chapter aims to provide a study of current 6G research, ambitions, and technological advancements while offering a vision of what 6G may look like in the future Japan. The first generation of network was started in 1980 and currently we are utilizing the fifth generation with an expectation of the sixth generation somewhere in 2023. Figure 2.1 shows all six network generations.

What we are expecting from 6G networks? A high speed is not the only expectation. Figure 2.2 represents the major expectations from 6G networks that include ultra-high speed, ultra-low latency, high reliability, massive capacity, terahertz frequencies, AI-driven networks, holographic communication, energy-efficient operations, etc.

The major contributions presented in this include (i) key features and significance of THz frequency in 6G development, (ii) government regulations and MoUs, (iii) promising applications, and (iv) security and possible open research issues. Such issues are deeply examined to provide a specific and targeted conclusion. Therefore, the presented study can be considered as an understanding of current practices and a catalyst in exploring new directions for future networks of the 2030s [1].

As an update in January 2022, 6G technology in Japan is in the early stages of research. The 6G implementation is proposed to be the next generation of wireless technology after 5G, to promise a much faster data speed, lower latency, and felicitating emerging technologies like holographic


DOI: 10.1201/9781003583127-2

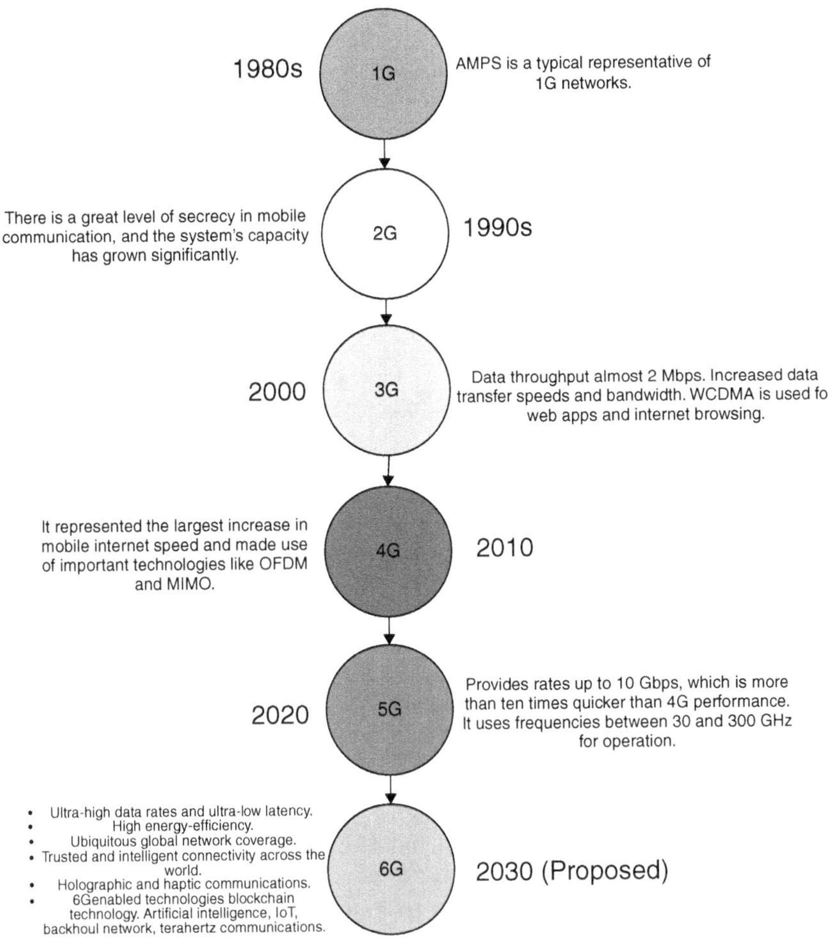


*Figure 2.1* Network generations (first to sixth).

communications, advanced artificial intelligence, the Internet of Things (IoT), and many more.

Some developing countries like Japan have been pioneers in telecommunications technology. Japan with companies like NTT Docomo, SoftBank, and KDDI is leading the way in telecom research. These organizations, along with academia and governmental agencies, are involved in early explorations and infrastructure for 6G. However, a realistic implementation of 6G networks is still years away and is expected in 2030. Continuous efforts are in progress to define the infrastructural requirements, supported technologies, and use cases for 6G, but extensive deployment is not expected until well into 2030 [2].

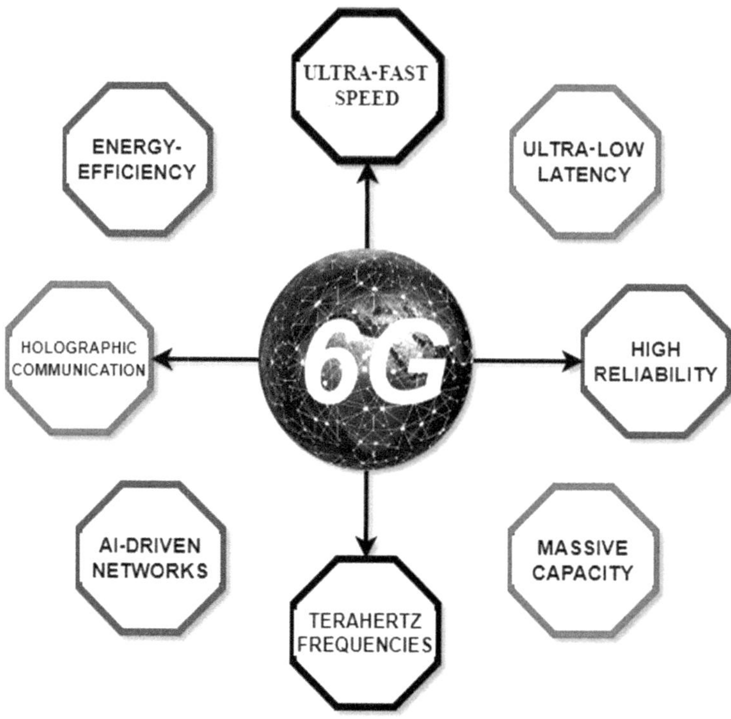


*Figure 2.2* Promising areas of 6G networks.

Japan's ambitious plan to implement 6G by 2030 focuses on speed, capacity, and coverage. Speed, they want to be able to communicate ten times quicker than 5G, which will allow vast amounts of data to be transferred almost instantly. Capacity, in order to manage the increase in connected devices anticipated in the upcoming 10 years, the 6G agenda of Japan will make use of new, high-frequency bands (such as terahertz waves). Coverage, to enable continuous connectivity everywhere, the Japanese networks will cover the skies, seas, and even space in addition to the ground [3]. With capabilities like self-healing networks and extremely low latency for real-time applications like remote surgery, 6G strives to be more intelligent.

This cutting-edge network is considered to be a major force behind "Society 5.0," Japan's concept of a hyperconnected, human-centered future that seamlessly blends digital and physical realms. They aspire to dominate the world market for 6G technology by 2030, holding important patents and a large market share.

Japan is a leader in innovation, having made significant investments in 6G research and development (R&D). Businesses like NTT are achieving important advances, such as data transmission in the 300 GHz range, which might form the basis of future 6G systems [3]. In order to speed up development, the government is also promoting R&D by cutting rules for important companies like NTT.

As the particular uses, Japan sees 6G enabling special uses that go beyond speed and capacity. This could include ultra-realistic telepresence for distant work, education, and healthcare; utilizing real-time data analysis and control in smart factories to revolutionize the production process; sophisticated self-driving cars that depend on extremely dependable, low-latency communication to navigate safely.

## 2.2 RELATED WORK

This section discusses the anticipation of the 6G network as the next major development in the telecommunications industry in recent years. It emphasizes the need and practices for advanced technology to enable immersive virtual experiences, addressing the increased demand for real-time communication, healthcare, business, and education in the post-pandemic era. The focus is on the future megatrends driving 6G, including automation in industry, robotics, and self-driving technology, as well as key requirements such as meeting performance criteria, UN sustainability goals, business models, edge intelligence, digital divide, and machine-learning trends. The studies also discuss the global acceleration of the technology and industrial revolution, supported by the widespread application of state-of-the-art information and communication technologies such as artificial intelligence, augmented and virtual reality, the Internet of Things, and blockchain technology, leading to the development of the 6G communication systems [4–6]. The development of 6G is seen to have noteworthy implications for communication development with intelligent connectivity. The discussions are also focused on the standardization of 5G communications and the commencement of visioning and planning for 6G communications to accommodate the needs anticipated for the year 2030. It emphasizes the importance of user-focused cellular telecommunication and highlights the need for high security, secrecy, and privacy in 6G [7, 8]. The paragraph also outlines potential application scenarios, key features, and required communication technologies for 6G while acknowledging non-technical issues that could impact its research and deployment.

The studies [9, 10] are based on the ongoing commercialization of 5G technology and the anticipated evolution into 6G networks, which are expected to prioritize energy efficiency and high-quality service. The study outlines the need for significant advancements in mobile network architecture to accommodate future applications and mentions the integration of global mobile networks with satellite, aerial, and underwater networks in providing ubiquitous access. They majorly focused on architectural changes corresponding to 6G networks, which also include ubiquitous 3D coverage, pervasive artificial intelligence, and improved network protocols.

A study [11] discusses the rapid development of brilliant terminus and the surge in wireless networking detailed flow, highlighting those current cellular networks, including 5G, may not fully meet emerging technical demands. It introduces the vision of 6G wireless communication, which is anticipated to be deployed between 2027 and 2030, incorporating artificial intelligence support. This paper thoroughly examines the concept of 6G wireless communication, delving into its envisioned network architecture, technical hurdles, potential remedies, transmission protocols at the physical layer, network configuration, and approaches to security [12, 13].

This research [14] outlines several state-of-the-art technologies like artificial intelligence and machine learning (AIML), quantum computing and assisted learning techniques, decentralized ledger technology, THZ and mm-wave data transmission, microcell network, multi-user interface cancellation technique/NOMA, distributed computing at the network edge and more as important components of the 6G network. The chapter also covers potential applications, use cases, performance indicators, challenges, and future research opportunities related to 5G (B5G) and sixth-generation wireless networks. Dynamic energy harvesting and the need to reduce human intervention have led to the recognition of artificial intelligence (AI) as the key solution. The integration of artificial intelligence procedures, which has systems that include automated learning, algorithmic learning, and cognitive computing, with conventional AI methods and mathematical models is seen as essential for minimizing system complications and improving the structure in 6G technology. The study also discusses the growing demand for visual, auditory, and tactile communication over 6G cellular networks, highlighting the potential of Tactile Internet (TI) in enabling real-time haptic interaction. TI is expected to have diverse applications and impact various industry sectors. However, it also presents unorthodox interaction problems, including communication delay, accessibility, dependability, and safety. An artificial intelligence-driven virtual model, constructed on the edge cloud component, is suggested as a potential remedy to tackle these network obstacles.

The studies [15, 16] discuss the potential of the IoT extensively in the forthcoming era of 6G, the landscape of communication will undergo a complete transformation, particularly focusing on its convergence with optical radio transmission (ORT) technology. It emphasizes the transition from 5G to 6G IoT, highlighting the high-density, high-capacity, and robust system architecture of 6G IoT systems, powered by the automated intelligence-driven intelligent mechanism of the system. The potential of ORT technology in allowing optimization and proficiency in the sixth generation of technology in sensor systems is also highlighted. The paragraph further mentions the expanding ORT market, related technologies, ongoing research, and the challenges and emerging technology solutions for radio-over-fiber (RoF)-supported 6G IoT systems.

The studies [17,18] present the shift from 5G to the upcoming 6G wireless communications and highlights a lack of collaboration between industry and academia in its development. The authors present a consolidated view of the vision, values, use cases, and enabling technologies for 6G, representing expertise from various aspects of mobile communications. They also revisit concurrent initiatives on 6G and identify six key research challenges, including intelligence empowerment, network evolution, extreme experiences, sustainability, trustworthiness, and global service coverage. Additionally, they specify a set of jointly defined use cases for 6G and discuss anticipated developments in radio, network, and management domains. The authors [19] also discussed the ongoing shift from 5G to the potential future of 6G mobile communication systems. It emphasizes a holistic approach to 6G, beginning with societal and lifestyle drivers for cutting-edge systems and delving into the technological prerequisites for sixth-generation wireless technology. The need for a significant increase in spectrum resource maximization, including ranges from a hundred gigahertz to one terahertz, is highlighted. Additionally, it outlines hurdles as well as prospects for structural strategies throughout all the tiers of the OSI model, such as core network redesign and new modulation and coding methods. The paragraph emphasizes the practical considerations and research challenges involved in realizing 6G technologies over the coming decade.

The study [19] outlines the disadvantages of recent times of the 5G telecommunication systems in the face of cutting-edge innovations like the Internet of All, cooperative androids, holographic presence, and tourism in outer space and ocean depths. It discusses the anticipated development of the 6G communications to consider the disadvantages. It highlights the growing interest in 6G research and development in both academia and the industry, emphasizing the societal and technological trends driving this progression. Additionally, it emphasizes the need for 6G to meet the demands of emerging applications and outlines the requirements and enabling technologies for 6G networks. The paragraph also mentions ongoing research projects, standardization efforts, and technical challenges in the pursuit of 6G development.

A study [20] discusses the future vision of the IoT in the context of 6G technology and its convergence with the RoF system. It highlights how 6G IoT will involve compact disparate gadgets with exceptionally substantial potential, sturdy system framework, and artificial intelligence algorithmic processes. The significance of optic fiber and RoF technologies for 6G IoT systems is underscored, alongside the burgeoning RoF market and complementary technologies. The paragraph concludes by addressing the challenges and emerging technology solutions for RoF-supported 6G IoT systems. The study provides an overview of the 6G global landscape, detailing both private and public initiatives that have resulted in significant capital placements on state-of-the-art intelligence and exchange systems as well as software assistance.

The study [21] discusses the early exploration of 6G wireless networks, focusing on security and privacy considerations. It outlines the envisioned impact of security on 6G systems, potential challenges related to 6G technologies, and proposed solutions. The research also touches on security key performance indicators (KPIs) and the expected risk terrain built upon the state-of-the-art system infrastructure. Lastly, it highlights protection aspects associated with facilitating innovations examples including decentralized ledger technology (DLT), tangible stratum safety, decentralized artificial intelligence/machine learning (AI/ML), perceptible illumination communication (VLC), terahertz (THz), and quantum computation.

The study [22] discusses the increasing energy costs associated with the growing network infrastructure and connected terminals in the 5G and upcoming 6G eras. The study emphasizes the need for green communications to address these challenges and highlights the role of AIML in improving energy efficiency and managing energy harvesting. The presented study delves into the applications of artificial intelligence algorithms, encompassing machine and deep learning, to decrease algorithm intricacy and augment precision within 6G networks. It also outlines the existing issues and potential research areas for AI models in achieving green 6G communications.

A study [23] discusses the crucial importance of healthcare infrastructure in light of the UN's sustainable development goal for good health and well-being. It highlights the strain on global health systems due to the aging population and limited resources, underscoring the potential of the Internet of Medical Things (IoMT) [24] to address these challenges. It acknowledges the deployment of 5G and the future vision for 6G, emphasizing the role of 6G to comprehend the full potential of the Internet of Medical Things. The study addresses the shortcomings of 5G while introducing an open, "6G enabled IoMT" architecture in incorporating a variety of medical services and applications. Additionally, it outlines the challenges and open issues related to the proposed architecture, providing a brief overview of 6G technology.

## 2.3 THE 6G INFRASTRUCTURE DEVELOPMENT IN JAPAN

The 6G infrastructure development is still in the early stages globally. However, several key trends and efforts are emerging as follows:

*Research Initiatives:* Academia, research organizations, and telecommunications companies across the globe are conducting research into 6G technologies. These efforts are aimed to identify the potential requirements, challenges, and opportunities for 6G networks.

*Standardized Efforts:* Standardized bodies such as the International Telecommunication Union (ITU) and industry consortia have started the groundwork for 6G standards and protocols. These efforts involve defining technical specifications to shape the future 6G networks.

*Technological Focus:* 6G is expected to shape the advancements of 5G technology by introducing new technologies, such as terahertz frequency bands, intelligent reflecting surfaces, and AI-driven optimizations.

*Global Collaborations:* Collaboration among the countries, companies, and research institutions is crucial for advancing 6G developments.

*Exploring Use Cases:* Contributions are being made for the identification of potential use cases and applications for 6G beyond traditional telecommunications. This includes fields like augmented and virtual reality, holographic communication, autonomous vehicles, and the Internet of Things.

*Regulatory Considerations:* Discussions around the spectrum allocation, regulations, and policy frameworks for 6G have been started by emerging governments and regulatory authorities to consider the implications of 6G deployment and how to facilitate their rollout while addressing those potential challenges.

It is essential to consider that the landscape of 6G development is continually evolving, and new technological advancements and initiatives may emerge in the near future. Several companies and research institutes in Japan are actively participating in research and developing 6G technologies. Some of these include the following:

*NEC Corporation:* NEC Corporation is a Japan-based global technological company that is involved in various aspects of 6G research and development. It has been working on massive Multiple Input Multiple Output and beamforming technology, which is expected to be the key mechanism of 6G networks. Figure 2.3 represents the NEC Corporation's initiatives for 6G in Japan.

*NTT Docomo:* NTT Docomo, one of Japan's leading telecommunications companies at the forefront of 6G research, has been involved in various initiatives in exploring the potential technologies, use cases, and infrastructure requirements for 6G networks across the country. Figure 2.4 highlights the major initiatives by NTT DOCOMO's for 6G development.

*SoftBank:* SoftBank is another major telecommunication company in Japan that is also investing in 6G research and developments by participating in

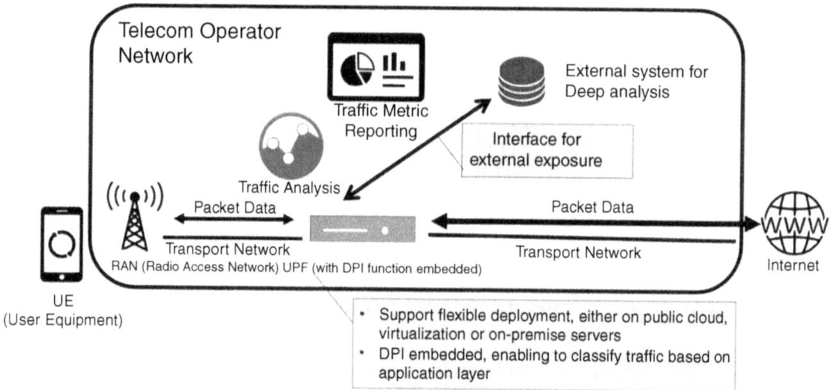


*Figure 2.3* NEC Corporation's initiatives for 6G.

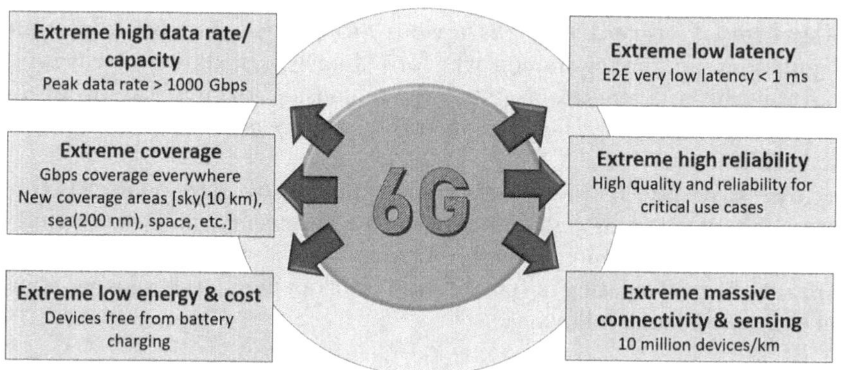


*Figure 2.4* NTT DOCOMO's initiatives for 6G.

collaborative efforts with other industry players and research institutions. Figure 2.5 highlights the major challenges with SoftBank in a highly efficient and reliable 6G network.

*Fujitsu:* Fujitsu is a leading Japanese telecom company with expertise in network infrastructure. It is conducting research into advanced wireless technologies and is likely involved in 6G development exertions. Figure 2.6 highlights the major initiatives by Fujitsu for 6G research and development.

*Panasonic:* The company Panasonic is known for its consumer electronics and is also involved in telecommunications research and development.

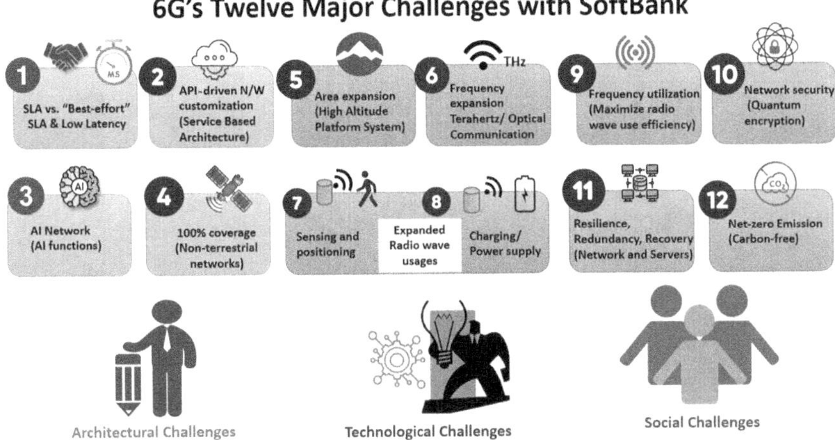


*Figure 2.5* 6G challenges with SoftBank.

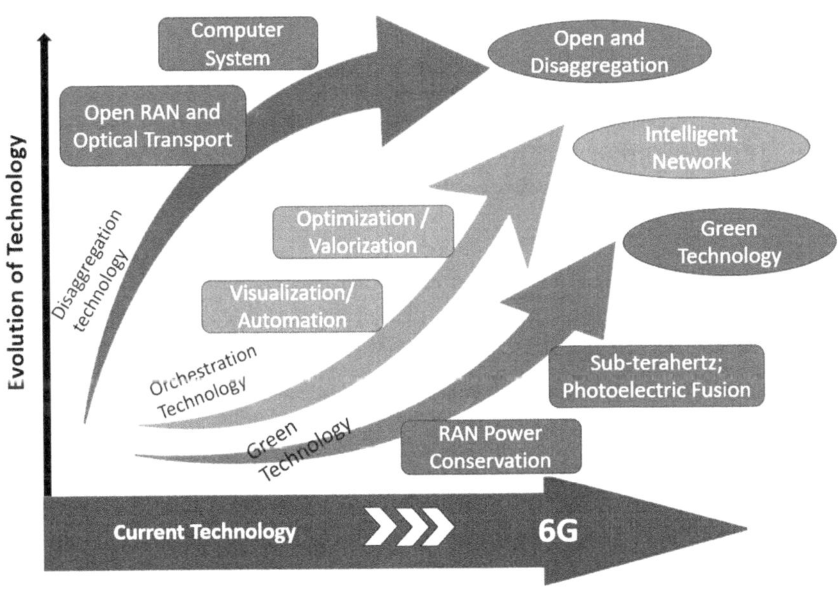


*Figure 2.6* Fujitsu's initiatives for 6G.

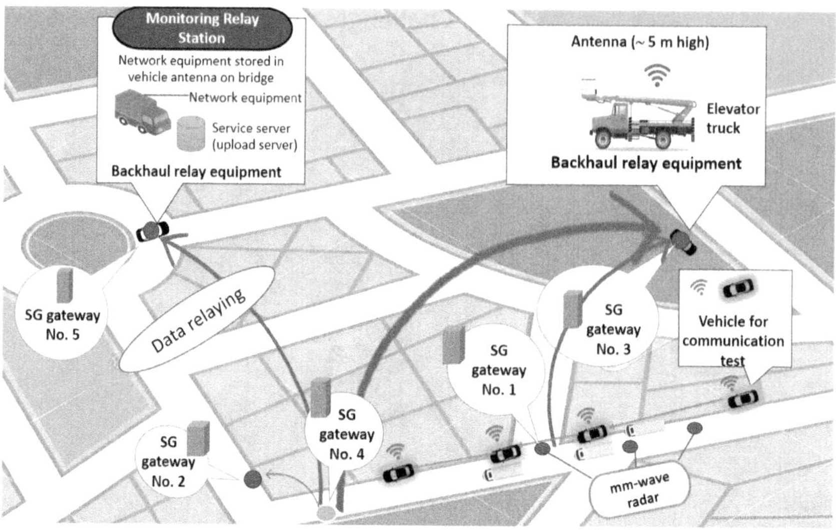


*Figure 2.7* Panasonic's initiatives for 6G infrastructure.

They have been exploring technologies, such as terahertz communications, which is expected to play a key role in 6G deployment. Panasonic-Japan has connected Yokohama city in testing the world's first beyond-5G network demonstration proposed for autonomous driving in a 6G environment. Figure 6.7 represents a roadmap of the 6G plan by Panasonic.

*KDDI Corporation:* KDDI Corporation is another significant company in Japan in the telecommunication industry that is engaged in 6G research initiatives. It is exploring potential use cases and technological innovations that could shape the forthcoming 6G networks. Figure 6.8 presents the information security initiatives by KDDI Corporation for the 6G network.

*Universities and Research Institutions*: Several universities and research institutions in Japan, such as the University of Tokyo, Tokyo Institute of Technology, Kyoto University, and many more are conducting research into 6G technologies. These bodies play a crucial role in advancing the state of the art in 6G wireless communication.

The above-mentioned companies and institutes are just a few examples of the many organizations in Japan that are actively involved in 6G research and development. Possible collaboration among industry players, research institutes, and government agencies is crucial for driving innovations and shaping the future of 6G technology.

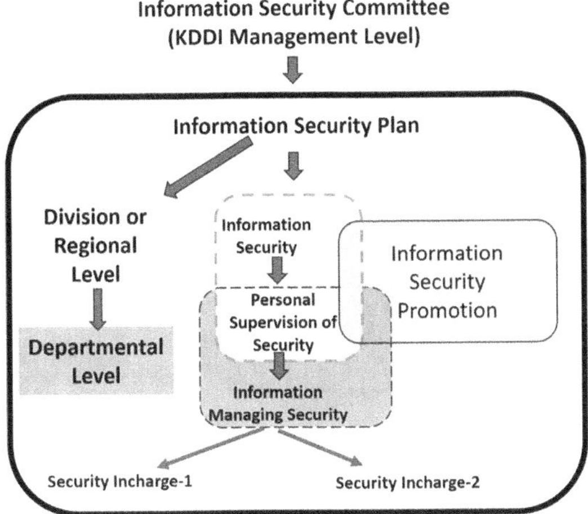


*Figure 2.8* KDDI Corporation's initiatives for information security in the 6G network.

Concerning Japan's 6G vision as a global leader in technology, Japan has once again taken the lead in shaping the future. The Japanese government, in collaboration with leading tech companies and academic institutions, has launched ambitious plans to develop and deploy 6G technology by 2030.

The following are major key aspects of Japan's 6G initiatives:

*Supercharged Speeds:* Imagine downloading an entire 4K movie in mere seconds or streaming high-definition content without any buffering. With 6G, these lightning-fast speeds will become a reality, transforming the way we consume media and access information.

*Massive IoT Connectivity:* 6G will enable seamless connectivity for an unprecedented number of devices, allowing for the expansion of IoT and the realization of smart cities [25], intelligent transportation systems [26], and more.

*Immersive Experiences:* Virtual reality (VR) and augmented reality (AR) applications will reach new heights with 6G, providing users with immersive experiences that blur the lines between the physical and digital worlds.

*Enhanced AI Capabilities:* With 6G, AI will become even more powerful, enabling advanced applications, such as autonomous vehicles, precision healthcare, and smart robotics.

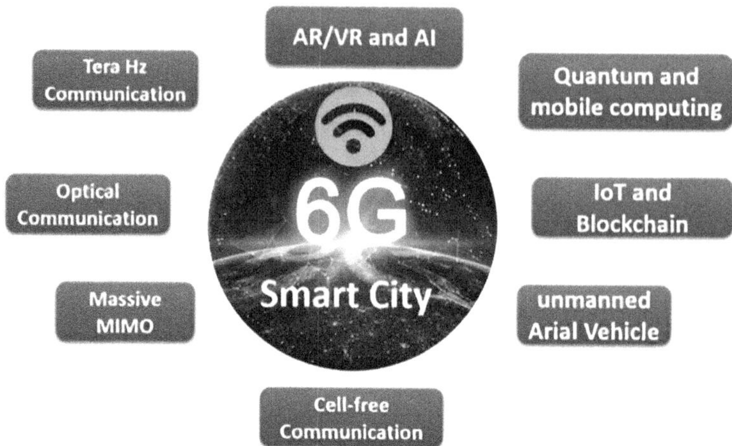


*Figure 2.9* Supported technological requirements for a 6G smart city.

## 2.4 6G MOMENTUM WORLDWIDE

Although 6G is likely to go live around 2030, a large number of 6G initiatives underway at the global level and agreeing investments are offering fascinating prospects in the next five years. The requirements likely demanded by 6G include yet unfulfilled 5G use cases and more advanced scenarios emerging for the next-generation 6G network. Examples of such emerging scenarios include terahertz frequencies, haptic communications, ubiquitous services, imaging, and remote sensing. Some examples and their corresponding technology requirements are shown in Figure 6.9.

What for other countries looking at the 6G role? South Korea has delineated a 6G research and development plan with $15 million worth of investment in the first phase in progress till the end of 2025. The Japanese government is planning to create a comprehensive plan to introduce 6G by 2030, set to be 10 times faster than 5G, according to the Internal Affairs and Communications Ministry. A panel including university researchers and private sector officials will be created by January's end to discuss its implementation. 6G networks will have the ability to connect multiple devices simultaneously. The government of Japan established a government-civilian research society in January 2020 with NTT Docomo and Toshiba to summarize inclusive strategies like 6G performance goals and support policies. Japan is endorsing the growth of 6G technology in the governmental budget every year after 2020. The major developments in 6G have also been recognized and are in progress in China to support connectivity, sensing, and AI [27].

As of August 2023, Japanese mobile operator namely SoftBank announced its plan of action for 6G. This consideration appears as an unforeseen head start, as Japan only started its 5G rollout in 2020. As per quarter 1 of 2021, 5G had a penetration rate of 3% among mobile subscribers, which was relatively low compared to most developed countries. The start of 6G will change all previous statistics.

The demand for faster, more reliable, and sustainable communication networks is on the rise in a world that is becoming increasingly interconnected. Large-scale adoption of the 5G technology has already transformed the communications industry and, as a domino effect, the telecom sector. With the initiation of 6G, the future will hold greater promise as the technology can be vibrant towards building a more sustainable setup to connect the world. Services that seemed futuristic, such as eHealth, autonomous vehicles, smart traffic systems, and advanced mobile cloud gaming, are enhancing the day-to-day experiences of societies. The sixth-generation telecom network is supposed to use higher frequencies in providing substantially higher volume and much lower latency with speeds almost a hundred times faster than 5G.

## 2.5 INFRASTRUCTURE AND CHALLENGES

### 2.5.1 Equipment

Japanese telecommunication rivals NTT and KDDI will collaborate in developing technology for ultra-energy-efficient networks that use light to transmit signals from communication lines to servers and semiconductors. By 2030, or later, they intend for their infrastructure to consume one-hundredth of the amount of power that today's communication networks and data centers use and for it to become a global standard. The telcos are eyeing their infrastructure for use in 6G networks; the next-generation standard for mobile communications is expected to be put into practical use around 2030. The implementation of 6G is expected to be 10 times faster than 5G. Still, if the communication infrastructure remains the same as today, power consumption will increase and transmission capacity could prove insufficient. The following are major considerations by the Japanese telecom sector in the development of the 6G network:

- API-driven network customization
- Use of artificial intelligence algorithms
- Complete coverage
- Area expansion
- Terahertz and optical communication
- Sensing and positioning

- Charging stations/power supply
- Maximum frequency utilization and efficiency
- Network security
- Resilience, redundancy, recovery
- Carbon-free and net-zero

### 2.5.2 Spectrum allocation

The term 6G spectrum denotes the new and unused bandwidth being considered for a 6G wireless network. 6G is predictable to discover higher bandwidth, possibly to include sub-terahertz bands.

The frequency bandwidth in the 6G spectrum is projected to cover a wider range than current cellular frequencies. Precisely, the spectrum is supposed to encompass the mid-band frequencies 7–15 GHz, including distinct bandwidths like 7.125 to 8.5 GHz, 10.7 to 13.25 GHz, and 14 to 15.35 GHz. Such mid-bandwidth is selected for its potential to strike a balance between capacity and coverage, which is crucial for robustness and efficiency in a 6G network.

### 2.5.3 Network deployment

In preparation for the launch of future 6G systems, Docomo and NTT said they are already engaged in the verification of various mobile communications. Docomo and NTT have collaborated with Fujitsu, NEC, and Nokia on trials in 2022 but are not yet standardized for 6G network.

The Japanese companies have recently agreed with Ericsson to plan the testing of 6G wireless interfaces for mid-band frequencies, as well as sub-terahertz 100 GHz bands. Also, they agreed with Keysight Technologies to test radio propagation for ultra-wideband communication.

## 2.6 DISCUSSION

As 6G is under development and moving to the prototype development phase, research institutes across the globe have started to pay attention to 6G infrastructure, which is aimed to be positioned somewhere in 2030. Green 6G is anticipated to improve the performance of information transmission up to 1 TBPS and ultra-low latency in µs. It is aimed to provide one thousand times higher capacity than 5G networks. One of the goals of 6G is to achieve ubiquitous connectivity through the integration of satellite communication networks and underwater sea communications to provide global coverage. Energy-efficient mechanisms and the use of nanomaterials may greatly improve the system's energy efficiency and realize sustainable green networks [28].

AI-assisted 6G networks are being planned to be introduced, with new scopes between the years 2027 and 2030. The forthcoming 6G technology is optimistic and exciting, however, there are many obstacles in the way of the development and implementation of 6G networks. The major obstacles include the volume of the system, consistency, security, latency, baud rate, energy efficiency, unstoppable smart connectivity, etc. In this study, various 6G-enabled technologies are discussed from the perspective of Japan like the use of AI, terahertz communications, wireless optical, augmented and virtual reality (AR/VR), and blockchain along with connected challenges. The research showcased the underpinning of 6G including the performance of the 6G network, and key empowering technologies in upgrading the forthcoming communication era.

## 2.7 THE FUTURE OF 6G TECHNOLOGY

Unprecedented speed and bandwidth in 6G will offer unprecedented data speeds and bandwidth, making it possible to transmit and receive vast amounts of information in real-time. This enhanced capability is crucial for applications such as AR/VR, the IoT, blockchain, etc. to open up new possibilities for innovation and efficiency [29].

Low latency for instantaneous communication will be one of the critical features of 6G with ultra-low latency, ensuring almost instantaneous communication between devices. This is particularly important for applications like autonomous vehicles, remote surgery, and critical infrastructure management, where real-time responsiveness is essential for safety and reliability [30].

Energy-efficiency and sustainability to build the lessons learned from previous generations, 6G aims to be more energy-efficient and environmentally friendly. By optimizing the use of resources and minimizing energy consumption, 6G can play a significant role in reducing the carbon footprint associated with global communication networks.

The key technical objectives for 6G development are as follows:

- Use of ultra-high data transmission speed (up to 1 terabyte per second) and use of ultra-low latency.
- Providing energy-efficient resources for connected devices.
- Ensuring coverage of ubiquitous global networks.
- Ensuring to guarantee reliable and smart connectivity across the network.

## 2.8 CONCLUSION

The presented research explored the potential impact of 6G technologies on Japan's society and economy in the 2030s, considering challenges such as an aging population and natural disasters. It outlines a design process using secondary research, workshops, and storytelling to develop scenarios for Japan's future. The methodologies generated four distinct future scenarios illustrating the active use of 6G technologies. The paper aims to stimulate discussion on the implications of novel technologies and the role of design-led research in driving technological innovation.

The discussion on the transition from 5G to 6G in the wireless technology cycle, emphasized how 6G enables automated and intelligent networks to support connected intelligence. By highlighting the variety of applications and use cases, it draws attention to the international efforts underway to establish the groundwork for the communication needs of 2030. It highlighted how AI is contributing to the development and enhancement of 6G designs, standards, protocols, and processes, with an emphasis on enabling ubiquitous AI services from the core of the network to its endpoints.

This chapter provided an extensive impression of the present scenario and growth in wireless communication technologies over the past few decades. It investigated the advancements from 5G to the ongoing development of 6G as well as innovations in ubiquitous connectivity, integrating sensing and artificial intelligence with communication functionality. The chapter also explored the challenges in the implementation of 6G and the global standardization efforts being done globally. Furthermore, the study showcased initiatives at the end of individual and collaborative efforts by Japanese companies. The case study emphasized how important ultra-reliable low-latency communications are for applications looking for ultra-low latency and reliability in a 6G network.

The study has thrown light on the key aspects of 6G by visualizing the development in mobile traffic and service subscriptions by envisioning the potential use cases and applications. The study also discussed key performance parameters in supporting the 6G use cases that are recognized quantitatively.

## REFERENCES

[1] C. D. Alwis *et al.*, "Survey on 6G Frontiers: Trends, Applications, Requirements, Technologies and Future Research," in *IEEE Open Journal of the Communications Society*, vol. 2, pp. 836–886, 2021, doi: 10.1109/OJCOMS.2021.3071496

[2] C.-X. Wang *et al.*, "On the Road to 6G: Visions, Requirements, Key Technologies, and Testbeds," in *IEEE Communications Surveys & Tutorials*, vol. 25, no. 2, pp. 905–974, Secondquarter 2023, doi: 10.1109/COMST.2023.3249835

[3] E. Udin, "6G Technology: Japan Plans to Achieve 6G by 2030," Gizchina.com, January 19, 2020, gizchina.com/2020/01/19/6g-technology-japan-plans-to-achieve-6g-by-2030/

[4] R. Kumar, *Information and Communication Technologies*, University Science Press, 2008, ISBN: 978-81-381-0519-0.

[5] R. Kumar, V. Jain, G. T. W. Han and A. Touzene, *Immersive Virtual and Augmented Reality in Healthcare: An IoT and Blockchain Perspective*, Taylor & Francis, CRC Press, 2023, ISBN: 9781032372617.

[6] R. Kumar, V. Jain, A. A. Elngar and A. Al-Haraizah, *Augmented and Virtual Reality in Social Learning: Technological Impacts and Challenges*, De Gruyter, 2023, ISBN: 978-3-11-099492-6

[7] R. Kumar, A. Anand, M. Fukui and V. Jain, "Integrated Cellular and Adhoc Relay System: A Performance Analysis," in *AIP Conference Proceedings*, vol. 2954, p. 020015, 2023.

[8] B. Sharma, R. Kumar, A. Kumar, M. Chhabra and S. Chaturvedi, "A Systematic Review of IoT Malware Detection using Machine Learning," *2023 10th International Conference on Computing for Sustainable Global Development (INDIACom)*, New Delhi, India, 2023, pp. 91–96.

[9] T. Huang, W. Yang, J. Wu, J. Ma, X. Zhang and D. Zhang, "A Survey on Green 6G Network: Architecture and Technologies," in *IEEE Access*, vol. 7, pp. 175758–175768, 2019, doi: 10.1109/ACCESS.2019.2957648

[10] M. W. Akhtar, S. A. Hassan, R. Ghaffar, H. Jung, S. Garg and M. S. Hossain, "The Shift to 6G Communications: Vision and Requirements," in *Human-Centric Computing and Information Sciences*, vol. 10, no. 1, pp. 1–27, 2020, https://doi.org/10.1186/s13673-020-00258-2

[11] J. R. Bhat and S. A. Alqahtani, "6G Ecosystem: Current Status and Future Perspective," in *IEEE Access*, vol. 9, pp. 43134–43167, 2021, doi: 10.1109/ACCESS.2021.3054833

[12] R. Rana and R. Kumar, "Performance Analysis of AODV in Presence of Malicious Node", in *Acta Electronica Malaysia (AEM)*, vol. 3, no. 1, pp. 1–5, 2019.

[13] R. Kumar, A. Anand, N. Singh, D. Singh and P. J. Singh, "An Efficient Machine Learning Approach for Detecting Malwares in IoT Environment," *2023 3rd International Conference on Technological Advancements in Computational Sciences (ICTACS)*, Tashkent, Uzbekistan, 2023, pp. 475–481, doi: 10.1109/ICTACS59847.2023.10390194

[14] G. Minopoulos, K. Psannis, G. Kokkonis and Y. Ishibashi, "An Artificial Intelligence-Based Tactile Internet System for Enhanced Security over 6G Networks," *2022 5th World Symposium on Communication Engineering (WSCE)*, Nagoya, Japan, 2022, pp. 114–118, doi: 10.1109/WSCE56210.2022.9916036

[15] Y. Lu and X. Zheng, "6G: A Survey on Technologies, Scenarios, Challenges, and the Related Issues," in *Journal of Industrial Information Integration*, vol. 19, p. 100158, 2020. https://doi.org/10.1016/j.jii.2020.100158

[16] N. Chen and M. Okada, "Toward 6G Internet of Things and the Convergence with RoF System," in *IEEE Internet of Things Journal*, vol. 8, no. 11, pp. 8719–8733, June1, 2021, doi: 10.1109/JIOT.2020.3047613

[17] M. A. Uusitalo *et al.*, "6G Vision, Value, Use Cases and Technologies from European 6G Flagship Project Hexa-X," in *IEEE Access*, vol. 9, pp. 160004–160020, 2021, doi: 10.1109/ACCESS.2021.3130030
[18] M. Banafaa, I. Shayea, J. Din, M. H. Azmi, A. Alashbi, Y. I. Daradkeh and A. Alhammadi, "6G Mobile Communication Technology: Requirements, Targets, Applications, Challenges, Advantages, and Opportunities," in *Alexandria Engineering Journal*, vol. 64, pp. 245–274, 2023, https://doi.org/10.1016/j.aej.2022.08.017
[19] R. Singh, A. Kaushik, W. Shin, D. R. Marco, V. Sciancalepore, D. Lee, H. Sasaki, A. Shojaeifard and O. A. Dobre, "Towards 6G Evolution: Three Enhancements, Three Innovations, and Three Major Challenges," in arXiv.org, February 16, 2024, https://arxiv.org/abs/2402.10781
[20] S. Sarmiento *et al.*, "High Capacity Converged Passive Optical Network and RoF-Based 5G+ Fronthaul Using 4-PAM and NOMA-CAP Signals," in *Journal of Lightwave Technology*, vol. 39, no. 2, pp. 372–380, January 15, 2021, doi: 10.1109/JLT.2020.3028492
[21] P. Porambage, G. Gür, D. P. M. Osorio, M. Liyanage, A. Gurtov and M. Ylianttila, "The Roadmap to 6G Security and Privacy," in *IEEE Open Journal of the Communications Society*, vol. 2, pp. 1094–1122, 2021, doi: 10.1109/OJCOMS.2021.3078081
[22] B. Mao, F. Tang, Y. Kawamoto and N. Kato, "AI Models for Green Communications Towards 6G," in *IEEE Communications Surveys & Tutorials*, vol. 24, no. 1, pp. 210–247, Firstquarter 2022, doi: 10.1109/COMST.2021.3130901
[23] N. Kim, G. Kim, S. Shim, S. Jang, J. Song and B. Lee, "Key technologies for 6G-Enabled Smart Sustainable City," in *Electronics*, vol. 13, no. 2, p. 268, 2024, https://doi.org/10.3390/electronics13020268
[24] S. S. Dhanda, B. Singh, P. Jindal, T. K. Sharma and D. Panwar, "6G-Enabled Internet of Medical Things," in *Expert Systems*, vol. 41, no. 1, pp. 1–24, 2023, https://doi.org/10.1111/exsy.13472
[25] R. Kumar, V. Jain, L. W. Yie and S. Teyarachakul, *Convergence of IoT, Blockchain and Computational Intelligence in Smart Cities*, Taylor & Francis, CRC Press, 2023, ISBN: 9781032404240.
[26] W. Y. Leong and R. Kumar, "5G Intelligent Transportation Systems for Smart Cities," book chapter in Rajendra Kumar, Vishal Jain, Leong Wai Yie and Sunantha Teyarachakul, editors, *Convergence of IoT, Blockchain, and Computational Intelligence in Smart Cities*, CRC Press, ISBN 9781032404240.
[27] S. Kamath, S. Anand, S. Buchke and K. Agnihotri, "A Review of Recent Developments in 6G Communications Systems," in *Engineering Proceedings*, vol. 59, p. 167, 2023, https://doi.org/10.3390/engproc2023059167
[28] R. Jain, S. Chaudhary and R. Kumar, "Green Approach for Next Generation Computing: A Survey", in *International Journal of Advanced Engineering Research and Science (IJAERS)*, vol. 4, no. 1, pp. 78–82, January 2017.

[29] A. Anand, R. Kumar, P. Pachauri, V. Jain and K. T. Ng, "6G and IoT-Supported Augmented and Virtual Reality–Enabled Simulation, Environment," book chapter in Kumar, Rajendra, Jain, Vishal, A. Elngar, Ahmed and Al-Haraizah, Ahed., editors, *Augmented and Virtual Reality in Social Learning: Technological Impacts and Challenges*, De Gruyter, 2023, pp. 203–2017, https://doi.org/10.1515/9783110981445-012

[30] S. Foster, "Japan's NTT Vision for the 6G Future," *Asia Times*, 2023, May 13, https://asiatimes.com/2023/05/japans-ntt-has-a-vision-for-the-6g-future/

Chapter 3

# Embedded communication system design, implementation of the MB-OFDM transceiver and its optimization for short-range IoT in 5G/6G applications

*Sanket N Dessai, Hemant Patidar, Satyanarayan Dubey, Prashant Upadhyay, Prasanna Kumar Singh, and Muralidharan R*

## 3.1 INTRODUCTION

Short-range communications will be the hotspot of all technological gadgets in the future. Such short-range communication necessitates high data speeds for many digital data streams, cheap costs for widespread consumer adoption, and low power consumption. Ultra-wide band (UWB) multiband-OFDM (MB-OFDM) offers a potential approach for meeting these requirements, making it an appealing choice for future wireless networks. The MB-OFDM system may maintain the same transmit power as though it were using the whole bandwidth by interleaving symbols across distinct sub-bands [1–3]. Because the bandwidth offered is 528 MHz, the data rate rises according to Shannon channel capacity as the bandwidth increases from 40 MHz to 500 MHz. The band spectrum is unlicensed; unlicensed UWB users may operate in the same bands as licensed and other unlicensed activities, providing multiband operates at such low spectral density levels that the transmitted power density is below the noise floor.

### 3.1.1 Multiband communication system

Figure 3.1 depicts the block diagram for the MB-OFDM communication system. The data bits are encoded with the forward error-coding technique, and the coded data is modulated to form the complex conjugated symbols, which are further OFDM modulated, i.e., the pilot sequences and guards are added, and the inverse fast Fourier transformer (IFFT) is implemented to generate the OFDM frame, which is interleaved over a specified band using a time-frequency kernel. At the receiver, the received data from different bands is OFDM de-modulated to remove data such as pilots and guards and demodulated and decoded using a specific algorithm in order to reconstruct the data bits.

 DOI: 10.1201/9781003583127-3

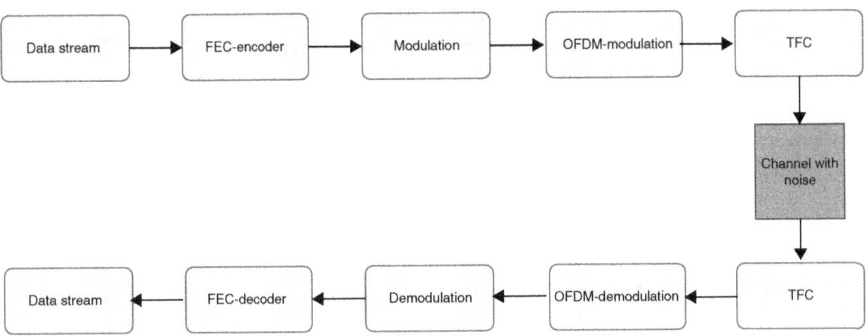


*Figure 3.1* MB-OFDM communication block diagram.

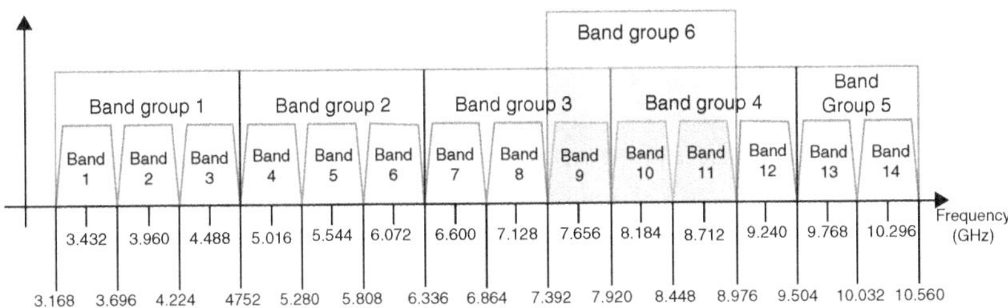


*Figure 3.2* Band planning for MB-OFDM.

### 3.1.2 Multiband technique

The multiband OFDM system is a short-range communication OFDM solution [2]. The accessible UWB spectrum ranges from 3.1 to 10.6 GHz and is broken into many sub-bands with lesser bandwidth. According to Federal Communications Commission (FCC) standards, there are four sets of three-band systems to enable four separate user accesses.

Figure 3.2 shows the spectrum band planning regulated by the FCC, which consists of five band groups and has three sub-bands within them except the fifth band group, which has two sub-bands. Figure 3.2 also shows the center frequencies of each sub-band where the data are interleaved.

### 3.1.3 Need for MB-OFDM

MB-OFDM is necessary for the deployment of high-speed data transmission in the ultra-wide band spectrum by efficiently partitioning the available spectrum into numerous bands and controlling the trade-offs. MB-OFDM offers great data throughput for short-range communications.

### 3.1.4 Motivation

The main motivation and idea behind implementing MB-OFDM is the data speed they provide in short-range communication with unlicensed bandwidth by effectively utilizing the bandwidth. The idea of utilizing an unlicensed spectrum and obtaining high data rates motivated and provided keen interest in designing, simulating, and implementing MB-OFDM transmitter and receiver blocks and optimizing the embedded platform.

## 3.2 LITERATURE REVIEW

The transmission of the MB-OFDM is performed with the transmitting antennas, while the simulations are performed with the additive white Gaussian noise (AWGN) noise. The receiver absorbs the signal and down-converts it into digital form using an analog-to-digital converter. The carrier frequency synchronization and symbol timing synchronization are performed by channel estimation and packed synchronization. The forward error correction (FEC) encoder is used to de-map the data and de-interleave it to recover the data at a higher rate [4–6].

A computing architecture comparison was performed between the Cortex A8 and Cortex A-53, and it was observed that the ARM Cortex A-53 has a quad-core and advanced pipeline [7, 8–18]. Code optimization helps improve the code execution time [19]. The field programmable gate array (FPGA) development of the MB-OFDM achieved a data rate of 640 Mbps [20], and hence the more suitable architectures for FPGA-based processing are Zynq and Arria, with the ARM Cortex A-53 architecture supporting the signal processing and communication blocks [21, 22]. The choice of the FPGA-based Cortex architecture depends on the applications of the Internet of Things (IoT) and its infrastructure for machine learning, along with the wireless communication based on it [21, 22].

The application of MB-OFDM is most suitable in the area of the wireless body area network (WBAN) with IoT and automotive short-range communications with 5G. In the body of the network, it is required to have a higher data rate for WBAN (499 MHz) and provide an open area for research. 5G for short-range communication will impact the lives of humans [23–27].

## 3.3 DEVELOPMENT OF A SIMULATION MODEL FOR A MULTIBAND TRANSCEIVER SYSTEM

This chapter provides details about the development of individual blocks of the developed MB-OFDM. Starting with the design of basic modulation blocks and then advancing with the design of the digital synthesizer, physical layer frame, and implementation/simulation details are covered extensively in this chapter. A block diagram of the MB-OFDM system is arrived at after extensive background study and literature survey. This chapter discusses the overall design and simulation aspects of the multiband UWB physical layer.

### 3.3.1 MB-OFDM system requirements

Based on the literature review, the following are the requirements specifications of the system:

(1) The multiband system needs a convolution encoder, modulator, interleaver, Viterbi decoder, FFT, and IFFT.
(2) Requires a frequency synthesizer for hopping frames at different frequencies.
(3) It requires a base estimation sequence for all time frequency coding (TFC) and a cover sequence.
(4) It requires MATLAB, Code Composer Studio, Real View Development Studio, Xilinx Vivado, and Vitis for the tool for the Zynq platform for implementation.

In this study, the channel is considered static, and it is precisely known at the receiver for all systems under development.

### 3.3.2 Design and development of the MB-OFDM system

To make a comparative study of multiband systems with conventional systems, it is essential to build the entire system from the base system. Starting with the basic communication system with encoding techniques, modulation, and frequency diversity, then advancing to build multiband systems implementing different center frequencies. The block diagram of the design is shown in Figure 3.3, which implements a scrambler, a convolution encoder, and a puncture in order to vary the coding rates. Different modulation and demodulation techniques, such as Quadrature phase-shift keying (QPSK), dual carrier modulation (DCM), 16-quadrature amplitude modulation (QAM), and 64-QAM, are implemented to achieve a higher data rate, along with designing the preamble and preamble header. The Viterbi decoder is designed for convolution decoding.

#### *3.3.2.1 Modulation*

In this project, system design and its data rates depend on modulation technique because utilizing the band effectively is done by the different modulation techniques. QPSK, 16-QAM, DCM, and 64-QAM are the digital modulation techniques being used in this project to design a system and

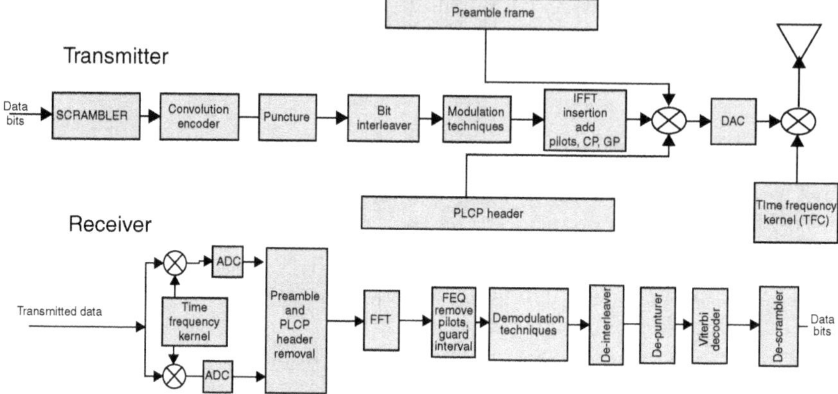


*Figure 3.3* Design of MB-OFDM block diagram.

compare these techniques on MB-OFDM for higher data rates. QPSK is better than binary phase shifting key (BPSK) in terms of data rate and bandwidth and better than QAM in terms of bit error rates, but the data rate is lower compared to 16- and 64-QAM and DCM. An input image, which is in the form of a matrix of varying intensity values (pixels), is converted into a bit stream of 0's and 1's. This data bit stream is then modulated using a different modulation scheme and passed through the AWGN channel. The data received is then demodulated to the original bit stream of 0's and 1's for image reconstruction.

#### *3.3.2.2 Scrambler*

Figure 3.4 shows the flow diagram of the scrambler. The seeder, which is 15-bit, is initialized into each delay block according to the value of the seed identifiers, as shown in Table 3.1. The input bits from the image 0's and 1's data are given as input to the scrambler. Initially, the seed values are initialized in all 15 registers. Register 15 and register 14 are XOR, and the result is stored as a temporary variable. The data is shifted to register 1, and the values in register 1 are shifted to register 2, and so on. The output is obtained by XOR input data with a temp or reg 1 value, and the process repeats in a loop to perform scrambling on all input data bits.

At the receiver side, the same operation is performed since descrambled data can be obtained again by performing the same scrambling operation, and hence the same is performed at the receiver side. The seed values used at the transmitter should be used at the receiver. The receiver receives the seed values of the transmitter through a physical header.

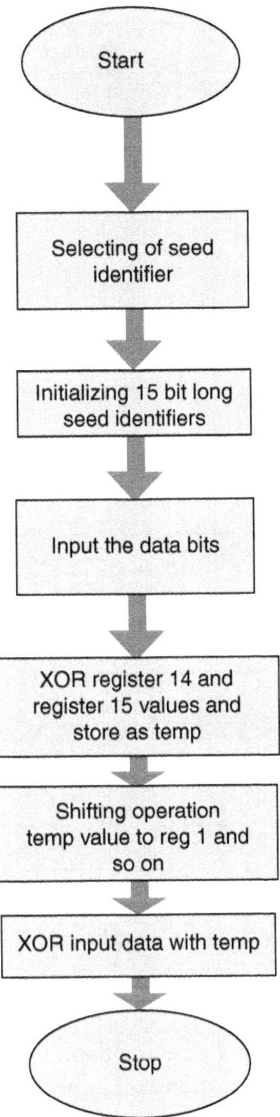


*Figure 3.4* Flow chart of scrambler implementation.

#### *3.3.2.3 Convolution encoder*

The scrambled data is sent into the second block of the system convolutional encoder. This block adds redundancy patterns to the data to increase the SNR ratio and enable more accurate decoding at the receiver. The mother code, which corresponds to a coding rate of 1/3, was designed using the

*Table 3.1* The values of seed identifiers

| *Seed identifier (b1, b0)* | *Seed values* |
|---|---|
| 0, 0 | 0011 111111111 111 |
| 0, 1 | 0111 111111111 111 |
| 1, 0 | 1011 111111111 111 |
| 1, 1 | 1111 111111111 111 |

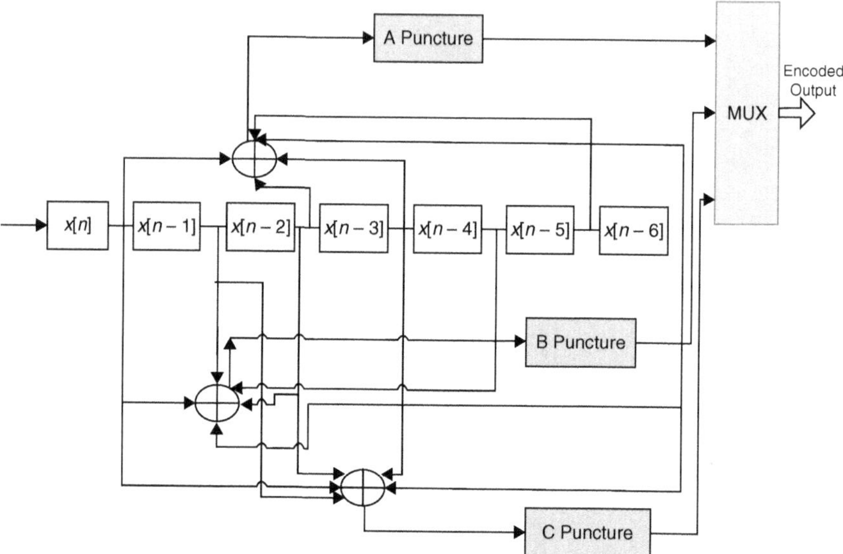


*Figure 3.5* Implementation of convolution encoder.

generator polynomials $g_0$ = [133], $g_1$ = [165], and $g_2$ = [171] (in octet form), and the constraint length was $K$ = 7. The convolutional code is built on the code polynomials 133, 165, and 171, or the binary patterns 1011011, 1110101, and 1111001.

As demonstrated in Figure 3.5, binary patterns define the weights in a feed-forward shift register. At the start of encoding, the shift register's initial state is always set to zero. Figure 3.6 shows the flow chart of the convolution encoder. First, the generator matrix binary values are taken, and outputs are drawn as shown in Figure 3.3. The register elements are 7 because the constraint length is 7. All the register elements are initialized to zero, and the binary bits from the scrambler are fed into the encoder. Three

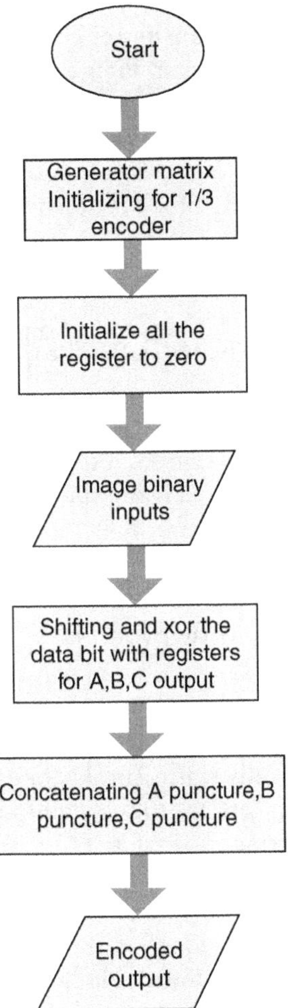


*Figure 3.6* Flow chart of convolution encoder.

outputs are taken for a single input by shifting the data and/or the values in register elements, and finally, three outputs are concatenated.

The mother code yields a coding rate of 11/32. The encoder accepts eleven bits from x0 to x10 and outputs 33 bits from A0 to C10. By removing bit C10, we achieve a coding rate of 11/32. (i.e., 11 input bits generate 32 output bits).

At the receiver, a dummy bit is inserted in position C10, and the sequence is recovered using the Viterbi decoder, yielding 11 decoded bits. Similarly, puncturing patterns for coding rates 1/2, 5/8, and 3/4 can be performed.

#### 3.3.2.4 Bit interleaver

The next block in the system transmitter is the bit interleaver, which is used to avoid bursts of errors that occur. Figure 3.7 shows the flow chart for the bit interleaver, in which the input bits from the convolution encoder are given to the bit interleaver. All the variables are initialized based on data rate, based on the data rate $N_{cbps}$, $N_{cbp6s}$, $N_{cyc}$ are initialized with different values for different data rates. Block interleaving is performed by applying the following Equation (3.1).

$$S(i) = U\left\{\text{Floor}\left(\frac{i}{N_{\text{cbps}}}\right) + 6\,\text{Mod}(i, \text{NCBPS})\right\} \tag{3.1}$$

where $S(i)$ is the output of the symbol interleaver, which is input to the tone interleaver, which is implemented using the tone interleaver Equation (3.1), which further randomizes the data.

$$T(i) = S\left\{\text{Floor}\left(\frac{i}{N_{\text{tint}}}\right) + 10\,\text{Mod}(i, N_{\text{tint}})\right\} \tag{3.2}$$

$T(i)$ of Equation (3.2) is the output from the tone interleaver and the tone interleaved data bits cyclically shifted by 33 for each frame and by 66 for the next frame. Now the data bits are totally randomized, and there is a reduction in burst errors occurring.

#### 3.3.2.5 Modulation

The constellation mapper is the fourth block in the transmitter system architecture, and it modulates OFDM subcarriers using various ways. Different modulation techniques such as QPSK, DCM, 16-QAM, and 64-QAM are implemented.

##### QPSK

Gray-coded constellation mapping converts an input binary sequence into a complex-valued sequence. In-phase and quadrature values, abbreviated as $I$ and $Q$, are calculated from a pair of input bits. Figure 3.8 shows the flow chart for the implementation of QPSK. Equation (3.3) yields complex-valued sequences $d$ for the $I$ and $Q$ values.

$$d = (\text{I} + j\,\text{Q})\,\text{KMOD} \tag{3.3}$$

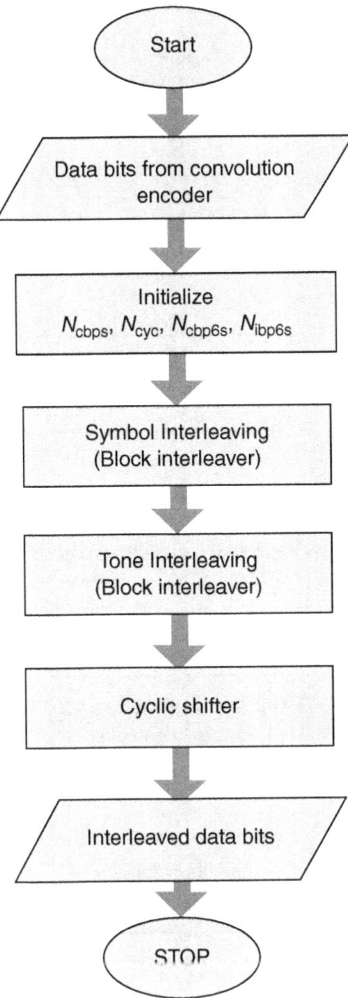


*Figure 3.7* Flow chart for the interleaver block.

where $j = \sqrt{1}$ is an imaginary number and KMOD = $1/\sqrt{2}$ normalizes $d$ to unit energy. The flow charts for implementing QPSK modulation are shown in Figure 3.8. The input bits are divided into odd and even bits, and then the odd bits and even bits are mapped in the constellation mapper with the below formula, Equation (3.4).

$$d = \frac{1}{\sqrt{2}}(I + j \cdot Q) \tag{3.4}$$

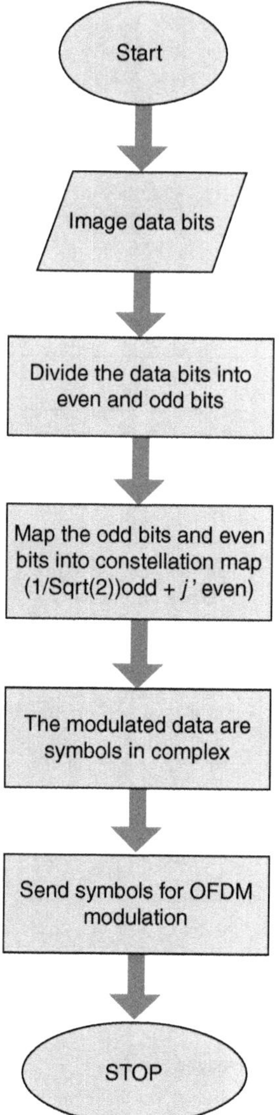


*Figure 3.8* Flow chart for implementation of QPSK.

### *QAM-16*

Figure 3.9 shows the flow chart for implementing 16-QAM modulation, which is a repetitive 8-PSK modulation technique. The bits are divided into in-phase and quadrature bits, and they are mapped in a 16-point

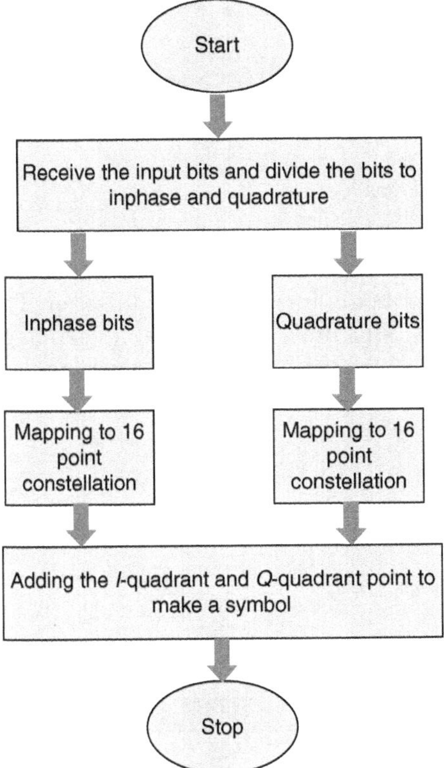


*Figure 3.9* Flow chart for 16-QAM and 64-QAM.

constellation diagram. The QAM16 has a KMOD value of 10. In 16-QAM, four bits are mapped as one symbol, and hence the data rate will be high but the probability of error will be high. The equation for 16-QAM is given by Equation (3.5).

$$d = \frac{1}{\sqrt{10}}(I + j \cdot Q) \tag{3.5}$$

*64-QAM modulation*

Figure 3.9 shows the flow chart for the implementation of 64-QAM in which the number of bits mapped is 6 bits to form one symbol, however, the same procedure as 16-QAM is followed by dividing the bits into odd and even bits and then mapped to 64-point constellation diagram. 64-QAM modulation with the KMOD value is given in Equation (3.6).

$$d = \frac{1}{\sqrt{42}}(I + j \cdot Q) \tag{3.6}$$

#### 3.3.2.6 OFDM modulation

The flow chart for OFDM modulation is illustrated in Figure 3.10. The modulation approach generates a complex-valued sequence, indicating that the symbols are ready for OFDM modulation. The sequence added with pilots, guards, and nulls is taken, and carrier generation for the payload is done. Each frame is done by inserting pilots' data tones and guards and finally, the frame of OFDM is generated, which is of 128

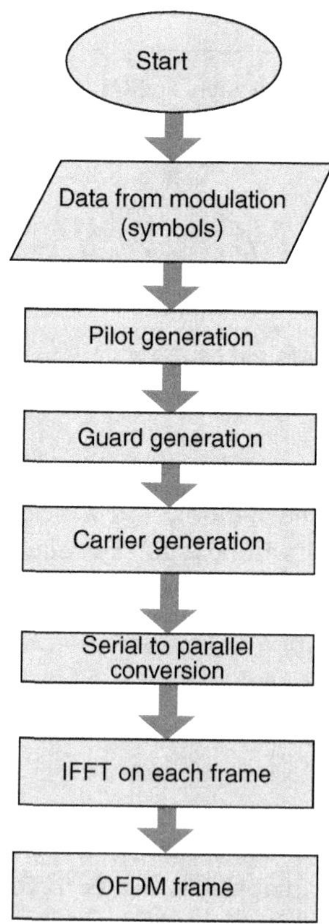


*Figure 3.10* Flow chart for OFDM.

*Table 3.2* MB-OFDM design parameters with their value

| *Parameters* | *Values* |
|---|---|
| Count of OFDM subcarriers | 128 |
| Count of data subcarriers | 100 |
| Count of specified pilot subcarriers | 12 |
| The number of guard subcarriers | 10 |
| Frequency spacing (f) of the subcarriers. | 4.125 MHz (=528 MHz/128) |
| $T_{FFT}$: IFFT/FFT interval | 242.42 ns (1/f) |
| $T_{cp}$: Duration of a cyclic prefix | 60.61 ns (=32/528 MHz) |
| $T_{GI}$: Duration of guard intervals | 9.47 ns (=5/528 MHz) |
| $T_{sys}$: Duration of the symbol | **312.5 ns** $T_{FFT+}T_{cp+}T_{GI}$ $+T_{sys}$ |

long symbols. Pilots are added at $n$ = −5, −15, −25, −35, −45, and −55. The multiband OFDM frame sequence in series is now converted to parallel, and IFFT is implemented and the OFDM frame is obtained and sent to a digital-to-analog converter. However, the reverse of the operation is applied to demodulating it. Table 3.2 shows the design parameters.

#### *3.3.2.7 PLCP header and preamble*

The flow diagram of PLCP header is shown in Figure 3.11 its specifics before it is convolutionally encoded. The first frame is a preamble and is designed to simplify the receiver's task for packet detection, timing estimation, phase estimation, and amplitude estimation. Figure 3.12 depicts the implementation of the PLCP preamble, which consists of two segments: the synchronization sequence and the channel estimation sequence.

#### *3.3.2.8 Multiband operating range*

Figure 3.13 shows the multiband hopping technique in which the symbols that are obtained, the mapping preamble, the PLCP header, and the payload, i.e., a frame of 165 samples long, are taken, and each frame is hopped at a different frequency with the provided frequency and band relation. In this system design, only the single band group is utilized, hence three frequencies are obtained from the center frequency equation from the variable $n$. The $n$ value is changed to 1, 2, and 3 to obtain the center frequency value. After the center frequency value is obtained, sine and cos waves are generated by $\exp(j_2 \times \pi \times f)$, the real part of the DAC. Figure 3.13 also shows the flow chart of multiband hopping where the frames are taken, i.e., a frame of 165 samples long is taken, and each frame is hopped at a different frequency with the provided frequency and band relation.

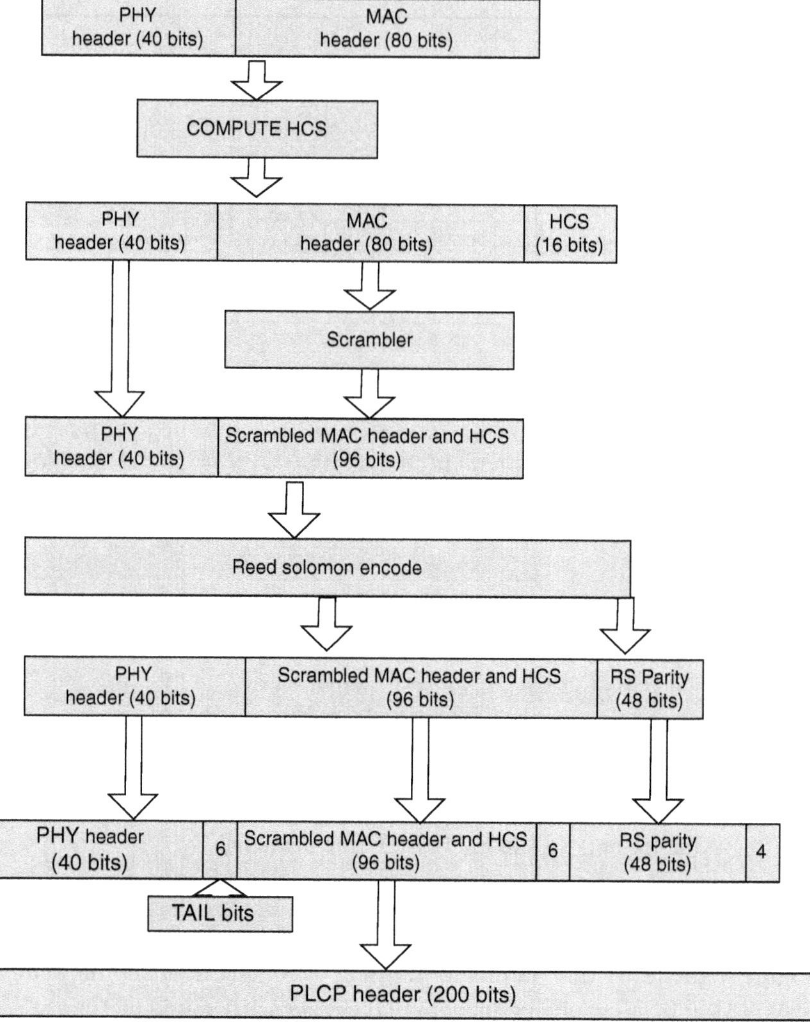


*Figure 3.11* Flow diagram for the PLCP header.

At the receiver side, the data is received from different frequencies hopping and processing are carried out.

### 3.3.3 MB-OFDM demodulation

A flow diagram for OFDM de-modulation is shown in Figure 3.14. The transmitted output is passed through AWGN noise, and the noised data is

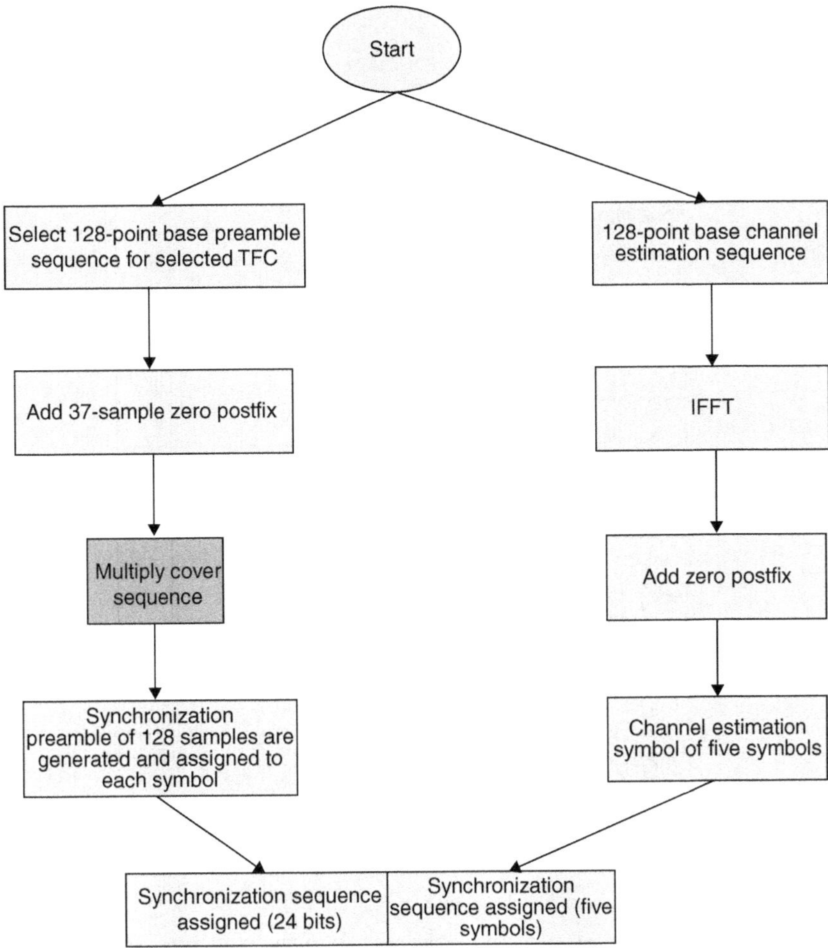


*Figure 3.12* Flow diagram for the PLCP preamble.

FFT operated back. They are converted from parallel to serial, and reshaping is done. The pilot tones added to OFDM frames are removed, and the guard tones are also removed and left with payload data. This complex conjugated payload data is further processed to obtain the data bits.

### 3.3.4 De-interleaver

The de-interleaver operation is implemented by the inverse process of the interleaver. The flow diagram of the de-interleaver is shown in Figure 3.15, in which the demodulated bits are obtained and all the parameters are

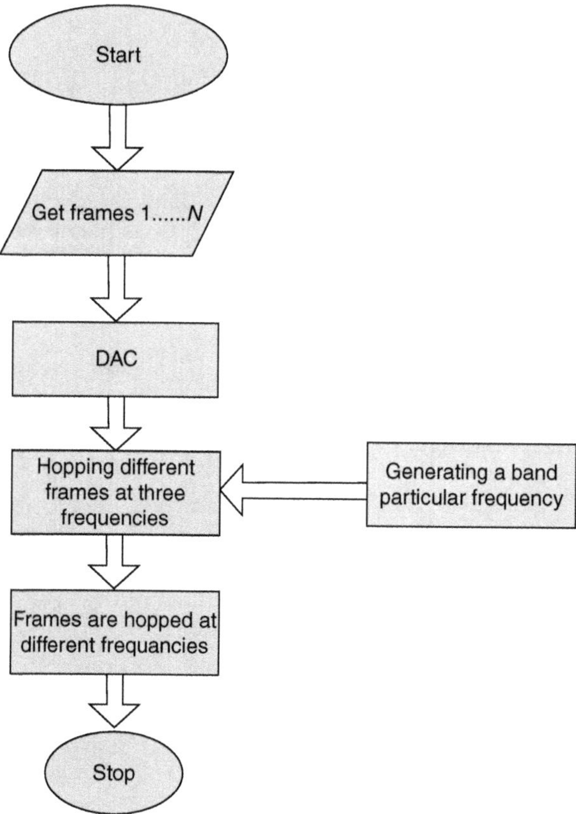


*Figure 3.13* Flow diagram for the digital synthesizer.

initialized. The parameters initialized depend on the data rate, but the same values are initialized and they are re-shifted by the same factor. The shifted data is sent to the de-tone interleaver, which performs the inverse operation of the tone interleaver. In the same way, the de-symbol interleaving is done, which is inverse to the symbol interleaver, and after the symbol de-interleaving, the de-interleaved bits are obtained back.

### 3.3.5 Demodulation techniques

In the receiver-side system design, the demodulation of symbols is done and they are converted into bits by mapping to an appropriate point based on the channel noise. Different de-modulation techniques are also used at the receiver and their de-modulation processes are explained as follows:

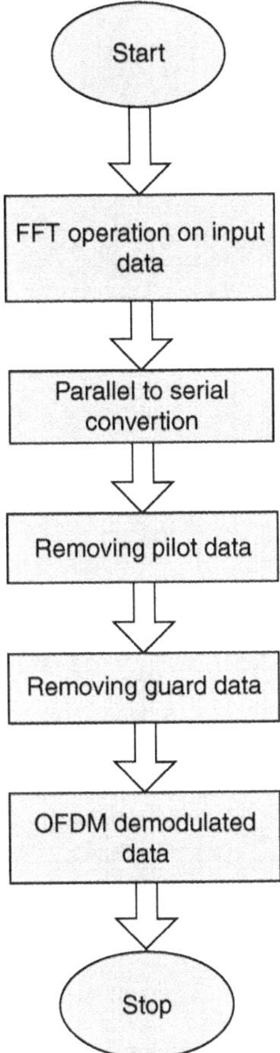


*Figure 3.14* Flow diagram of MB-OFDM demodulation.

#### *3.3.5.1 QPSK demodulation*

The flow chart for QPSK demodulation is as shown in Figure 3.16(b). The modulated bits are obtained, four-point gray-coded constellations are initialized, the distance of each symbol of the noised symbols that are received is calculated, and the minimum distance near the constellation points is mapped to obtain the binary data bits.

#### *3.3.5.2 16-QAM demodulation*

The flow chart for 16-QAM demodulation is shown in Figure 3.16(c). The modulated bits are obtained, 16-point gray-coded constellations are

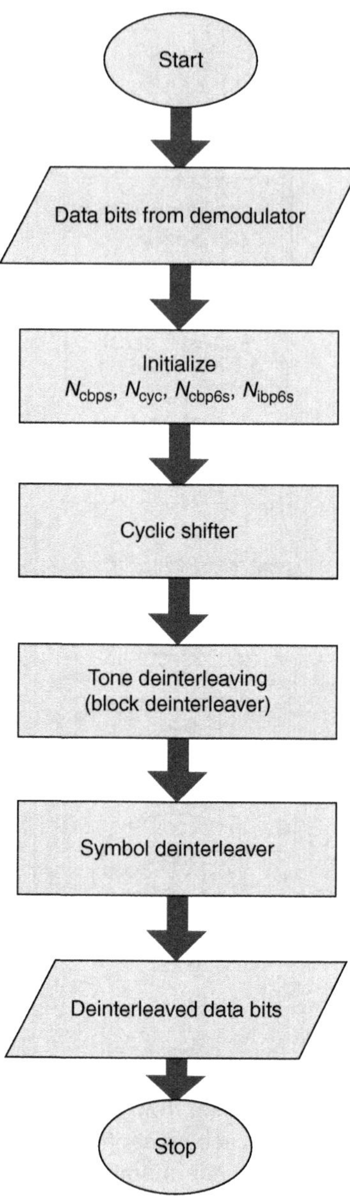


*Figure 3.15* Flow diagram of de-interleaver.

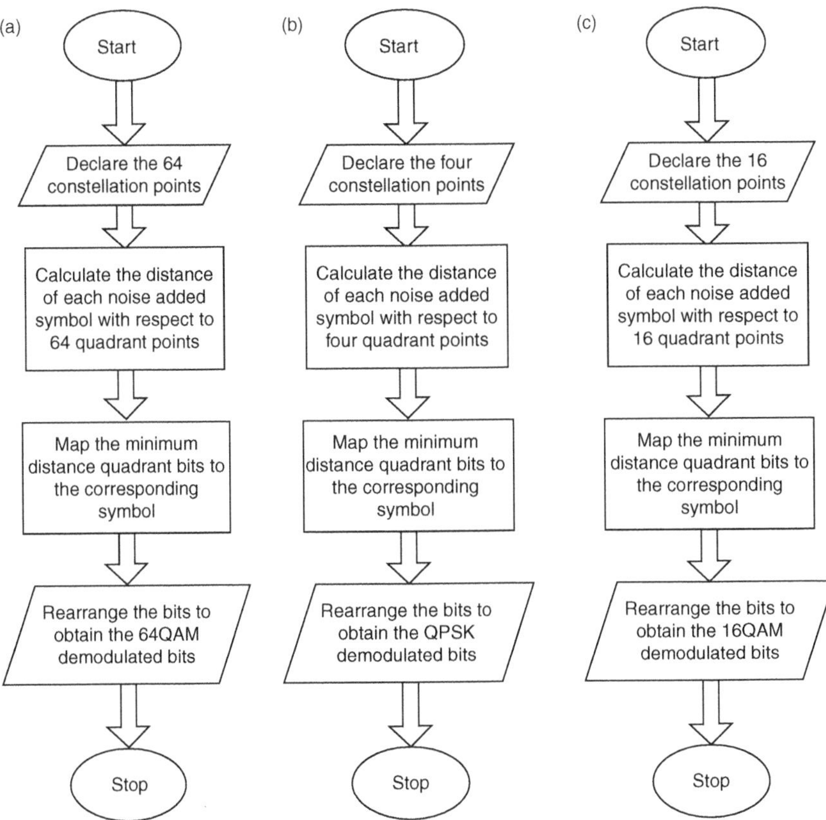


*Figure 3.16* QPSK, 16-QAM, 64-QAM demodulation flow diagram.

initialized, the distance of each symbol of the noised symbols that are received is calculated, and the minimum distance near the constellation points is mapped to obtain the binary data bits. However, if the minimum distance of the original symbols is lost, then it is mapped to a different point, and hence it is more likely to have errors compared to QPSK.

### *3.3.5.3 64-QAM demodulation*

The flow chart for 64-QAM demodulation is shown in Figure 3.16(a). The modulated bits are obtained, 64-point gray-coded constellations are initialized, the distance of each symbol of the noised symbols that are received is calculated, and the minimum distance near the constellation points is mapped to obtain the binary data bits. However, if the minimum distance of the original symbols is lost, then it is mapped to a different point, and hence it is more likely to have errors compared to QPSK and 16-QAM.

### 3.3.6 Viterbi decoder

A decoder must utilize some form of decoding rule to locate a way across the trellis to decode a convolutional code, where each path in the trellis defines a unique decoder output sequence. Figure 3.17 depicts the flow chart for the Viterbi decoder. The initialization phase of the Viterbi decoding algorithm determines the types of variables utilized by the algorithm and their starting values.

## 3.4 IMPLEMENTATION OF MULTIBAND OFDM IN C

This section documents the C implementation of the MATLAB model of the MB-OFDM.

### 3.4.1 Encoder implementation

The MATLAB code implemented for the convolution encoder of MATLAB is taken and C is programmed in the same method. In the C code function Conv_encoder, all seven registers are declared and initialized to zero. A temporary register is declared to store the XOR-ed data. The data bits are inserted, and each bit of data is stored in the first register. All the register values are shifted as the data flows in and XOR operation is performed for the register values as implemented in MATLAB.

### 3.4.2 Modulation and de-modulation C implementation

The modulation and de-modulation are implemented in C as in MATLAB, where the inputs are taken and complex conjugated to form the symbol. The code is implemented using the structure concept, with the structure having real and imaginary structures assigned, and with if condition checks for the combination value of even and odd bits and assigns the particular real and complex values to the output.

The implementation of QPSK modulation is done by dividing the bits into odd and even, and then constellation mapping is done, and the two bits are demodulated by replacing –0.707 by 0 and 0.707 by 1. The implementation of 16-QAM is done by dividing the bits into odd and even, and then constellation mapping is done, and the four bits are demodulated by replacing mapped constellation points in 1's and 0's.

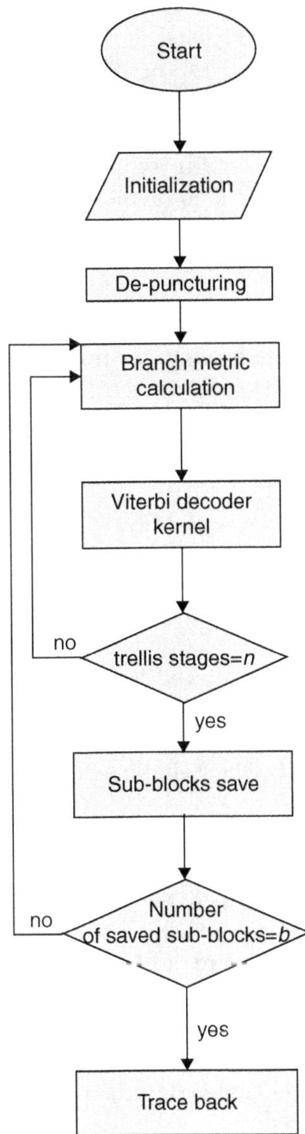


*Figure 3.17* Viterbi decoder flow diagram.

### 3.4.3 Scrambler and descrambler implementation in C

The scrambler is implemented with the function scrambler. Initially, all the seed values are initialized to the 15-register, considered a pseudo-random sequence. The data bit is stored in register 1, and the value of reg 1 is shifted, and the process repeats till 15 registers, and the 14 and 15 registers are XOR-ed and stored in the temp variable, which is feedback to the input. The descrambler is implemented by again calling the scrambler function.

### 3.4.4 Interleaver and de-interleaver implementation in C

The interleaving and de-interleaving are implemented in C by calling the functions Inter_leaver and De_interleaver. Interleaver is implemented in C by taking the data bits and applying permutation to the received data bits so that the data bits are randomized, and the inverse is done for de-interleaver, where the input bits are de-permuted.

### 3.4.5 IFFT and FFT implementation in C

The implementation of FFT and IFFT is done by using structures so that the imaginary and real can be accessed. The implementation of C code is done by calling the function FFT, and for IFFT, the data is conjugated and applied to the FFT function itself. First, the size of the data is taken and compared with 2, and the data bits are divided into odd and even, and the even number is multiplied by the twiddle factor.

### 3.4.6 Guard addition and removal implementation in C

The guard bits are to be added to the OFDM symbol. i.e., the zeros are inserted, and at the receiver, the guard is removed. The implementation in C is done by using the function guard_add, and removal is implemented using the function guard_remove.

In this chapter, test cases are implemented for the multiband OFDM transceiver system, which tests the functionality of the system. These are mainly to validate the developed system blocks and their appropriate code for their complete functionality.

## 3.5 TEST PLAN AND VERIFICATIONS

### 3.5.1 Test plan

Performance analysis of the developed multiband OFDM system is based on different parameters. BER vs. SNR curves highlight the system performance graphically. Input data: this type of input is among the critical data any system requires to process. The developed multiband OFDM system has to process any type of input data as its application is high-end data transmission. In this project, the developed system is tested with all kinds of images—binary, grayscale, and color images.

Figure 3.18 shows (a) a binary image, (b) a grayscale image, and (c) a color image being tested. The processing time of any system depends on the size of the input data. The larger the size, the longer the time to obtain the result. Processing a binary image of size 32 × 32 (total of 1 × 1024 binary bits) or 64 × 64 (1 × 4096) or 128 × 128 (1 × 16384) or 256 × 256 (1 × 65536) is faster when compared to grayscale and even faster when compared color images. A grayscale image of size 32 × 32 has a total of 32 × 32 × 8 = 8192 bits to be processed. The color image of size 32 × 32 has a total of 32 × 32 × 8 × 3 = 24576 bits to be processed.

**MSE:** Mean squared error (MSE) is a measure used to assess the quality of a reconstructed picture. Equation (3.7) is used to compute the MSE, which is the cumulative squared error between the original image and the rebuilt image. A lower MSE number indicates a smaller error.

$$\text{MSE} = \frac{\sum_{M,N}\left[I_1(m,n) - I_2(m,n)\right]^2}{M \times N} \tag{3.7}$$

where $M$ and $N$ are the input image's rows and columns, respectively. However, MSE is a dimensionless metric.

**PSNR**: Another error statistic used to compare the quality of reconstructed images is the peak SNR ratio. The reconstructed image quality improves as the PSNR increases. The PSNR is calculated using Equation (3.8):

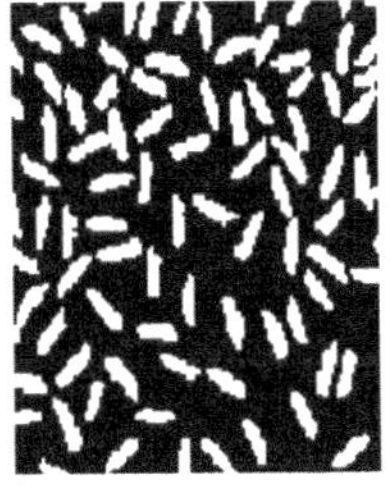

(a) Binary image

(b) Grayscale image

(c) Color image

*Figure 3.18* Images for test cases.

$$\text{PSNR} = 10\log_{10}\left(\frac{R^2}{\text{MSE}}\right)\text{Decibels} \tag{3.8}$$

where $R$ is the greatest fluctuation in the supplied picture data type. For example, if the input picture contains a double-precision floating-point data type, $R$ equals 1. If the data type is an 8-bit unsigned integer, the value is 255.

### 3.5.2 Test cases and validation results

System design starts with incorporating the functionalities of a digital modulator/demodulator. The data rates vary based on the modulation techniques, and hence 64-QAM will have the highest data rate if implemented with a high bit error rate, hence normally higher performance (lower BER) is achieved using lower transfer rate techniques like QPSK and BPSK compared to 16-QAM or 64-QAM in the same SNR environment. However, by using 16-QAM, moderate performance is achieved, and the data rate will also be improved. Validation of the MB-OFDM with various modulation techniques will be performed in the next sessions. Table 3.3 shows the MSE and PSNR values for QPSK and QAM.

Theoretically and practically, the QPSK modulation technique yields better results. Hence, the QPSK modulation scheme is chosen for the validation of the developed system blocks of MB-OFDM. Table 3.4 shows the MSE and PSNR values obtained for a binary image, grayscale image, and color image subjected to QPSK modulation for SNR varying in the range from –5 dB to 10 dB. It is observed that as the MSE values tend to 0, the corresponding PSNR tends to infinity. The lower the MSE or higher the PSNR, the better the reconstructed image.

To visualize the amount of noise in the reconstructed images, Table 3.5 shows the results obtained for binary, grayscale, and color images subjected to 16-QAM modulation at varying SNRs.

*Table 3.3* MSE and PSNR values for QPSK and 16-QAM

| *SNR (in dB)* | *MSE_QPSK* | *PSNR_QPSK (in dB)* | *MSE_16QAM* | *PSNR_16QAM (in dB)* |
|---|---|---|---|---|
| -5 | 60.7073 | 30.0213 | 82.3781 | 28.9727 |
| -3 | 40.2540 | 31.4531 | 72.5802 | 29.5226 |
| 0 | 22.0957 | 33.6506 | 57.3872 | 30.5427 |
| 2 | 12.8321 | 36.7219 | 41.8437 | 31.9145 |
| 5 | 2.9098 | 41.2202 | 25.2718 | 34.1044 |
| 10 | 0.00002 | 95.1205 | 0.3691 | 52.4594 |

*Table 3.4* MSE and PSNR for different images

| *SNR (in dB)* | *MSE QPSK (binary image)* | *PSNR QPSK (binary image) (in dB)* | *MSE QPSK (grayscale image)* | *PSNR QPSK (grayscale image) (in dB)* | *MSE QPSK (color image)* | *PSNR QPSK (color image) (in dB)* |
|---|---|---|---|---|---|---|
| -5 | 0.2841 | 5.4650 | 85.8143 | 28.7952 | 97.4936 | 28.2410 |
| -3 | 0.2366 | 6.2604 | 74.3021 | 29.4208 | 87.8124 | 28.6952 |
| 0 | 0.1587 | 7.9945 | 57.7470 | 30.5155 | 68.1400 | 29.7968 |
| 2 | 0.1047 | 9.8005 | 42.9599 | 31.8002 | 48.214 | 31.2991 |
| 5 | 0.0377 | 14.2366 | 16.7084 | 35.9015 | 19.9072 | 35.1407 |
| 10 | 0.0012 | 29.3567 | 0.5767 | 50.5211 | 0.5209 | 50.9633 |
| 15 | 0 | Inf | 0 | Inf | 0 | Inf |
| 20 | 0 | Inf | 0 | Inf | 0 | Inf |

The amount of noise in the reconstructed images is observed, Table 3.6 shows the results obtained for binary, grayscale, and color images subjected to QPSK modulation at varying SNRs.

### 3.5.3 Test case and validation for sub-blocks of MB-OFDM transceiver

In this section, each block of a designed OFDM transceiver is considered and verified with MATLAB's built-in functions. Table 3.7 shows the results of test cases for the OFDM transceiver. In the test case, convolution encoders (16-QAM, QPSK, DCM, and 64-QAM), OFDM modulation and demodulation operations, and PHY headers and preamble frames are tested with developed code and built MATLAB functions.

Table 3.8 shows the results of test cases for the OFDM transceiver in the simulation tool of the ARM Cortex-A53. In the test case of the convolution encoder, QPSK, OFDM transmitter, and OFDM receiver, scrambler operation is tested with developed C code in the tool, and results are verified with MATLAB functions. The C code had been developed for different sub-blocks of OFDM symbols, which include the convolution encoder, QPSK modulation and demodulation, and OFDM transmitter, OFDM receiver, and scrambler.

### 3.5.4 Test case validation for video processing

The test plan is implemented for video with an SNR value of 4 dB using the QPSK modulation technique, which is shown in Figure 3.19 for the color video. The video frames are taken and processed on the developed system model. The test case considered is for 10 frames of the video, which can be

*Table 3.5* Validation results for the 16-QAM modulation technique at varying SNRs

| *Results for 16-QAM modulation technique at SNR 0 dB* | | |
|---|---|---|
| *BINARY IMAGE* | *GRAYSCALE IMAGE* | *COLOR IMAGE* |
| |  | 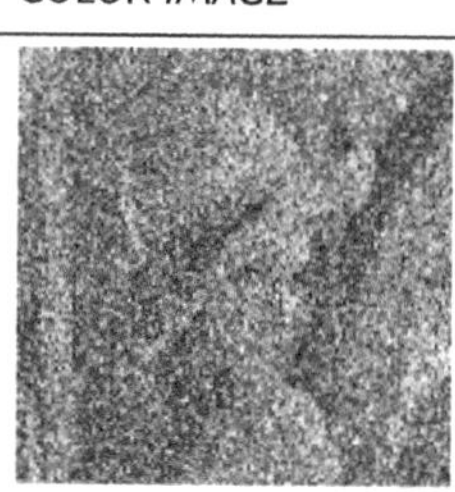 |
| MSE_QPSK = 0.2841<br>PSNR_QPSK = 51.099 dB | MSE_QPSK = 85.8143<br>PSNR_QPSK= 28.7952 dB | MSE_QPSK = 97.4936<br>PSNR_QPSK= 28.2410 dB |
| *Results for 16-QAM modulation at SNR 4 dB* | | |
| BINARY IMAGE | GRAYSCALE IMAGE | COLOR IMAGE |
|  |  | 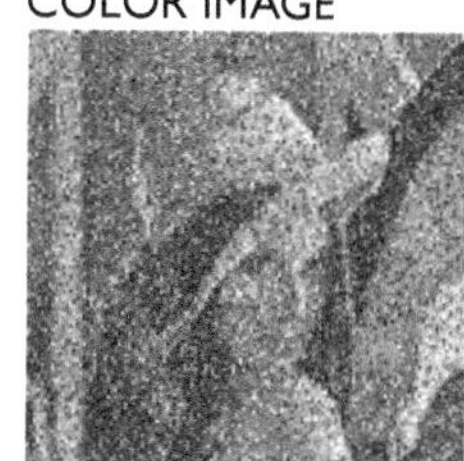 |
| MSE_QPSK = 0.1587<br>PSNR_QPSK = 52.365 dB | MSE_QPSK = 57.7470<br>PSNR_QPSK= 30.5155 dB | MSE_QPSK = 68.1400<br>PSNR_QPSK= 29.7968 dB |
| *Results for 16-QAM modulation at SNR 6 dB* | | |
| BINARY IMAGE | GRAYSCALE IMAGE | COLOR IMAGE 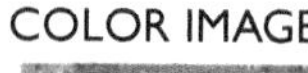 |
|  |  |  |
| MSE_QPSK = 0.0377<br>PSNR_QPSK = 65.0898 dB | MSE_QPSK = 16.7084<br>PSNR_QPSK = 35.9015 dB | MSE_QPSK = 19.9072<br>PSNR_QPSK = 35.1407 dB |
| *Results for 16-QAM modulation at SNR 10 dB* | | |
| BINARY IMAGE | GRAYSCALE IMAGE | COLOR IMAGE |
|  |  |  |
| MSE_QPSK = 0.0012<br>PSNR_QPSK = 76.295dB | MSE_QPSK = 0.5767<br>PSNR_QPSK = 50.5211 dB | MSE_QPSK = 0.5209<br>PSNR_QPSK = 50.9633 dB |

*Table 3.6* Validation results for QPSK modulation technique at varying SNRs

| *Results for QPSK modulation technique at SNR 0 dB* | | |
|---|---|---|
| *BINARY IMAGE* | *GRAYSCALE IMAGE* | *COLOR IMAGE* |
| 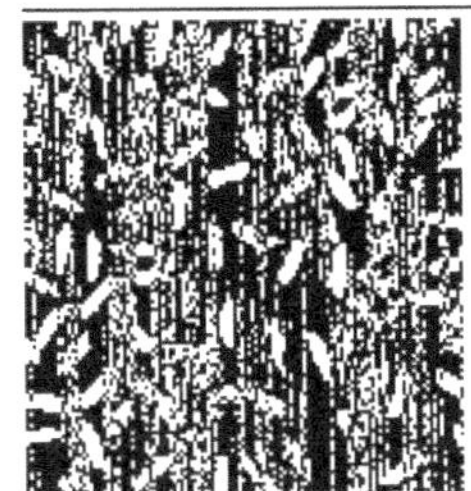 | 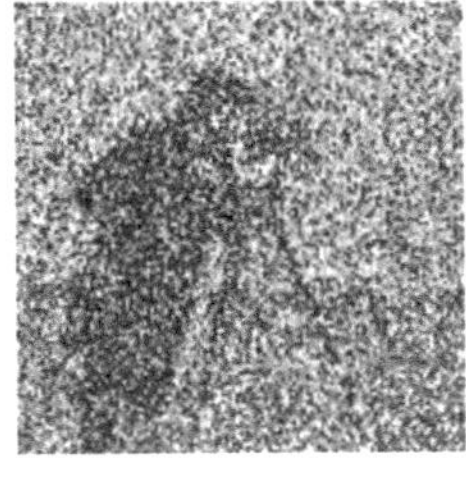 | 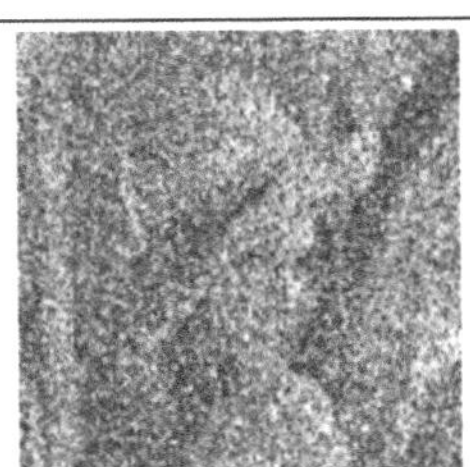 |
| MSE_QPSK = 0.2841<br>PSNR_QPSK = 51.099 dB | MSE_QPSK = 85.8143<br>PSNR_QPSK= 28.7952 dB | MSE_QPSK = 97.4936<br>PSNR_QPSK= 28.2410 dB |
| *Results for QPSK modulation technique at SNR 4 dB* | | |
| BINARY IMAGE | GRAYSCALE IMAGE | COLOR IMAGE |
| 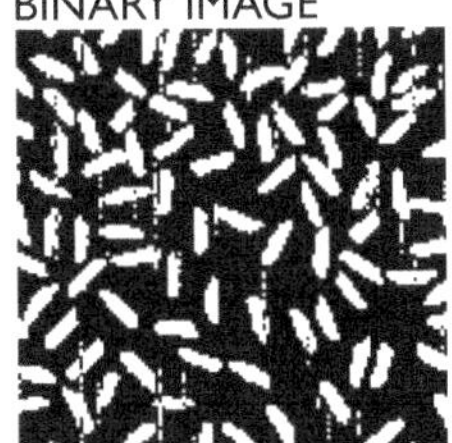 | | 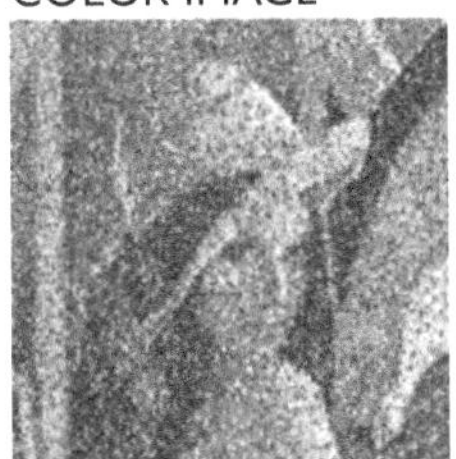 |
| MSE_QPSK = 0.1587<br>PSNR_QPSK = 52.365 dB | MSE_QPSK = 57.7470<br>PSNR_QPSK= 30.5155 dB | MSE_QPSK = 68.1400<br>PSNR_QPSK= 29.7968 dB |
| *Results for QPSK modulation technique at SNR 6 dB* | | |
| BINARY IMAGE | GRAYSCALE IMAGE | COLOR IMAGE |
|  |  |  |
| MSE_QPSK = 0.0377<br>PSNR_QPSK = 65.0898dB | MSE_QPSK = 16.7084<br>PSNR_QPSK = 35.9015 dB | MSE_QPSK = 19.9072<br>PSNR_QPSK = 35.1407 dB |
| *Results for QPSK Modulation Technique at SNR 10 dB* | | |
| BINARY IMAGE | GRAYSCALE IMAGE | COLOR IMAGE |
|  |  |  |
| MSE_QPSK = 0.0012<br>PSNR_QPSK = 76.295dB | MSE_QPSK = 0.5767<br>PSNR_QPSK = 50.5211 dB | MSE_QPSK = 0.5209<br>PSNR_QPSK = 50.9633 dB |

*Table 3.7* Test cases of sub-blocks of the OFDM transceiver in MATLAB

| *Test case no.* | *Sub-block* | *Test data* | *Expected results* | *Actual result* | *Status* |
|---|---|---|---|---|---|
| TS_1 | PHY Header | Input bits | PHY header is modulated and the frame is frames | PHY header is developed | Pass |
| TS_2 | Preamble | Input sequence | The preamble sequence is generated | Preamble sequences are developed | Pass |
| TS_3 | Convolution encoder | Input bits | Each bit is XOR to get 2 extra bits | convolution encoder 1/3 working | Pass |
| TS_4 | Convolution encoder Puncturer | Input bits | Each bit is XOR to get 2 extra bits | Puncture with 5/8,3/4,1/2,1/3 | Pass |
| TS_5 | QPSK | Input from the convolution encoder | QPSK operation With constellation plot | Working | Pass |
| TS_6 | DCM | Input from the convolution encoder | DCM operation With constellation plot | Working | Pass |
| TS_7 | 16-QAM | Input from the convolution encoder | 16-QAM operation With constellation plot | Working | Pass |
| TS_8 | 64-QAM | Input from the convolution encoder | 64-QAM operation With constellation plot | Working | Pass |
| TS_9 | OFDM Modulation | Modulated data | MB-OFDM frame Operation | Modulated frame | Pass |
| TS_10 | OFDM De-modulation | Modulated data | Complex modulated data | Working | Pass |
| TS_11 | Output data | Input bits | Original reconstructed image | Obtained | Pass |

*Table 3.8* Test cases and validation of sub-blocks of the OFDM transceiver in C

| *Test-case no.* | *Sub-block* | *Test data in C* | *Expected results* | *Actual result* | *Status* |
|---|---|---|---|---|---|
| TS_1 | Convolution encoder | Input bits | Each bit is XOR to get 2 extra bits | convolution encoder 1/3 working | Pass |
| TS_2 | QPSK | Input from convolution encoder | QPSK operation | Mapping of signals | Pass |
| TS_3 | OFDM transmitter | Modulated data | Modulated complex data | Working | Pass |
| TS_4 | OFDM Receiver | Modulated data | Binary data | Working | Pass |

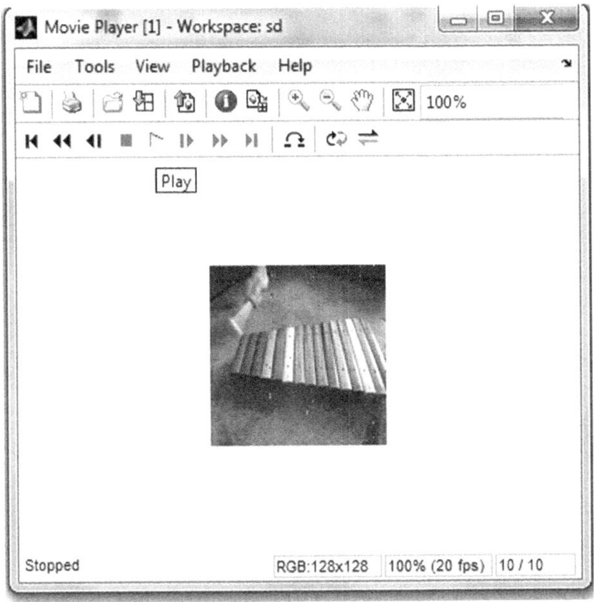


*Figure 3.19* Model testing for video processing.

observed on the right-side bottom of Figure 3.19, and the test was successful with some minor noise.

### 3.5.5 Testing results obtained in C with MATLAB

The model that was developed in MATLAB was implemented in the C language in an ARM-based tool, and the test cases were developed and tested by comparing the results of MATLAB and the ARM Cortex A-53 tool studio, Xilinx Vitis. The below session provides a few block test cases. Figure 3.20 and Figure 3.21 provide the results obtained in the MATLAB and ARM tools for modulation technique, and comparing the values of the MATLAB and ARM Cortex A-53 tools, almost all the values are the same and verified.

Figures 3.22 and 3.23 provide the results obtained in the MATLAB and Xilinx Vitis tools for the Descrambler technique, and comparing the values of MATLAB and Xilinx Vitis, almost all the values are the same and verified. Since Descrambler is verified, Scrambler will also work the same.

```
Columns 1 through 7

 0.7071 + 0.7071i   0.7071 + 0.7071i   0.7071 - 0.7071i  -0.7071 + 0.7071i  -0.7071 + 0.7071i   0.7071 + 0.7071i  -0.7071 - 0.7071i

Columns 8 through 14

-0.7071 - 0.7071i   0.7071 - 0.7071i   0.7071 - 0.7071i  -0.7071 + 0.7071i  -0.7071 + 0.7071i  -0.7071 - 0.7071i  -0.7071 + 0.7071i

Columns 15 through 21

 0.7071 - 0.7071i  -0.7071 + 0.7071i  -0.7071 + 0.7071i  -0.7071 - 0.7071i   0.7071 - 0.7071i  -0.7071 - 0.7071i  -0.7071 - 0.7071i

Columns 22 through 28

 0.7071 + 0.7071i  -0.7071 - 0.7071i   0.7071 + 0.7071i   0.7071 + 0.7071i  -0.7071 - 0.7071i   0.7071 + 0.7071i   0.7071 + 0.7071i

Columns 29 through 35

-0.7071 - 0.7071i  -0.7071 - 0.7071i  -0.7071 + 0.7071i   0.7071 - 0.7071i  -0.7071 - 0.7071i   0.7071 + 0.7071i  -0.7071 - 0.7071i

Columns 36 through 42

-0.7071 + 0.7071i   0.7071 - 0.7071i   0.7071 + 0.7071i   0.7071 + 0.7071i   0.7071 + 0.7071i  -0.7071 - 0.7071i   0.7071 + 0.7071i

Columns 43 through 49

-0.7071 - 0.7071i   0.7071 + 0.7071i   0.7071 + 0.7071i  -0.7071 + 0.7071i  -0.7071 + 0.7071i  -0.7071 + 0.7071i   0.7071 + 0.7071i
```

*Figure 3.20* Results obtained for modulation in MATLAB.

## 3.6 RESULTS AND DISCUSSIONS

This section deals with the performance analysis of the developed MB-OFDM system. The performance of the system is tested based on different performance metrics like BER, SNR, MSE, PSNR, and capacity limits. The performance of the MB-OFDM system is compared with various modulation techniques to select the best data rate according to application. Power spectral density is tested to satisfy the power of the generated MB-OFDM. The hopping of frames is done at different frequencies to generate different data rates. Data rate calculation can be done by implementing Equation (3.9)

$$C = B\log_2\left(1+\frac{S}{N}\right) \tag{3.9}$$

where $C$ is the channel's capacity in bits per second, $B$ is the signal's bandwidth, and $S/N$ is the SNR within that band. This is because the channel's capacity correlates with signal bandwidth, but only in logarithmic terms to the SNR ratio. Thus, increasing the bandwidth is the most successful method to boost the data rate, rendering UWB signals most effective at providing the highest data rate while maintaining a very low signal strength density. Table 3.9 lists the coded bits per symbol for various modulation techniques.

### 3.6.1 Result obtained for various data rates for different modulation techniques

Different modulation techniques provide different data rates since the number of frames sent changes for different modulation techniques the

```
Modulated output is : 0.7071 + 0.7071i
Modulated output is : 0.7071 + 0.7071i
Modulated output is : 0.7071 - 0.7071i
Modulated output is :-0.7071 + 0.7071i
Modulated output is :-0.7071 + 0.7071i
Modulated output is : 0.7071 + 0.7071i
Modulated output is :-0.7071 - 0.7071i
Modulated output is :-0.7071 - 0.7071i
Modulated output is : 0.7071 - 0.7071i
Modulated output is : 0.7071 - 0.7071i
Modulated output is :-0.7071 + 0.7071i
Modulated output is :-0.7071 + 0.7071i
Modulated output is :-0.7071 - 0.7071i
Modulated output is :-0.7071 + 0.7071i
Modulated output is : 0.7071 - 0.7071i
Modulated output is :-0.7071 + 0.7071i
Modulated output is :-0.7071 + 0.7071i
Modulated output is :-0.7071 - 0.7071i
Modulated output is : 0.7071 - 0.7071i
Modulated output is :-0.7071 - 0.7071i
Modulated output is :-0.7071 - 0.7071i
Modulated output is : 0.7071 + 0.7071i
Modulated output is :-0.7071 - 0.7071i
Modulated output is : 0.7071 + 0.7071i
Modulated output is : 0.7071 + 0.7071i
Modulated output is :-0.7071 - 0.7071i
Modulated output is : 0.7071 + 0.7071i
Modulated output is : 0.7071 + 0.7071i
```

*Figure 3.21* Results obtained for modulation in the ARM-based tool.

```
scrambler_mac_hcs =

  Columns 1 through 23

    1   0   0   0   1   1   1   0   1   0   1   0   1   1   0   1   1   0   1   1   0   0   0

  Columns 24 through 46

    0   1   1   1   0   0   0   1   1   0   1   1   1   1   1   0   0   0   1   1   1   0   0

  Columns 47 through 69

    0   0   0   1   0   0   1   1   0   1   1   1   0   0   1   1   0   0   0   0   0   1   0

  Columns 70 through 92

    1   1   0   1   0   0   1   0   0   1   1   1   1   1   1   0   1   1   1   1   0   1   1
```

*Figure 3.22* Results obtained for descrambler in MATLAB.

```
Descrambled output:1
Descrambled output:0
Descrambled output:0
Descrambled output:0
Descrambled output:1
Descrambled output:1
Descrambled output:1
Descrambled output:0
Descrambled output:1
Descrambled output:0
Descrambled output:1
Descrambled output:1
Descrambled output:1
Descrambled output:0
Descrambled output:1
Descrambled output:1
Descrambled output:0
Descrambled output:1
Descrambled output:1
```

*Figure 3.23* Results obtained for descrambler in ARM Cortex A-53.

*Table 3.9* Various data rate calculated

| *Data rate* | *Modulation* | *Coding rate* | *Frequency spreading gain* | *Time spreading gain* | *Overall gain* | *Coded bits per symbol* |
|---|---|---|---|---|---|---|
| 53.5 (Mbps) | QPSK | 1/3 | 2 | 2 | 4 | 100 |
| 55 (Mbps) | QPSK | 11/32 | 2 | 2 | 4 | 100 |
| 80 (Mbps) | QPSK | 1/2 | 2 | 2 | 4 | 100 |
| 106.7 (Mbps) | QPSK | 1/3 | 2 | 2 | 2 | 200 |
| 110 (Mbps) | QPSK | 11/32 | 1 | 2 | 2 | 200 |
| 160 (Mbps) | QPSK | 1/2 | 1 | 2 | 2 | 200 |
| 200 (Mbps) | QPSK | 5/8 | 1 | 2 | 2 | 200 |
| 320 (Mbps) | QPSK | 1/2 | 1 | 1 | 1 | 200 |
| 400 (Mbps) | QPSK | 5/8 | 1 | 1 | 1 | 200 |
| 480 (Mbps) | QPSK | 3/4 | 1 | 1 | 1 | 200 |
| 2.56(Gbps) | 16QAM | 1/2 | 1 | 1 | 1 | 400 |
| 3.2 (Gbps) | 16QAM | 5/8 | 1 | 1 | 1 | 400 |
| 3.84 (Gbps) | 16QAM | 3/4 | 1 | 1 | 1 | 400 |
| 5.6 (Gbps) | 64QAM | 1/2 | 1 | 1 | 1 | 600 |
| 7.2 (Gbps) | 64QAM | 5/8 | 1 | 1 | 1 | 600 |
| 8.64 (Gbps) | 64QAM | 3/4 | 1 | 1 | 1 | 600 |

below session provides the results for data rates varying for different modulation techniques.

#### *3.6.1.1 Results obtained for different data rates for QPSK modulation + MB-OFDM*

Multiband OFDM utilizes different modulation techniques so that there will be a reduction in the number of frames sent; the higher-order modulation technique gives fewer frames. Hence, the data rate varies, and their SNR and BER graphs are shown in Figure 3.24. Different data rates are 160 Mbps, 200 Mbps, 320 Mbps, and 400 Mbps obtained by implementing the DCM modulation technique. The BER vs. SNR graph shown in Figure 3.24 clarifies that the higher the data rate, the more prone to error. For 160 Mbps with an SNR of 2 dB, the number of bit errors is $10^{-5}$, which is negligible, but for data rates of 160 Mbps and 320 Mbps, the BER is $10^{-3,}$ and for 400 Mbps and 480 Mbps, the BER is $10^{-1}$ and $10^{0}$, respectively, and hence it clarifies that the higher the data rate, the higher the BER. The DCM technique provides better performance than the QPSK modulation technique.

#### *3.6.1.2 Results obtained for data rates for DCM + MB-OFDM*

Multiband OFDM is utilized in short-range communications, and since they have a large bandwidth of 528 MHz, the data rates are high, as shown in Figure 3.25. Different data rates are 106.7 Mbps, 160 Mbps, 200 Mbps, 320 Mbps, and 400 Mbps obtained by implementing the DCM modulation technique. The BER vs. SNR graph shown in Figure 3.25 clarifies that the

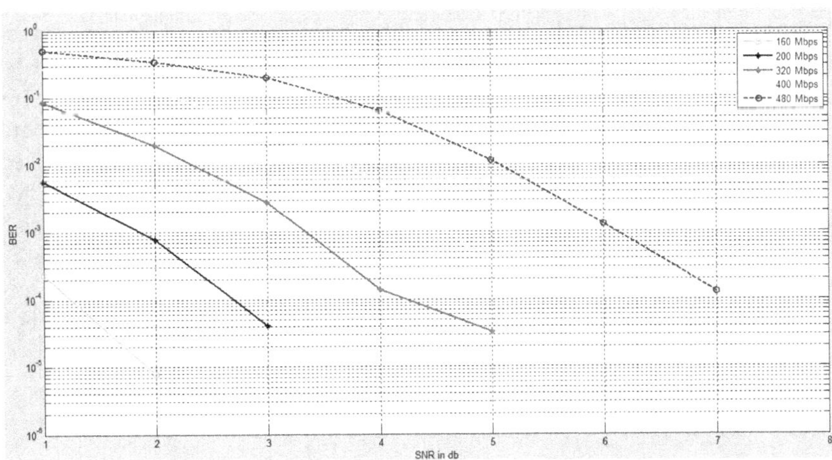


*Figure 3.24* Shows BER v/s SNR curves for different data rate for QPSK modulation.

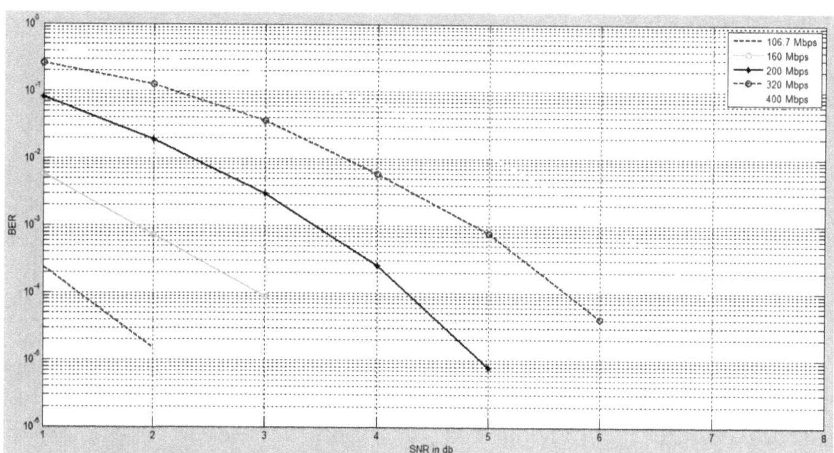


*Figure 3.25* BER v/s SNR curves for different data rates for DCM.

higher the data rate, the more likely it is to error. For 106.7 Mbps with an SNR of 2 dB, the number of bit errors is $10^{-5}$, which is negligible, but for data rates of 160 Mbps, the BER is $10^{-3}$, for 200 Mbps, the BER is $10^{-2}$, and for 300 and 480 Mbps, the BER is $10^{-1}$, hence it clarifies that the higher the data rate, the higher the BER. The DCM technique provides better performance than the QPSK modulation technique.

#### *3.6.1.3 Results obtained for various data rates for MB-OFDM + 16QAM*

The mapping of the number of bits to symbols increases for 16-QAM compared to QPSK and DCM, and hence the number of frames transmitted decreases. In order to achieve a higher data rate, 16-QAM can be implemented. Figure 3.26 shows the MB-OFDM implemented using 16-QAM modulation techniques, which yield different data rates. The data rates applied are 2.6 Gbps, 3.2 Gbps, and 3.84 Gbps. It can be observed that at low SNR values, the BER is very high compared to that of DCM, and hence the image noise is higher. At higher SNR values, BER is also high compared to the other two modulation techniques. However, there is a drastic increase in data rate from 480 Mbps to 3.84 Mbps, which is achieved without frequency spreading, as shown in the graph. 2.56 Gbps have a good BER compared to 3.2 Gbps and 3.8 Gbps.

#### *3.6.1.4 Results obtained for various data rates for MB-OFDM + 64-QAM*

The mapping of the number of bits to symbols increases for 64-QAM compared to QPSK, DCM, and 16-QAM, and hence the number of

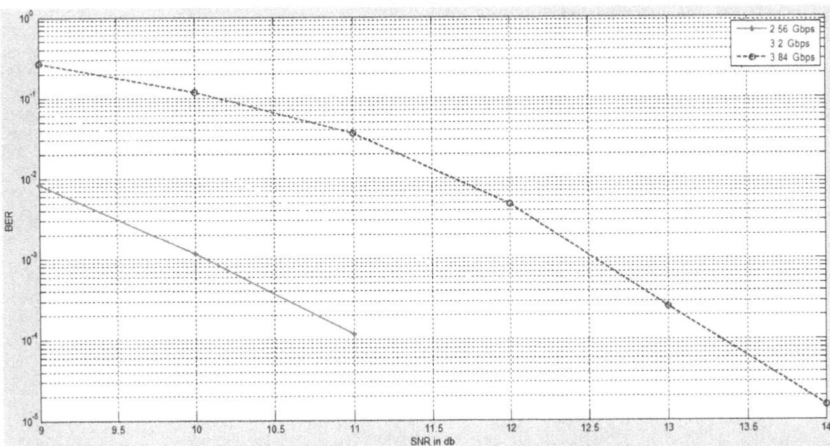


*Figure 3.26* BER v/s SNR curves for different data rates for 16-QAM.

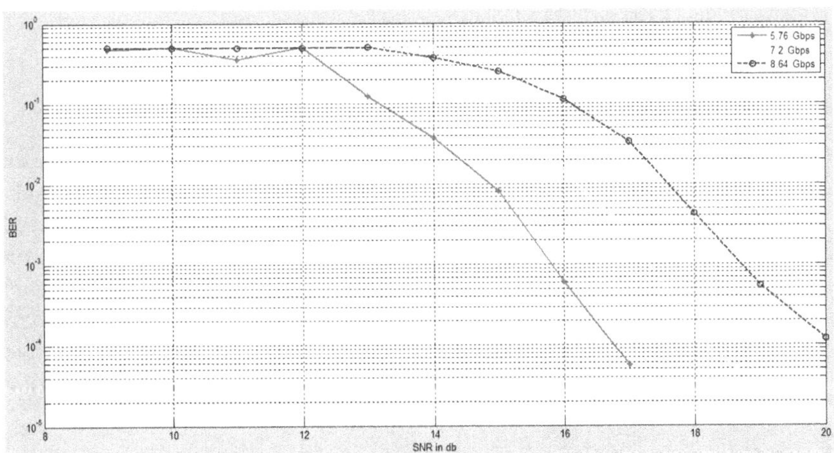


*Figure 3.27* BER v/s SNR curves for different data rates for 64-QAM.

frames transmitted decreases. In order to achieve a much higher data rate, 64-QAM can be implemented. Figure 3.27 shows the MB-OFDM implemented using 64-QAM modulation techniques, which yield different data rates. The data rates applied are 5.76 Gbps, 7.2 Gbps, and 8.64 Gbps. It can be observed that at low SNR values, the BER is very high compared to that of DCM, and hence the image noise is higher. At higher SNR values, BER is also high compared to the other two modulation

```
-0.0221 - 0.0221i   0.0110 + 0.0884i  -0.0221 - 0.1105i   0.0221 + 0.0552i   0.0331 - 0.0000i
-0.1079 - 0.0006i  -0.0390 - 0.0288i  -0.0515 - 0.0809i   0.0335 - 0.0723i   0.0410 - 0.0126i
 0.1069 + 0.0370i   0.0251 + 0.0318i  -0.0538 + 0.0100i  -0.0525 + 0.0133i   0.0772 - 0.0744i
-0.0592 - 0.0524i   0.1169 - 0.0248i  -0.0608 - 0.0513i   0.0460 - 0.0013i   0.0504 + 0.1425i
 0.0009 + 0.1099i  -0.0673 - 0.0053i   0.0125 + 0.0483i  -0.0362 - 0.0126i   0.0352 - 0.0400i
 0.0539 + 0.1343i  -0.0727 + 0.0152i   0.0219 + 0.0385i  -0.1615 + 0.0075i  -0.0487 + 0.1024i
-0.0620 + 0.0427i  -0.0301 - 0.0987i  -0.0013 - 0.0425i   0.0568 - 0.0241i  -0.0526 + 0.0826i
-0.0560 + 0.0016i   0.0472 - 0.0655i  -0.0317 - 0.0116i   0.0105 + 0.0318i  -0.0378 - 0.0068i
-0.0143 + 0.0290i   0.1104 + 0.0198i  -0.0267 - 0.1495i   0.0442 + 0.0416i  -0.0189 + 0.0255i
-0.0353 + 0.0387i   0.0039 - 0.0112i   0.0124 - 0.0107i   0.0769 + 0.0095i   0.0455 - 0.0920i
 0.0143 - 0.0437i   0.0325 - 0.0598i  -0.0183 + 0.0403i  -0.0027 - 0.0407i   0.1233 - 0.0933i
 0.0785 - 0.0345i   0.0788 - 0.0021i   0.0351 - 0.0365i   0.0143 + 0.0317i   0.0091 + 0.1456i
-0.0273 + 0.0389i  -0.0190 + 0.0569i   0.0199 + 0.0367i   0.0684 - 0.0260i  -0.1253 + 0.0022i
-0.0079 + 0.0705i  -0.0476 + 0.0410i  -0.1069 + 0.0525i   0.0352 + 0.0400i   0.0008 - 0.0704i
 0.0246 + 0.1565i  -0.0732 - 0.0649i   0.0402 + 0.0281i  -0.0203 + 0.0219i   0.0563 + 0.0567i
-0.0045 + 0.1126i  -0.0580 - 0.0431i   0.0505 - 0.0633i  -0.0447 + 0.0919i   0.0778 + 0.0369i
```

*Figure 3.28* Transmitted data result in MATLAB.

techniques. At an SNR of 16 dB, the BER is $10^{-3}$ for the data rate of 5.76 Gbps and $10^{-2}$ for 7.2 Gbps and $10^{-1}$ for 8.64 Gbps but there is a drastic increase in the data rate from 3.4 Gbps to 8.64 Gbps, which is achieved without frequency spreading, as shown in the graph. However, low data rates can be obtained with good BER by performing frequency and time domain spreading.

### 3.6.2 Results comparison of C with MATLAB

The model developed in MATLAB was implemented in an ARM-based tool, but the payload of the physical layer is implemented, and the results obtained in MATLAB and on ARM-based Xilinx tools are compared, Figure 3.28 shows the MATLAB output obtained for the transmitted sequence, and Figure 3.29 shows the ARM-based tool output of the transmitted output. The MATLAB outputs are written column-wise, and one can observe the similar result obtained. Figure 3.30 shows the received data bits in MATLAB and C.

### 3.6.3 Embedded implementation with computing architecture core analysis and its parameters

ARM processing is powerful and has been used in communication systems since the beginning. The ARM architecture was developed to support a wider range of technological applications. A higher performance of 60% is provided by Cortex-A17 over the Cortex-A9 core, with a reduction of 20% in power consumption for the same workload execution. ARM Cortex-A53 is a 64-bit-width bus and CPU lines that help index faster executions of the system. So, most design is carried out with the architecture of ARM-Cortex-A9 and Cortex-A53. The details of the computing architecture and

```
IFFT output of transmitter-0.0221 - 0.1105i
IFFT output of transmitter 0.0221 + 0.0552i
IFFT output of transmitter 0.0331 - 0.0000i
IFFT output of transmitter -0.1079 - 0.0006i
IFFT output of transmitter-0.0390 - 0.0288i
IFFT output of transmitter-0.0515 - 0.0809i
IFFT output of transmitter0.0335 - 0.0723i
IFFT output of transmitter0.0410 - 0.0126i
IFFT output of transmitter 0.1069 + 0.0370i
IFFT output of transmitter 0.0251 + 0.0318i
IFFT output of transmitter -0.0538 + 0.0100i
IFFT output of transmitter -0.0525 + 0.0133i
IFFT output of transmitter0.0772 - 0.0744i
IFFT output of transmitter-0.0592 - 0.0524i
IFFT output of transmitter0.1169 - 0.0248i
IFFT output of transmitter-0.0608 - 0.0513i
IFFT output of transmitter 0.0460 - 0.0013i
IFFT output of transmitter0.0504 + 0.1425i
IFFT output of transmitter0.0009 + 0.1099i
IFFT output of transmitter-0.0673 - 0.0053i
IFFT output of transmitter 0.0125 + 0.0483i
IFFT output of transmitter -0.0362 - 0.0126i
IFFT output of transmitter0.0352 - 0.0400i
IFFT output of transmitter0.0539 + 0.1343i
IFFT output of transmitter -0.0727 + 0.0152i
IFFT output of transmitter0.0219 + 0.0385i
IFFT output of transmitter-0.1615 + 0.0075i
IFFT output of transmitter-0.0487 + 0.1024i
IFFT output of transmitter-0.0620 + 0.0427i
IFFT output of transmitter-0.0301 - 0.0987i
IFFT output of transmitter-0.0013 - 0.0425i
```

*Figure 3.29* Transmitted data result in ARM-based tool.

its parameters are shown in Table 3.10. The processing performance of the architectures depends on the frequency of operations; the number of execution units used for the applications is shown in Equation (3.10). The processing speed is measured in millions of operations per second. The FPGA processing architecture and Raspberry Pi board are used for the experimental studies. The optimization of the code has been performed as per guidelines for the optimizations of helping the compiler, dead code eliminations, and other looping constructs perform better functions of code execution and code coverage. The implementation of MB-OFDM is shown in Figure 3.31.

*Figure 3.30* Received data bits in MATLAB and C.

The execution units include the data processing operation, load/store operations, address generation operation, and multicore operation. The computing efficiency 'PP' is measured in thousands of information manipulations per second (MdOp/s) for a machine's design with a particular number of processing units and operating clock frequency. The PP unit is MdOP/s, as indicated in Equation (3.10).

$$PP_{\text{ARM Cortex-A9}} = \text{number of processing cells in a design} \times \text{the operating clock frequency.} \quad (3.10)$$

In the case of the ARM Cortex-A9 with 2-cores, there are three execution units in each core, which makes six execution units and a clock frequency of 1 GHz.

$$PP_{\text{ARM Cortex-A8}} = 6 \times 1000000000 \text{ dOp/s}$$

$$PP_{\text{ARM Cortex-A8}} = 6000 \text{ MdOp/s}$$

From Table 3.10, it is concluded that as the processing performance improves, the system executes using computer architectures with many multicores, such as the ARM Cortex-A53 Architecture and the ARM Cortex A-17. The Xilinx UltraScale with RF SoC is used with transceivers ranging from 0.5 to 28 Gbps to achieve system performance.

### 3.6.4 Comparative examination of data rates and execution timings for the proposed approach and comparable methods

Table 3.11 shows the comparative figures acquired for data rates, while Table 3.12 shows the results obtained for execution time. The suggested

*Table 3.10* Computing architecture core analysis with its parameters for higher data rate with 64-QAM based MB-OFDM

| *Architecture and its parameters* | *ARM Cortex A8* | *ARM Cortex -A9* | *ARM Cortex-A15* | *ARM Cortex-A17* | *ARM Cortex-A 53* |
|---|---|---|---|---|---|
| **Number of cores** | 1 | 2 | 4 | 4 | 4 |
| **ARM architecture versions** | Av7 | Av7 | Av7 | Av7 | Av8 |
| **DMIPS/MHz** | 1 x 2.0 | 2 x 2.5 | 4 x 3.5 | 4 x 3.5 | 4 x 2.4 |
| **Frequency GHz** | 1 | 1 | 2 | 2 | 1.5 |
| **Processing CPU bit** | 32 | 32 | 32 | 32 | 32-64 |
| **Stages of pipelines** | 13 stages | 8 stages | 15 stages | 10-12 stages | 8-stages with 2-way superscalar |
| **Order of pipeline execution** | In-order | Out-order | Out-order | Out-order | In-order |
| **Vector float point (VFP)** | VFPv3 | VFPv4 | VFPv4 | VFPv4 | VFPv3 |
| **VFP register** | 32 x 64 bit | 32 x 64 bit | 32 x 64 bit | 32 x 64 bit | 32 x 64 bit |
| **DSP and NEON SIMD** | 128 bits wide | 128 bits wide | 128 bits wide | 128 bits wide | 128 bits wide |
| **Number of execution** | 1 CPU + 1VPU + 1 Neon = 3 | 2 x 3 (1 CPU + 1VPU + 1 Neon) =6 | 4 x 3 (1 CPU + 1VPU + 1 Neon) =12 | 4 x 3 (1 CPU + 1VPU + 1 Neon) =12 | 4 x 4 (1 CPU + 1 VPU + 1 Neon + 1 64 bitwidth Indexing) =16 |
| **Processing performance (MdOp/s)** | 3000 | 6000 | 24,000 | 24,000 | 24,000 |
| **Widely used** | Lower | Higher | Moderate | Moderate | Higher |
| **Number of instructions** | 6,326,663 | 6,346,353 | 6,452,345 | 6,448,763 | 6,463,742 |
| **Number of instructions (with optimizations)** | 4,466,866 | 4,442,636 | 4,455,426 | 4,454,678 | 4,446,843 |
| **Average instructions/sec** | 4,896,643 | 2,537,687 | 1,182,430 | 1,182,363 | 1,187,243 |
| **Average instructions/sec (with optimizations)** | 2,633,826 | 1,589,872 | 0,943,246 | 0,926,742 | 0,923,647 |
| **Clocks** | 17,636,686 | 08,673,458 | 04,435,332 | 04,402,392 | 04,321,742 |
| **Clocks (with optimizations)** | 08,662,663 | 05,537,866 | 02,348,325 | 02,373,432 | 02,362,473 |
| **Throughput computing η (execution time) $m_{sec}$** | 503 | 387 | 216 | 198 | 186 |
| **Throughput computing η (execution time with optimizations) $m_{sec}$** | 486 | 239 | 156 | 149 | 146 |

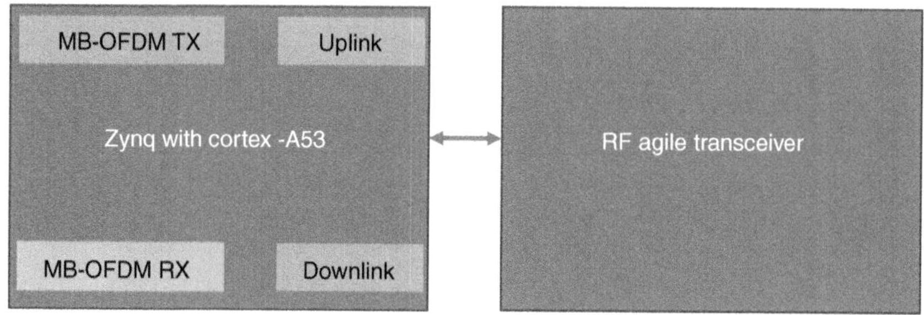


*Figure 3.31* MB-OFDM node block implementation with Zynq with Cortex-A53 and RF agile transceiver block diagram.

*Table 3.11* A comparison of data rates from this study and others in the literature

| *Authors* | *MB-OFDM/OFDM data rate throughput* |
|---|---|
| The Proposed Method | 10.9 Gbps |
| [28] Rani et.al. | 10 Gbps |
| [29] Agarwal et al. | 1 Gbps |
| [30] Suryawanshi et al. | 1 Gbps |
| [31] Venkataramanan & Lakshmi | 100 Mbps |
| [32] Kim et al. | 18 Mbps |

*Table 3.12* A comparison of execution time comes from this study and previous publications

| *Authors* | *SoC architecture* | *MB-OFDM execution time ($m_{sec}$)* | *Comparison of implemented sub-components* |
|---|---|---|---|
| The Proposed Method | Zynq UltraScale with ARM Cortex-A53 | 186 | Full MB-OFDM with programmability and reconfigurability |
| [33] Singh et.al. | ARM Cortex A9 | 700 | 16-QAM, and 64-FFT |
| [34] Dou & Zhang | STM32F407 Cortex-M4 | 2.88 | 64-FFT |
| [35] Zitouni et al. | ARM Cortex-A9 | -- | -- |
| [36] Fadhil | FPGA | -- | -- |

architecture obtained a data rate of 10.9 Gbps, compared to Rani et al. with 10 Gbps [28], Agarwal et al. and Suryawanshi et al. with 1 Gbps [29, 30].

This study used Zynq ARM-Cortex-A53 processors to implement integrated MB-OFDM. There is no thorough implementation of MB-OFDM in any research that compares it to other architectural platforms; rather, representations of its various parts and subcomponents are available. The proposed work focuses on comprehensive solutions of MB-OFDM/OFDM on systems with the configurations shown in Table 3.12. Singh et al. created OFDM using 16 QAM and 64 FFT and investigated its reconfigurability, estimating its running time at 700 ms [33]. Dou and Zhang presented the FFT and IFFT of 256 at a computation time of 2.88 using a floating point and implemented OFDM on the STM32F407, a microcontroller based on the ARM Cortex-M4 [34]. Zitouni et al. disclosed an FPGA-based version of the Zynq ARM Cortex-A9. Embedded applications have already used the Cortex-A9 [35].

## 3.7 CONCLUSIONS

This chapter deals with the design and implementation of MB-OFDM for 5G and 6G short-range applications. The BER performance of the MB-OFDM was analyzed, and it was observed that the 64-QAM-based MB-OFDM provides a higher data rate of 8.69 Gbps for the applications developed. Also, there is a provision for selecting the different data rates based on the modulation schemes: a higher data rate based on 16-QAM and 64-QAM-based MB-OFDM and a lower data rate based on QPSK and DCM. As the higher data rate modulation schemes are used, the BER also increases with lower SNR and decreases with higher SNR (15 dB for 16-QAM-based MB-OFDM and 21 dB for 64-QAM-based MB-OFDM). The chapter explains the different data rate achievements through modeling and architecture implementation on ARM processor Cortex-A (8, 9, 15, 17, 53) families. The implementation of the computing C-code is designed and optimized for code optimizations. As processing performance increases, the period of execution decreases with advanced quad-core organizations and their pipeline stages.

## ABBREVIATIONS

| | |
|---|---|
| Fifth generation/sixth generation | 5G/6G |
| Advanced RISC machine | ARM |
| Additive white Gaussian noise | AWGN |
| Bit error rate | BER |
| Binary phase shifting key | BPSK |
| Dual carrier modulation **DCM** | |
| Ultra-wide band | UWB |
| Federal Communications Commission | FCC |
| Fast Fourier transformer/inverse fast Fourier transformer | FFT/IFFT |
| Forward error correction | FEC |
| Field programmable gate array | FPGA |

| | |
|---|---|
| Internet of Things | **IoT** |
| Multiband-orthogonal frequency division multiplexing | **MB-OFDM** |
| Mean square error | **MSE** |
| Orthogonal frequency division multiplexing | **OFDM** |
| Peak signal to noise ratio | **PSNR** |
| Quadrature amplitude modulation | **QAM** |
| Quadrature phase shifting key | **QPSK** |
| Signal to noise ratio | **SNR** |
| Time frequency coding | **TFC** |

## REFERENCES

[1] G. Heidari, *WiMedia UWB Technology of Choice for Wireless USB and Bluetooth*, West Sussex: John Wiley and Sons, 1st Edition, 2008.

[2] W. Siriwongpairat and K.J.R. Liu, *Ultra-Wideband Communications Systems Multiband OFDM Approach*, New Jersey: John Wiley and Sons, 1st Edition, 2007.

[3] I. Djordjevic, H. Batshon, L. Xu and T. Wang, "Four-Dimensional Optical Multiband-OFDM for Beyond 1.4 Tb/S Serial Optical Transmission," *Optics Express*, Vol. 19, No. 2, pp. 876–882, 2011.

[4] A. Sharma, Z.I.J. Garcia, R. Martnez, A. Perallos and J.C. Batchelor, "Channel-Based Antenna Synthesis for Improved In-Vehicle UWB MB-OFDM Communications," *IET Microwaves, Antennas and Propagation*, Vol. 13, No. 9, pp. 1358–1367, 2019.

[5] A. Chandra, A. Prokeš, T. Mikulášek, J. Blumenstein, P. Kukolev, T. Zemen and C. F. Mecklenbräuker, "Frequency-Domain In-Vehicle UWB Channel Modelling," *IEEE Transactions on Vehicular Technology*, Vol. 65, No. 6, pp. 3929–3940, June 2016.

[6] M.S.G. Varna, "Performance Evaluation of Fixed and Variable Zero Padding length in Ultra-Wideband Receiver Design using MB-OFDM Based on ECMA-368 standard", MS Thesis, Department of Electrical Engineering, Northern Illinois University, Dekalb, Illinois, 2014.

[7] E. Briggs, "OFDM Physical layer Architecture and Real-Time Multi-Path Fading Channel Emulation for the 3GPP Long Term Evolution Downlink", PhD Thesis, Faculty of Texas Tech University, Texas, US, 2012.

[8] ARM Cortex-A8 Technical Reference Manual, 2010, [Online]. Available: https://developer.arm.com/documentation/ddi0344/latest, ddi0344, ARM Ltd, 2010. [Accessed on 26 October 2022].

[9] ARM Cortex-A9 Technical Reference Manual, 2012, [Online]. Available: https://developer.arm.com/documentation/ddi0388/latest, ddi0388, ARM Ltd, 2008, [Accessed on 26 October 2022].

[10] ARM Cortex-A9 MPCore Technical Reference Manual, 2012, [Online]. Available: https://developer.arm.com/documentation/100486/0401/, 100486, ARM Ltd, 2016, [Accessed on 06 December 2022].

[11] ARM Cortex-A9 Performance, 2012, [Online]. Available: www.arm.com/products/processors/cortex-a/cortex-a9.php, Cortex-a9, ARM Ltd, 2016, [Accessed on 10 November 2022].

[12] ARM Cortex-A15 MP-Core Processor Technical Reference Manual, 2013, [Online]. Available: https://developer.arm.com/documentation/ddi0438/i, ddi0438, ARM Ltd, 2013, [Accessed on 10 November 2022].
[13] N. Kirsch, "ARM Cortex-A17 To Have 60% More Performance Than Cortex-A9 Processor," [Online]. Available: www.legitreviews.com/arm-cortex-a17-60-performance-cortex-a9-processor_135639, February 11, 2014, [Accessed on 10 November 2022].
[14] "ARM Cortex-A17 Processor", [Online]. Available: https://developer.arm.com/Processors/Cortex-A17, Cortex-a17, ARM Ltd, 2014, [Accessed on 10 November 2022].
[15] ARM Cortex-A53 MP-Core Processor Technical Reference Manual, 2018, [Online]. Available: https://developer.arm.com/documentation/ddi0500/latest/, Ddi0500, ARM Ltd, 2018, [Accessed on 10 November 2022].
[16] A.L. Shimpi, "The ARM Diaries, Part 2: Understanding the Cortex-12", July 17, 2013, [Online]. Available: www.anandtech.com/show/7126/the-arm-diaries-part-2-understanding-the-cortex-a12/2, [Accessed on 10 November 2022].
[17] A.L. Shimpi, "ARM Cortex A17: An Evolved Cortex 12 for the mainstream in 2015," February 11, 2014, [Online]. Available: www.anandtech.com/show/7739/arm-cortex-a17, [Accessed on 10 November 2022].
[18] S. Rosinger, "ARM Cortex-A17/Cortex-A12 Processor Update," September 10, 2020, [Online]. Available: https://community.arm.com/arm-community-blogs/b/architectures-and-processors-blog/posts/arm-cortex-a17-cortex-a12-processor-update, [Accessed on 10 November 2022].
[19] A. Madhu, "DSP Implementation of OFDM Acoustic Modem," MTech Thesis in Digital Signal Processing, Department of Electronics and Communication Engineering, National Institute of Technology Rourkela, Rourkela, Odisha, India, 2008.
[20] P. Hedaoo, U.M. Gokhale and S. Thakur, "VLSI Architecture for MB-OFDM Transmitter," *IOSR Journal of VLSI and Signal Processing (IOSR-JVPS)*, Vol. 4, No. 3(II), pp. 2319–4197, May–June 2014.
[21] K.W. Berlee, "Design and Implementation of Real-Time Cognitive Dynamic Spectrum Radio Targeting the FM Radio Band with PHYDYAS-FS-FBMC," PhD Thesis, University of Strathclyde, Glasgow, UK, 2021.
[22] S. Bailey, Rapid ASIC Design for Digital Signal Processors," PhD Thesis, Department of Electrical Engineering and Computer Sciences, University of California, Berkely, 2018.
[23] L. Wu, "Performance of Pulsed OFDM Systems for UWB Communication," *International Conference on Multimedia and Signal Processing, 14–15 May*, IEEE, Guilin, Guangxi, pp. 83–86, 2011.
[24] J. Wang, "Broadband Wireless Communications: 3G, 4G and Wireless LAN," *Boston: Kluwer Academic Publisher*, 1st Edition, 2001
[25] 3GPP 5G, 5G NR User Equipment (UE) Radio Access Capabilities. 3GPP Technical Specification 38.306., 2018.
[26] SNS Telecom and IT, "The Public Safety LTE and 5G Market: 2020-2030-Opportunities, Challenges, Strategies and Forecasts. Research and Market," October 2022, [Online]. Available: www.snstelecom.com/public-safety-lte, [Accessed on 16 November 2022].

[27] S.K. Rao and R. Prasad, "Impact of 5G Technologies on Industry 4.0," *Wireless Personal Communications*, Vol. 100, No. 1, pp. 145–159, 2018.
[28] A. Rani, M. S. Bhamrah and S. Dewra, "CO-OFDM System with 16-QAM Subcarrier Modulation using Reconfigurable Optical Add Drop Multiplexer," *Journal of Optical Communications*, Vol. 43, No.1, pp. 25–28, 2022.
[29] A. Agarwal, V.K. Sinha, R. Palisetty, P. Kumar, K.C. Ray, K. Kumar and T. Pandey, "Performance Analysis and FPGA Prototype of Variable Rate GO-OFDMA baseband Transmission Scheme," *Wireless Personal Communications*, Vol. 108, pp. 785–809, 2019.
[30] S. Suryawanshi, L. P. Thakare and A. Y. Deshmukh, "Implementation of MIMO-OFDM Transceiver Architecture Design with Simulink," *International Journal of Engineering Research and General Science*, Vol. 3, No. 2, pp. 834–841, 2015.
[31] V. Venkataramanan and S. Lakshmi, "Performance Analysis of LTE Physical Layer using Hardware Co-Simulation Techniques and Implementation on FPGA for Communication Systems," *International Journal of Communications Systems*, Vol. 35, No. 2, 2022.
[32] Y. Kim, L. Kwon and E.C. Park, "OFDMA Backoff Control Scheme for Improving Channel Efficiency in the Dynamic Network Environment of IEEE802.11ax WLANs," *Sensors*, Vol. 22, No. 15, p. 5111, 2021.
[33] N. Singh, S.V.S. Santosh and S.J. Darak, "Towards Intelligent Reconfigurable Wireless Physical Layer (PHY)," *IEEE Open Journal of Circuits and Systems*, Vol. 2, pp. 226–240, 2021.
[34] Y. Dou and F. Zhang, "Design and Implementation of OFDM Communication System Based on ARM," *10th International Conference on Intelligent Human-Machine Systems and Cybernetics (IHMSC), 25–26 August 2018*, Hangzhou, China, IEEE, 2018, pp. 362–365.
[35] R. Zitouni, H. Bouaroua and B. Senouci, "Hardware-Software Co-Design for Software Defined Radio: IEEE 802.11p Receiver Case Study," *11th ACM Conference on Recommender Systems, RecSys'17, 27–31 August 2017*, Como, Italy, New York, NY, Association for Computing Machinery, 2017.
[36] M.J. Fadhil, "High-Data Processing System of 6x6 MIMO-OFDM Using FPGA Technique with Spatial Algorithm," *Engineering and Technology Journal*, Vol. 36, No. 7A, pp. 723–732, 2018.

Chapter 4

# Blockchain and IoT for data privacy in the 6G network

## Challenges and solutions

*Bhupinder Singh, Vishal Jain, and Christian Kaunert*

## 4.1 INTRODUCTION

The imminent arrangement of 6G networks brings forth a new era of connectivity, which promises unprecedented data speeds and connectivity for a wide array of devices through the Internet of Things (IoT) [1, 2]. A peer-to-peer (P2P) network of nodes is responsible for maintaining a decentralized distributed ledger known as a blockchain [3, 4]. As the name implies, a blockchain is made up of a number of blocks of transactions that are chained or logically connected to one another to create a digital record [5, 6]. The verified transactions that fall under a specific time frame are then compiled into what is called a "block" [6, 7]. As such, there are a certain number of transactions in each block [8, 9]. These blocks are logically connected in the sequence of their production using cryptographic hashes [10, 11]. The exploration of these technologies not only addresses current concerns but also lays the groundwork for a resilient and trustworthy future for the Internet of Things in the upcoming 6G era [12, 13]. It explores the complex interplay between blockchain technology and the IoT in order to support data privacy in the context of the rapidly developing 6G networks [14, 15]. In light of the impending 6G era, which will bring with it previously unheard of levels of connection and data volume, combining blockchain with the IoT appears to be a viable solution for data privacy problems [16, 17]. It clarifies the intended function of this integration, which highlights the difficulties presented by the developing 6G environment and suggests creative ways to strengthen data privacy in the context of massive machine-type communication, ultra-reliable low-latency communication, and improved mobile broadband [18, 19].

Since around 1980, there has been a new generation of mobile phones almost every 10 years [20, 21]. First-generation (1G) phones evolved into the successful second-generation (2G) and then with the arrival of third-generation (3G) phones, auctions had mixed results [22]. Also, business-wise, 1G and 2G focused mostly on phone services before progressively adding data, as 3G was less successful however 4G saw substantial success. There

DOI: 10.1201/9781003583127-4

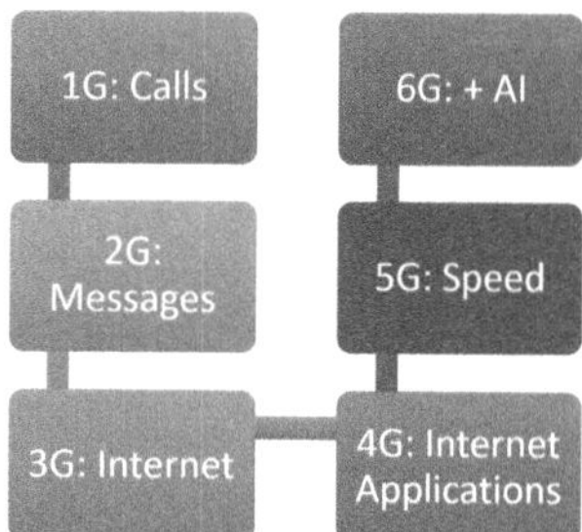


*Figure 4.1* Timeline of 1G to 6G.

is now much talk and expectation about what the fifth generation (5G) will be able to do [23, 24]. Some of the expectations include very high data rates (20 GB) very low latency (1 millisecond) and large capacities [25]. The next 6G networks ought to focus on improving coverage, throughput, and user-friendliness from any point in the world [26]. The thorough overview of the evolution of wireless communication networks from 1G to 6G by exploring prospective characteristics of 6G networks [27, 28] as shown in Figure 4.1.

### 4.1.1 Role of the IoT in the data-centric paradigm of 6G networks

The combination of 5G network technology with IoT applications has brought about a major transformation and as the IoT and 5G are expected to merge as 6G, it is expected to outperform its predecessor in a number of areas [29]. It is anticipated that this development will boost productivity in businesses as well as our daily lives [30]. 1G was introduced in the 1980s with the anticipated completion of 6G communications in the 2030s [31, 32]. The development of the wireless communications as seen in 6G is anticipated to provide every function offered by previous network generations in addition to a few new ones [33]. These comprise the development of wide-ranging IoT and AI-powered mobile apps, improved satellite communications, all-encompassing network coverage and totally autonomous system growth [34, 35].

### 4.1.2 Objectives of the chapter

This chapter has the following objectives:

- understanding the challenges to data privacy in 6G networks.
- exploring the integration of blockchain and the IoT for enhanced data privacy.

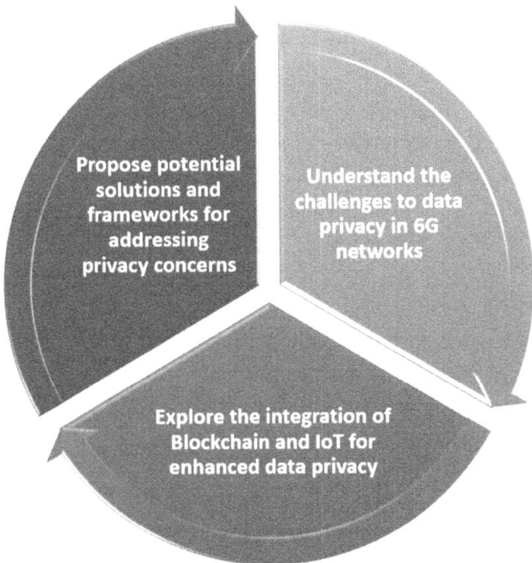


*Figure 4.2* Objectives of the study presented.

- proposing potential solutions and frameworks for addressing privacy concerns.

Figure 4.2 presents the major objectives of the chapter.

### 4.1.3 Structure of the chapter

The chapter comprehensively explores the various dynamics of blockchain and the IoT for data privacy in the 6G network: vision, challenges, and potential solutions. Section 4.2 elaborates the 6G networks: a data-centric paradigm. Section 4.3 expresses the IoT and data generation in 6G networks. Section 4.4 lays down the concept of blockchain technology: role of blockchain in enhancing data privacy in 6G networks. Section 4.5 highlights the challenges and concerns. Section 4.6 discusses the potential solutions. And, finally Section 4.7 concludes the Chapter with future scope.

## 4.2 6G NETWORKS: A DATA-CENTRIC PARADIGM

The increasing popularity of smart services based on the Internet of Everything (IoE) is driving the need for wireless networks, which calls for a reevaluation of existing networks with enhanced characteristics for the upcoming generation of communication systems [36]. Although 5G networks show great promise for IoE-based services, they are unable to fully satisfy the extensive needs of new smart application apps [37]. To overcome the main drawbacks of the current 5G network, there is an increasing need to imagine and create 6G wireless communication technologies [38]. The use of artificial intelligence in 6G has the potential to address complex problems associated with network optimization [39].

In order to augment the utility of upcoming 6G networks, scientists are investigating novel technologies including THz and quantum communications [40]. The rapidly changing 6G wireless communications environment needs strong support for large-scale data-driven applications and an expanding user base. The thorough comparison of 5G and 6G networks is essential and concentrates on the latest advancements and trends in 6G technology as well as network needs and key enabling technologies [41, 42]. The cutting-edge 6G connection options has a focus on smart community service, such as holographic beam-forming, edge computing, AI-enabled IoT networks, and backscatter communication, and also there are number of potential research avenues for the implementation of 6G-based IoT networks [43].

### 4.2.1 Key features of 6G

The center of the network changes with each new development in communication technology. The focus of the 2G and 3G eras was text and audio communication between people [44]. While the 5G era has focused on linking industrial automation systems and the IoT, the 4G era witnessed a dramatic move towards massive data consumption [45]. The extrasensory experiences will be possible in the 6G age due to the seamless merging of the digital, physical, and human domains. The concept of robust computing powers combined with intelligent information systems greatly improve human productivity, changing our way of life, work habits, and environmental care [46]. Key features of 6G are presented in Figure 4.3.

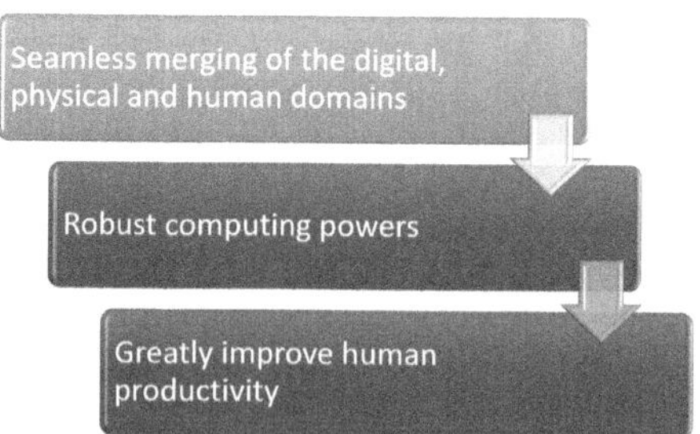


*Figure 4.3* Key features of 6G.

## 4.3 IOT AND DATA GENERATION IN 6G NETWORKS

The extensive deployment of the IoT has been facilitated by the widespread acceptance of wireless communications and the developments in smart device technology [47]. In order to connect millions of physical things to the Internet, sensing and computing capabilities must be integrated throughout society [48]. IoT has emerged as a crucial element of the Internet of the future in the current context garnering substantial interest from academic and industry circles because of its enormous potential to provide a wide range of client services across several aspects of modern life [49]. IoT makes it possible for various devices to automatically coordinate and communicate with each other without the need for human involvement [50]. This skill might lead to industry revolutions and provide game-changing solutions. The significant paradigm change is occurring in the current network environment with a shift away from protocol improvements and toward network behavior optimization [51]. There are many causes that have contributed to this change, such as the following:

**Nearing the limits of the available spectrum:** As the available spectrum reaches a saturation point and building new networks becomes less financially viable, it becomes more important than ever to schedule the currently allotted spectrum in a way that responds to subscriber and network demands, resulting in RAN-core convergence [52].

**Decoupling of software and hardware:** The system optimization has gained additional flexibility due to the labor split between the software and hardware may dynamically change the network design and functionality [53].

**Protocol-level optimizations:** There is currently little effort put into improving network behavior at the protocol level. The easy programmability and expanding network system accessibility to a wider spectrum of developers are becoming priorities [54].

**Cloud-based deployments and virtual networks:** As these technologies have developed, network management has become more all-encompassing. This entails setting up and deploying whole networks using configurations that are determined centrally [55]. The steps for network behavior optimization are presented in Figure 4.4.

### 4.3.1 IoT devices in 6G

The rapid growth of IoT networks and the impressive rise in the use of smart devices have made it difficult for 5G technology to keep up with the ever-increasing technical demands such as those for highly dynamic, autonomous, widely scalable, and intelligent services [56]. The capabilities of current 5G wireless systems are about to be surpassed by the fast development of automated and intelligent IoT networks [57]. The continuous

*Figure 4.4* Network behavior optimization.

improvements in current 5G systems are required to improve the quality of IoT service delivery and business outcomes due to the emergence of new IoT services and applications like aerial vehicles and remote robotic surgery. Smart cities, healthcare, and transportation are just a few of the industries that might benefit from the development of intelligent services made possible by the quickly developing IoT [58].

Sixth-generation (6G) networks are expected to be significantly impacted by the widespread deployment of IoT devices [59]. For the infrastructure underpinning 6G, which protects huge IoT networks and 6G from threats, especially new attacks, is a significant concern [60]. Innovative designs and paradigms that make use of software-based techniques, infrastructure virtualization, and intelligence are therefore desperately needed [61]. To achieve this goal, artificial intelligence must be integrated with important networking technologies like network functions' virtualization (NFV) and software-defined networking (SDN) in order to provide parts for creative and practical ways to improve IoT security [62].

The fields of computer science and engineering are hard at work utilizing deep-learning (DL) and machine-learning (ML) approaches to improve cyber threat identification. This includes identifying, forecasting, and mitigating abnormalities and unusual traffic patterns [63, 64]. Developing a scalable solution for a realistic infrastructure architecture that is suited for large Internet of Things applications is a difficulty when designing an intrusion detection system (IDS) [65]. It is evident that future solutions should be tailored to native virtualized infrastructure in order to handle the security concerns associated with big IoT applications, even if many aspects must be taken into account for an ideal intelligent solution [66].

### 4.3.2 Data privacy concerns in IoT

IoT typically refers to a broad category of physical objects that are online and collect, distribute, or use data [66]. The personal wearables like watches and glasses, home appliances like toasters and televisions, building amenities like lights and elevators, industrial equipment used in supply chains like forklifts and sprinklers, and components of urban infrastructure like trash cans and traffic lights all fall under this category [67, 68]. IoT device data has the potential to improve efficiency, convenience, and understanding of almost every element of our environment. IoT is presently providing several advantages to the public sector and has the potential to produce much more public value in the future [69].

### 4.3.3 Privacy implications of continuous monitoring and tracking

Emerging technologies have a wealth of available user data, and a space–air–ground integrated network environment is expected to present 6G mobile networks with a number of issues [70]. But for now 6G security and privacy issues are primarily theoretical [71]. It addresses privacy and security concerns for 6G technologies at the physical, connection, and service levels in a methodical manner [72]. It takes lessons from the flaws in current defensive mechanisms and security systems [73]. In addition to carrying over flaws from earlier generations, 6G brings fresh threat vectors from cutting-edge radio technology namely assaults directed at ubiquitous intelligence and the exposed placement of radio stripes in ultra-massive MIMO systems at Terahertz bands [74, 75]. The survey reveals promising methods to reduce the scope of attacks and breaches involving personal data [76]. These methods include quantum-safe communications, deep network slicing, physical layer protection, AI security, platform-agnostic security, real-time adaptive security, and novel data protection mechanisms like differential privacy and distributed ledgers [77].

## 4.4 BLOCKCHAIN TECHNOLOGY: ROLE OF BLOCKCHAIN IN ENHANCING DATA PRIVACY IN 6G NETWORKS

As compared to 5G networks, 6G networks are expected to have far faster network speeds, much lower communication latency, and deeper coverage [78, 79]. Terahertz, millimeter waves and light waves are just a few of the ultra-high frequency wireless spectrum resources that these networks will make considerable use of. In order to create a complete green network, 6G networks will integrate many technologies, including satellite Internet, microwave networks, and terrestrial mobile communications [80]. This network will enable real-time security evaluations, intelligent data perception, cooperative interactions between several entities, and coordinated coverage of both terrestrial and space domains [81].

There will be new application scenarios and higher-performance demands as the 6G era draws near. An air–space–ground integrated network communication system will be established as part of the 6G transition, with the goal of achieving ubiquitous network coverage in all circumstances [82, 83]. However, 6G networks must meet strict standards for bandwidth, latency, security, connection density, and flexibility due to the wide range of applications and communication scenarios, ultra-heterogeneous network connections, and the need for exceptional performance [84]. Among its many constituents are chain data structures, point-to-point transmission, distributed storage, consensus mechanisms, and encryption algorithms [85].

Blockchain is one of the most anticipated upcoming technologies and the system is exposed to a number of risks and challenges regarding aspects, such as system security, data privacy, sustainability, and scalability due to the demanding performance criteria of 6G networks, which include ultra-high peak rate, ultra-low latency, ultra-high reliability, ultra-low energy consumption, and seamless connectivity [86]. Blockchain technology with its distributed network design, intelligent node consensus, and smart contracts appears as an essential tool to tackle these issues [87]. The goal of 6G networks may be realized through the safe, intelligent, and efficient underlying technical basis provided by the combined use of blockchain and 6G networks [88]. In addition to stressing the need for an endogenous trust network in the 6G network, 6G needs the possible application of blockchain technology to tackle challenging new privacy issues. Blockchain is seen to be a workable way to protect 6G network privacy [89]. Blockchain provides 6G networks with a strong guarantee to create a distributed, safe, and reliable transaction environment [90].

## 4.5 CHALLENGES AND CONCERNS

The forthcoming sixth generation of the Internet of Things, or 6G-IoT, is expected to tackle significant issues pertaining to very dependable and ultra-low-latency communication, protocol optimization, and wide-spread geographic coverage for IoT networks [91]. The device capabilities have been further enhanced by the incorporation of AI, which allows devices to engage actively with the network instead of only sending data [92]. Because of this, the anticipated 6G-IoT framework is anticipated to provide efficient solutions for smart cities, industrial IoT, and distant applications with a significant influence on the development [93]. The major challenges and concerns of 6G are presented in Figure 4.5.

## 4.6 POTENTIAL SOLUTIONS

With the arrival of 6G, every improvement in network connection that 5G offers to end users will experience additional improvement. 6G is set to revolutionize several industries, including robots, agriculture, manufacturing, and smart cities [94]. 5G-Advanced is the planned standard enhancements for 5G that promise more efficiency and expanded capabilities, and an improved user experience is a major contributor to this breakthrough [95]. The wireless technology has always improved and expanded on the use cases of its predecessor while bringing new ones and this pattern is anticipated to continue [96]. With the utilizing 5G's technology advancements and use

*Figure 4.5* Challenges and concerns.

cases, 6G will promote wider adoption through cost- and optimization-saving measures. 6G will simultaneously provide new use cases [97].

The large-scale synchronous real-time updates along with the integration of sensors, artificial intelligence and machine learning (AI/ML) with digital twin models will provide a bridge between the physical and human worlds. Digital twin models are already in use with 5G and are anticipated to function on a much bigger scale with 6G [98]. Digital twin models are essential for assessing the physical environment, modeling results, predicting requirements, and executing productive activities. The network design will be greatly impacted by digital twins, which go beyond factories to include entire city networks and even digital portraits of people [99].

In the 6G age, smartphones will still be necessary, but new man–machine interfaces will improve control and information consumption [100]. The touchscreen typing will eventually give way to gesture and voice control, and gadgets could end up integrated in garments or change into skin patches. 24/7 monitoring of critical metrics would be beneficial for wearables especially in the healthcare industry [101]. As artificial intelligence and machine vision continue to advance and become more adept at identifying objects and people, wireless cameras will become ubiquitous sensors [102]. The digital currencies and keys may become widely used as a result of the ambient data gathered by a variety of sensing modalities including the acoustics. Even using brain sensors to communicate with machines would be a major advancement in user interface design [103, 104]. The viable solutions toward 6G challenges are presented using techniques shown in Figure 4.6.

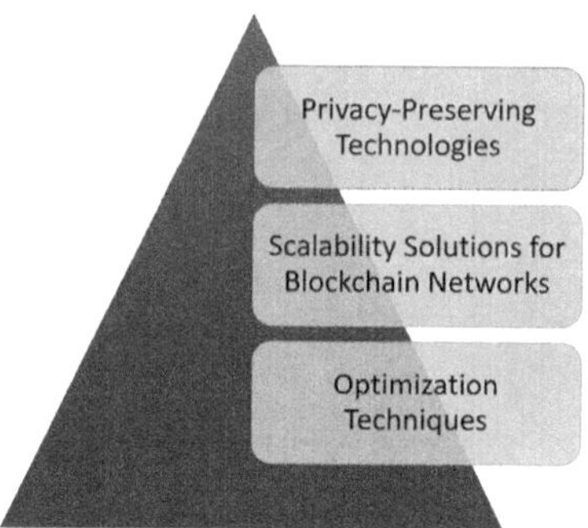


*Figure 4.6* Viable solutions.

## 4.7 CONCLUSION AND FUTURE SCOPE

The 6G wireless technology is a cellular technology that replaces 5G and has the capacity to use higher frequencies. This development makes it possible for 6G networks to have noticeably lower latency and much more capacity. The main goals of 6G internet is to enable communications with a latency of one microsecond, which is 1000 times quicker than 5G networks throughput of one millisecond or 1/1000th the latency. This combination of blockchain and Internet of Things technologies is an effort to add to the current conversation about improving data privacy in 6G networks. It aims to give insights for scholars, practitioners, and politicians interested in forming the future of communication networks by a detailed analysis of the problems and possible solutions. The 6G technology industry is expected to transform imaging, presence technology, and location awareness, bringing significant improvements. In conjunction with AI, 6G's computational infrastructure will be able to determine the best places to perform computing operations, including choices about data processing, sharing, and storage. The fact that 6G is not yet a working technology must be emphasized. Although several companies are contributing to the advancement of this next-generation wireless standard, it will likely take some years before industry standards for devices that support 6G-enabled networks become a reality.

## REFERENCES

[1] Li, W., Su, Z., Li, R., Zhang, K., & Wang, Y. (2020). Blockchain-based data security for artificial intelligence applications in 6G networks. *IEEE Network*, *34*(6), 31–37.

[2] Nguyen, T., Tran, N., Loven, L., Partala, J., Kechadi, M. T., & Pirttikangas, S. (2020). Privacy-aware blockchain innovation for 6G: Challenges and

opportunities. In *2020 2nd 6G Wireless Summit (6G SUMMIT)* IEEE (pp. 1–5).

[3] Pajooh, H. H., Demidenko, S., Aslam, S., & Harris, M. (2022). Blockchain and 6G-enabled IoT. *Inventions*, *7*(4), 109.

[4] Velliangiri, S., Manoharan, R., Ramachandran, S., & Rajasekar, V. (2021). Blockchain based privacy preserving framework for emerging 6G wireless communications. *IEEE Transactions on Industrial Informatics*, *18*(7), 4868–4874.

[5] Hewa, T., Gür, G., Kalla, A., Ylianttila, M., Bracken, A., & Liyanage, M. (2020). The role of blockchain in 6G: Challenges, opportunities and research directions. In *2020 2nd 6G Wireless Summit (6G SUMMIT)* IEEE (pp. 1–5).

[6] Jahid, A., Alsharif, M. H., & Hall, T. J. (2023). The convergence of blockchain, IoT and 6G: Potential, opportunities, challenges and research roadmap. *Journal of Network and Computer Applications*, *217*, 103677.

[7] Shen, X. S., Liu, D., Huang, C., Xue, L., Yin, H., Zhuang, W., ... & Ying, B. (2022). Blockchain for transparent data management toward 6G. *Engineering*, *8*, 74–85.

[8] Wang, J., Ling, X., Le, Y., Huang, Y., & You, X. (2021). Blockchain-enabled wireless communications: A new paradigm towards 6G. *National Science Review*, *8*(9), nwab069.

[9] Han, C., Kim, G. J., Alfarraj, O., Tolba, A., & Ren, Y. (2022). ZT-BDS: A secure blockchain-based zero-trust data storage scheme in 6G edge IoT. *Journal of Internet Technology*, *23*(2), 289–295.

[10] Mustafa Hilal, A., Alzahrani, J. S., Abunadi, I., Nemri, N., Al-Wesabi, F. N., Motwakel, A., ... & Sarwar Zamani, A. (2022). Intelligent deep learning model for privacy preserving IIoT on 6G environment. *Computers, Materials & Continua*, *72*(1), 333–348

[11] Zainuddin, A. A., Omar, N. F., Zakaria, N. N., & Camara, N. A. M. (2023). Privacy-preserving techniques for IoT data in 6G networks with blockchain integration: A review. *International Journal on Perceptive and Cognitive Computing*, *9*(2), 80–92.

[12] Xu, H., Klaine, P. V., Onireti, O., Cao, B., Imran, M., & Zhang, L. (2020). Blockchain-enabled resource management and sharing for 6G communications. *Digital Communications and Networks*, *6*(3), 261–269.

[13] Khan, A. H., Hassan, N. U., Yuen, C., Zhao, J., Niyato, D., Zhang, Y., & Poor, H. V. (2021). Blockchain and 6G: The future of secure and ubiquitous communication. *IEEE Wireless Communications*, *29*(1), 194–201.

[14] Sekaran, R., Patan, R., Raveendran, A., Al-Turjman, F., Ramachandran, M., & Mostarda, L. (2020). Survival study on blockchain based 6G-enabled mobile edge computation for IoT automation. *IEEE Access*, *8*, 143453–143463.

[15] Kumar, R., Jain, V., Yie, L. W., Teyarachakul, S. (2023). *Convergence of IoT, Blockchain, and Computational Intelligence in Smart Cities.*, USA: CRC Press. ISBN 9781032404240.

[16] Mao, B., Liu, J., Wu, Y., & Kato, N. (2023). Security and privacy on 6G network edge: A survey. *IEEE Communications Surveys & Tutorials*, 25(2), 1095–1127.

[17] Atlam, H. F., Azad, M. A., Altamimi, M., & Fadhel, N. (2022). Role of Blockchain and AI in Security and Privacy of 6G. In *AI and Blockchain Technology in 6G Wireless Network* (pp. 93–115). Singapore: Springer Nature Singapore.

[18] Lu, Y., Huang, X., Zhang, K., Maharjan, S., & Zhang, Y. (2020). Low-latency federated learning and blockchain for edge association in digital twin empowered 6G networks. *IEEE Transactions on Industrial Informatics*, *17*(7), 5098–5107.

[19] Shah, K., Chadotra, S., Tanwar, S., Gupta, R., & Kumar, N. (2022). Blockchain for IoV in 6G environment: Review solutions and challenges. *Cluster Computing*, *25*(3), 1927–1955.

[20] Ylianttila, M., Kantola, R., Gurtov, A., Mucchi, L., Oppermann, I., Yan, Z., ... & Röning, J. (2020). 6G white paper: Research challenges for trust, security and privacy. *arXiv preprint arXiv:2004.11665*.

[21] Aggarwal, S., Kumar, N., & Tanwar, S. (2020). Blockchain-envisioned UAV communication using 6G networks: Open issues, use cases, and future directions. *IEEE Internet of Things Journal*, *8*(7), 5416–5441.

[22] Mirtskhulava, L., Iavich, M., Razmadze, M., & Gulua, N. (2021, November). Securing medical data in 5G and 6G via multichain blockchain technology using post-quantum signatures. In *2021 IEEE International Conference on Information and Telecommunication Technologies and Radio Electronics (UkrMiCo)* (pp. 72–75). IEEE.

[23] Saini, R., Bera, A., Behera, B. K., Ahmed, E. A., Jamjoom, M., & Farouk, A. (2023). Designing quantum blockchain system integrated with 6G network. *Journal of King Saud University-Computer and Information Sciences*, *35*(10), 101847.

[24] Sun, Z., Liang, W., Qi, F., Dong, Z., & Cai, Y. (2021). Blockchain-based dynamic spectrum sharing for 6G UIoT networks. *IEEE Network*, *35*(5), 143–149.

[25] Nguyen, V. L., Lin, P. C., Cheng, B. C., Hwang, R. H., & Lin, Y. D. (2021). Security and privacy for 6G: A survey on prospective technologies and challenges. *IEEE Communications Surveys & Tutorials*, *23*(4), 2384–2428.

[26] Yadav, M., Agarwal, U., Rishiwal, V., Tanwar, S., Kumar, S., Alqahtani, F., & Tolba, A. (2023). Exploring synergy of blockchain and 6G network for industrial automation. *IEEE Access*, *11*, 137163–137187.

[27] Shahzad, K., Aseeri, A. O., & Shah, M. A. (2022). A blockchain-based authentication solution for 6G communication security in tactile networks. *Electronics*, *11*(9), 1374.

[28] Wang, M., Zhu, T., Zhang, T., Zhang, J., Yu, S., & Zhou, W. (2020). Security and privacy in 6G networks: New areas and new challenges. *Digital Communications and Networks*, *6*(3), 281–291.

[29] Peng, T., Gong, B., & Zhang, J. (2023). Towards privacy preserving in 6G networks: Verifiable searchable symmetric encryption based on blockchain. *Applied Sciences*, *13*(18), 10151.

[30] El Ghor, H., & Nakhal, B. (2023). A Blockchain-Enabled Approach for Secure Data Sharing in 6G-based Internet of Things Networks. In *Wireless*

*Networks: Cyber Security Threats and Countermeasures* (pp. 227–246). Cham: Springer International Publishing.

[31] Sun, Y., Liu, J., Wang, J., Cao, Y., & Kato, N. (2020). When machine learning meets privacy in 6G: A survey. *IEEE Communications Surveys & Tutorials*, *22*(4), 2694–2724.

[32] Nguyen, D. C., Ding, M., Pathirana, P. N., Seneviratne, A., Li, J., Niyato, D., ... & Poor, H. V. (2021). 6G internet of things: A comprehensive survey. *IEEE Internet of Things Journal*, *9*(1), 359–383.

[33] Gupta, R., Nair, A., Tanwar, S., & Kumar, N. (2021). Blockchain-assisted secure UAV communication in 6G environment: Architecture, opportunities, and challenges. *IET Communications*, *15*(10), 1352–1367.

[34] Bodkhe, U., Verma, A., Saraswat, D., Bhattacharya, P., & Tanwar, S. (2022). Adoption of blockchain for data privacy in 6G-envisioned augmented reality: Opportunities and challenges. *Emerging Technologies for Computing, Communication and Smart Cities: Proceedings of ETCCS, 2021*, 519–532.

[35] Wang, Z., Xu, Y., Liu, J., Li, Z., Li, Z., Jia, H., & Wang, D. (2022). An efficient data sharing scheme for privacy protection based on blockchain and edge intelligence in 6g-vanet. *Wireless Communications and Mobile Computing*, *2022*, 1–18.

[36] Osorio, D. P. M., Ahmad, I., Sánchez, J. D. V., Gurtov, A., Scholliers, J., Kutila, M., & Porambage, P. (2022). Towards 6G-enabled internet of vehicles: Security and privacy. *IEEE Open Journal of the Communications Society*, *3*, 82–105.

[37] Porambage, P., Gür, G., Osorio, D. P. M., Liyanage, M., Gurtov, A., & Ylianttila, M. (2021). The roadmap to 6G security and privacy. *IEEE Open Journal of the Communications Society*, *2*, 1094–1122.

[38] Lin, H., Garg, S., Hu, J., Kaddoum, G., Peng, M., & Hossain, M. S. (2020). A blockchain-based secure data aggregation strategy using sixth generation enabled network-in-box for industrial applications. *IEEE Transactions on Industrial Informatics*, *17*(10), 7204–7212.

[39] Kalla, A., De Alwis, C., Porambage, P., Gür, G., & Liyanage, M. (2022). A survey on the use of blockchain for future 6G: Technical aspects, use cases, challenges and research directions. *Journal of Industrial Information Integration*, *30*, 100404.

[40] Zuhair, M., Patel, F., Navapara, D., Bhattacharya, P., & Saraswat, D. (2021, April). BloCoV6: A blockchain-based 6G-assisted UAV contact tracing scheme for COVID-19 pandemic. In *2021 2nd International Conference on Intelligent Engineering and Management (ICIEM)* (pp. 271–276). IEEE.

[41] Khowaja, S. A., Khuwaja, P., Dev, K., Lee, I. H., Khan, W. U., Wang, W., ... & Magarini, M. (2022). A secure data sharing scheme in community segmented vehicular social networks for 6G. *IEEE Transactions on Industrial Informatics*, *19*(1), 890–899.

[42] Vuppula, R., & Pradhan, H. S. (2023). Blockchain-oriented location privacy preserving for cooperative spectrum sensing in 6G wireless networks. In *IET Blockchain. Wiley*.

[43] Maksymyuk, T., Gazda, J., Volosin, M., Bugar, G., Horvath, D., Klymash, M., & Dohler, M. (2020). Blockchain-empowered framework for decentralized network management in 6G. *IEEE Communications Magazine*, *58*(9), 86–92.
[44] Kathole, A. B., Katti, J., Dhabliya, D., Deshpande, V., Rajawat, A. S., Goyal, S. B., ... & Suciu, G. (2022). Energy-aware UAV based on blockchain model using IoE application in 6G network-driven cybertwin. *Energies*, *15*(21), 8304.
[45] Sharma, T., Prasad, S. K., & Sharma, V. (2022, February). Research challenges of blockchain in 6G Network. In *2022 IEEE Delhi Section Conference (DELCON)* (pp. 1–7). IEEE.
[46] Sun, W., Li, S., & Zhang, Y. (2021). Edge caching in blockchain empowered 6G. *China Communications*, *18*(1), 1–17.
[47] Srivastava, V., Mahara, T., & Yadav, P. (2021). An analysis of the ethical challenges of blockchain-enabled E-healthcare applications in 6G networks. *International Journal of Cognitive Computing in Engineering*, *2*, 171–179.
[48] Zhou, Z., Wang, M., Huang, J., Lin, S., & Lv, Z. (2021). Blockchain in big data security for intelligent transportation with 6G. *IEEE Transactions on Intelligent Transportation Systems*, *23*(7), 9736–9746.
[49] Guo, F., Yu, F. R., Zhang, H., Li, X., Ji, H., & Leung, V. C. (2021). Enabling massive IoT toward 6G: A comprehensive survey. *IEEE Internet of Things Journal*, *8*(15), 11891–11915.
[50] Li, M., Yu, F. R., Si, P., Zhang, Y., & Qian, Y. (2022). Intelligent resource optimization for blockchain-enabled IoT in 6G via collective reinforcement learning. *IEEE Network*, *36*(6), 175–182.
[51] Soni, M., & Singh, D. K. (2023). Blockchain-based group authentication scheme for 6G communication network. *Physical Communication*, *57*, 102005.
[52] Ozdogan, M. O., Carkacioglu, L., & Canberk, B. (2022, May). Digital twin driven blockchain based reliable and efficient 6G edge network. In *2022 18th International Conference on Distributed Computing in Sensor Systems (DCOSS)* (pp. 342–348). IEEE.
[53] Whig, P., Velu, A., & Naddikatu, R. R. (2022). The Economic Impact of AI-Enabled Blockchain in 6G-Based Industry. In *AI and Blockchain Technology in 6G Wireless Network* (pp. 205–224). Singapore: Springer Nature Singapore.
[54] Zhang, H., Leng, S., Wu, F., & Chai, H. (2021). A DAG blockchain-enhanced user-autonomy spectrum sharing framework for 6G-enabled IoT. *IEEE Internet of Things Journal*, *9*(11), 8012–8023.
[55] Bhattacharya, P., Saraswat, D., Dave, A., Acharya, M., Tanwar, S., Sharma, G., & Davidson, I. E. (2021). Coalition of 6G and blockchain in AR/VR space: Challenges and future directions. *IEEE Access*, *9*, 168455–168484.
[56] Janani, K., & Ramamoorthy, S. (2022). Defending IoT Security Infrastructure with the 6G Network, and Blockchain and Intelligent Learning Models for the Future Research Roadmap. In *Challenges and Risks Involved in Deploying 6G and NextGen Networks* (pp. 177–203). USA: IGI Global.

[57] Soleymani, S. A., Goudarzi, S., Anisi, M. H., Movahedi, Z., Jindal, A., & Kama, N. (2021). PACMAN: Privacy-preserving authentication scheme for managing cybertwin-based 6G networking. *IEEE Transactions on Industrial Informatics*, *18*(7), 4902–4911.

[58] Sandeepa, C., Siniarski, B., Kourtellis, N., Wang, S., & Liyanage, M. (2022). A survey on privacy for B5G/6G: New privacy challenges, and research directions. *Journal of Industrial Information Integration*, *30*, 100405.

[59] Nguyen, T., Lovén, L., Partala, J., & Pirttikangas, S. (2021). The Intersection of Blockchain and 6G Technologies. In *6G Mobile Wireless Networks* (pp. 393–417). Cham: Springer International Publishing.

[60] Tan, L., Yu, K., Yang, C., & Bashir, A. K. (2021, October). A blockchain-based Shamir's threshold cryptography for data protection in industrial internet of things of smart city. In *Proceedings of the 1st Workshop on Artificial Intelligence and Blockchain Technologies for Smart Cities with 6G* (pp. 13–18). ACM.

[61] Rathod, T., Jadav, N. K., Alshehri, M. D., Tanwar, S., Sharma, R., Felseghi, R. A., & Raboaca, M. S. (2022). Blockchain for future wireless networks: A decade survey. *Sensors*, *22*(11), 4182.

[62] Ayaz, F., Sheng, Z., Tian, D., Nekovee, M., & Saeed, N. (2022). Blockchain-empowered AI for 6G-enabled internet of vehicles. *Electronics*, *11*(20), 3339.

[63] Bodkhe, U., & Tanwar, S. (2023). Network management schemes for IoT environment towards 6G: A comprehensive review. *Microprocessors and Microsystems*, *103*, 104928.

[64] Lin, J. C. W., Srivastava, G., Zhang, Y., Djenouri, Y., & Aloqaily, M. (2020). Privacy-preserving multiobjective sanitization model in 6G IoT environments. *IEEE Internet of Things Journal*, *8*(7), 5340–5349.

[65] Manogaran, G., Rawal, B. S., Saravanan, V., Kumar, P. M., Martínez, O. S., Crespo, R. G., ... & Krishnamoorthy, S. (2020). Blockchain based integrated security measure for reliable service delegation in 6G communication environment. *Computer Communications*, *161*, 248–256.

[66] Šarac, M., Pavlović, N., Bacanin, N., Al-Turjman, F., & Adamović, S. (2021). Increasing privacy and security by integrating a blockchain secure interface into an IoT device security gateway architecture. *Energy Reports*, 7, 8075–8082.

[67] Ch, R., Srivastava, G., Gadekallu, T. R., Maddikunta, P. K. R., & Bhattacharya, S. (2020). Security and privacy of UAV data using blockchain technology. *Journal of Information security and Applications*, *55*, 102670.

[68] Wang, Y., Tian, Y., Hei, X., Zhu, L., & Ji, W. (2021). A novel IoV block-streaming service awareness and trusted verification scheme in 6G. *IEEE Transactions on Vehicular Technology*, *70*(6), 5197–5210.

[69] Luo, Y., Gong, B., Zhu, H., & Guo, C. (2023). A trusted federated incentive mechanism based on blockchain for 6G network data security. *Applied Sciences*, *13*(19), 10586.

[70] Qiao, L., Dang, S., Shihada, B., Alouini, M. S., Nowak, R., & Lv, Z. (2022). Can blockchain link the future?. *Digital Communications and Networks*, *8*(5), 687–694.

[71] Yu, D., Ren, J., Yang, Q., Tarkoma, S., Siddula, M., & Dressler, F. (2023). Guest editorial special issue on when blockchain meets 5G/6G—enabling endogenously secure IoT. *IEEE Internet of Things Journal*, *10*(14), 11946–11948.

[72] Manojkumar, V., Sastry, V. N., & Srinivasulu Reddy, U. (2022). Security, Privacy Challenges and Solutions for Various Applications in Blockchain Distributed Ledger for Wireless-Based Communication Networks. In *AI and Blockchain Technology in 6G Wireless Network* (pp. 117–133). Singapore: Springer Nature Singapore.

[73] Garg, S., Goyal, S., & Bhandari, A. (2022). Role of Blockchain in Security of 6G Networks. In *Challenges and Risks Involved in Deploying 6G and NextGen Networks* (pp. 106–129). USA: IGI Global.

[74] Hu, S., Liang, Y. C., Xiong, Z., & Niyato, D. (2021). Blockchain and artificial intelligence for dynamic resource sharing in 6G and beyond. *IEEE Wireless Communications*, *28*(4), 145–151.

[75] Dhar Dwivedi, A., Singh, R., Kaushik, K., Rao Mukkamala, R., & Alnumay, W. S. (2021). Blockchain and artificial intelligence for 5G-enabled internet of things: Challenges, opportunities, and solutions. *Transactions on Emerging Telecommunications Technologies*, *35*(4), e4329.

[76] Siriwardhana, Y., Porambage, P., Liyanage, M., & Ylianttila, M. (2021, June). AI and 6G security: Opportunities and challenges. In *2021 Joint European Conference on Networks and Communications & 6G Summit (EuCNC/6G Summit)* (pp. 616–621). IEEE.

[77] Zawish, M., Ashraf, N., Ansari, R. I., Davy, S., Qureshi, H. K., Aslam, N., & Hassan, S. A. (2022). Toward on-device AI and blockchain for 6G-enabled agricultural supply chain management. *IEEE Internet of Things Magazine*, *5*(2), 160–166.

[78] Porambage, P., Gür, G., Osorio, D. P. M., Livanage, M., & Ylianttila, M. (2021, June). 6G security challenges and potential solutions. In *2021 Joint European Conference on Networks and Communications & 6G Summit (EuCNC/6G Summit)* (pp. 622–627). IEEE.

[79] Gupta, M., Ved, C., & Kumari, M. (2022). Emergence of Blockchain Applications with the 6G-Enabled IoT-Based Smart City. In *Blockchain for 6G-Enabled Network-Based Applications* (pp. 213–235). USA: CRC Press.

[80] Gao, S., Li, G., Feng, L., Chen, Y., & Chen, Y. (2023). A secure data sharing system for 6G networks. *IEEE Access*, *11*, 133281–133293.

[81] Moosavi, N., & Taherdoost, H. (2022, August). Blockchain-enabled network for 6G wireless communication systems. In *International Conference on Intelligent Cyber Physical Systems and Internet of Things* (pp. 857–868). Springer International Publishing.

[82] Liu, L., Liang, W., Mang, G., & Dong, Z. (2021, June). Blockchain based spectrum sharing over 6g hybrid cloud. In *2021 International Wireless Communications and Mobile Computing (IWCMC)* (pp. 492–497). IEEE.

[83] Wang, M., Zhou, Z., & Ding, C. (2021). Blockchain-based decentralized reputation management system for internet of everything in 6G-enabled cybertwin architecture. *Journal of New Media*, *3*(4), 137–150.

[84] Shirwaikar, R. D., Faisal, A. M., Singh, A., & Shanbhag, D. D. (2021, December). A review on privacy and security in 6G networks. In *2021 International Conference on Forensics, Analytics, Big Data, Security (FABS)* (Vol. 1, pp. 1–6). IEEE.
[85] Kumar, P., Kumar, R., Islam, A. N., Garg, S., Kaddoum, G., & Han, Z. (2023). Distributed AI and blockchain for 6G-assisted terrestrial and non-terrestrial networks: Challenges and future directions. *IEEE Network*, *37*(2), 70–77.
[86] Putra, G. D., Dedeoglu, V., Kanhere, S. S., & Jurdak, R. (2022). Toward blockchain-based trust and reputation management for trustworthy 6G networks. *IEEE Network*, *36*(4), 112–119.
[87] Mohsan, S. A. H., Mazinani, A., Malik, W., Younas, I., Othman, N. Q. H., Amjad, H., & Mahmood, A. (2020). 6G: Envisioning the key technologies, applications and challenges. *International Journal of Advanced Computer Science and Applications*, *11*(9), 1097–1123.
[88] Wang, X., Zhu, H., Xiao, H., Zhou, Z., Yang, S., & Sun, L. (2023). Blockchain-enhanced trust management for mobile edge computing-enabled intelligent vehicular collaboration in the 6G era. *Transactions on Emerging Telecommunications Technologies*, *34*(7), e4791.
[89] Kalla, A., De Alwis, C., Gur, G., Gochhayat, S. P., Liyanage, M., & Porambage, P. (2022). Emerging directions for blockchainized 6G. *IEEE Consumer Electronics Magazine*, *13*(2), 42–51.
[90] Li, B., Deng, S., Yan, X., & Dustdar, S. (2022). The confluence of blockchain and 6G network: scenarios analysis and performance assessment. *arXiv preprint arXiv:2207.04744*.
[91] Asim, J., Khan, A. S., Saqib, R. M., Abdullah, J., Ahmad, Z., Honey, S., ... & Abbas, M. (2022). Blockchain-based multifactor authentication for future 6G cellular networks: A systematic review. *Applied Sciences*, *12*(7), 3551.
[92] Gupta, R., Shukla, A., & Tanwar, S. (2020). BATS: A blockchain and AI-empowered drone-assisted telesurgery system towards 6G. *IEEE Transactions on Network Science and Engineering*, *8*(4), 2958–2967.
[93] Yrjölä, S. (2020, June). How could blockchain transform 6G towards open ecosystemic business models?. In *2020 IEEE International Conference on Communications Workshops (ICC Workshops)* (pp. 1–6). IEEE.
[94] Friha, O., & Ferrag, M. A. (2021). Blockchain technology for 6G communication networks. *Cybersecurity Issues in Emerging Technologies*, 77.
[95] Khan, L. U., Yaqoob, I., Imran, M., Han, Z., & Hong, C. S. (2020). 6G wireless systems: A vision, architectural elements, and future directions. *IEEE Access*, *8*, 147029–147044.
[96] Pokhrel, S. R. (2020, September). Federated learning meets blockchain at 6G edge: A drone-assisted networking for disaster response. In *Proceedings of the 2nd ACM MobiCom Workshop on Drone Assisted Wireless Communications for 5G and Beyond* (pp. 49–54). ACM.
[97] Gupta, R., Jadav, N. K., Nair, A., Tanwar, S., & Shahinzadeh, H. (2022, September). Blockchain and AI-based secure onion routing framework for data dissemination in IoT environment underlying 6G networks. In *2022*

*Sixth International Conference on Smart Cities, Internet of Things and Applications (SCIoT)* (pp. 1–6). IEEE.

[98] Porambage, P., & Liyanage, M. (2023). *Security and Privacy Vision in 6G: A Comprehensive Guide.* John Wiley & Sons.

[99] Yan, Z., Yang, L. T., Li, T., Miche, Y., Yu, S., & Yau, S. S. (2022). Guest editorial: Trust, security and privacy of 6G. *IEEE Network*, *36*(4), 100–102.

[100] Ramkumar, G., Sreeja, B. P., Singh, D. P., Perwej, Y., Deepa, S., & Sakthivel, D. (2022, April). 6G–secure data cluster approach with blockchain. In *2022 3rd International Conference on Intelligent Engineering and Management (ICIEM)* (pp. 882–886). IEEE.

[101] Liu, M., Wu, Q., Hei, Y., & Li, D. (2023). Blockchain-based licensed spectrum fair distribution method towards 6G-envisioned communications. *Applied Sciences*, *13*(16), 9231.

[102] Kim, M., Oh, I., Yim, K., Sahlabadi, M., & Shukur, Z. (2023). Security of 6G enabled vehicle-to-everything communication in emerging federated learning and blockchain technologies. *IEEE Access*, *12*, 33972–34001

[103] Gera, B., Raghuvanshi, Y. S., Rawlley, O., Gupta, S., Dua, A., & Sharma, P. (2023). Leveraging AI-enabled 6G-driven IoT for sustainable smart cities. *International Journal of Communication Systems*, *36*(16), e5588.

[104] Kumar, R., Singh, R. C., Jain, V. (2023). *Modeling for Sustainable Development: A Multidisciplinary Approach.* USA: Nova Science Publishers, DOI: https://doi.org/10.52305/HAXA0362

Chapter 5

# Securing the future

## Navigating network security and data privacy challenges in the 6G era

*Pawan Whig, Balaram Yadav Kasula, Nikhitha Yathiraju, Anupriya Jain, and Seema Sharma*

### 5.1 INTRODUCTION

The appearance of the sixth era of remote correspondence, ordinarily known as 6G, heralds another period for the network and its innovative potential outcomes. As we stand at the cusp of this groundbreaking jump, it becomes simple to examine the perplexing transaction between network security and information protection in the 6G climate. This presentation makes way for a complete investigation of the effects and difficulties that emerge in establishing a tremendous and interconnected scene of 6G organizations.

To comprehend the meaning of 6G, we should initially follow the development of remote correspondence. Every age, from 1G to the current 5G, has achieved momentous changes in the manner in which we associate and impart. While 5G has established the groundwork for super quick and low-dormancy correspondence, 6G vows to push the limits much further. With expected highlights like terahertz frequencies, holographic correspondence, and consistent combination of man-made reasoning, 6G imagines a future where the network rises above the constraints of the present.

As we embrace the commitments of 6G, it becomes clear that the expanded network and the expansion of arising advancements come inseparably with uplifted security challenges. The complex trap of gadgets, fueled by man-made consciousness and flawlessly imparting through 6G organizations, presents a tremendous assault surface for noxious entertainers. Understanding and addressing these difficulties is fundamental to opening the maximum capacity of 6G while guaranteeing the honesty, privacy, and accessibility of information.

A main issue in the 6G story is the split in the difference of top-tier advances that reevaluate how we see association. Man-made scholarly ability, with its capacity to manage tremendous extents of information ceaselessly, stays as a foundation of 6G applications. The Catch of Things (IoT) further develops the affiliation's certificate, making a characteristic system where contraptions convey dependably. Edge figuring, with its decentralized dealing with limits, chips away at the ability of information managing and

DOI: 10.1201/9781003583127-5

diminishes inertia. In any case, these movements present its own strategy of prosperity assessments, requiring a careful and adaptable technique for overseeing network security.

As we leave on the 6G trip, it is key to perceive and comprehend the potential deficiencies that might be taken advantage of by noxious parts. The use of terahertz frequencies, for example, opens up new difficulties in guaranteeing the gathering of correspondence, as these frequencies have different causing credits veered from lower frequencies. The mix of PC-based information-driven applications presents the bet of antagonistic assaults, where an adroit situation can be controlled to seek after mistaken choices. Besides, the huge number of interconnected gadgets in the IoT scene makes a wandering randomly assault surface, which needs mindful ideas concerning security shows.

While network security spins around protecting the framework and correspondence channels, information confirmation arises as an identical worry in the 6G scene. The anticipated trade of information between gadgets, together with the specific considered reproduced understanding of applications, raises issues about the secret and commitment in regards to data crossing these affiliations. Finding some kind of congruity between taking care of the force of information for creative applications and safeguarding individual and different evened out security changes into a central test.

Past the mechanical viewpoints, moral contemplations come to the front in the 6G time period. The skillful new turn of events and plan of PC-based information, the moral use of information, and contemplations of critical worth and inclusivity become essential bits of the conversation. Besides, examining the administrative scene changes into a stunning task, as the fast speed of imaginative levels of progress means a large part of the time overpowers the organizing of extensive administrative plans. This portion hops into the moral and legitimate assessments that help the capable progress of 6G affiliations. This presentation gives a brief look into the complex scene of 6G, where the commitments of network and mechanical development are unpredictably interlaced with the difficulties of getting networks and saving information protection. Resulting sections will dig further into explicit perspectives, investigating inventive arrangements, best practices, and the cooperative endeavors expected to guarantee a solid and versatile 6G climate. As we leave on this excursion, the mission for an amicable conjunction of state-of-the-art innovation and strong safety efforts turns into the core value for molding the eventual fate of remote correspondence.

## 5.2 LITERATURE REVIEW

Porambage and Liyanage (2023) contribute essentially with an extensive manual for security and protection dreams in the 6G period, offering a guide for exploring these vital viewpoints [1, 2]. Tending to the particular security necessities and difficulties of 6G innovations and applications, Abdel Hakeem et al. (2022) give bits of knowledge vital to building a powerful security structure [3]. Sandeepa et al. (2022) turn towards protection, introducing an overview that explains arising difficulties and examination headings in the B5G/6G scene [4]. Expanding the talk, Zaman et al. (2023) offer an exhaustive outline of safety and protection contemplations, diving into the subtleties of the 6G time [5]. Furthermore, Bhat and Alqahtani (2021) give bits of knowledge into the ongoing status and future points of view of the 6G environment, giving an all-encompassing perspective on the mechanical scene [6]. Investigating the convergence of safety, protection, and open doors for aeronautical organizations, Lopez et al. (2022) direct an exploratory review, recognizing new hindrances and open doors inside 6G systems administration [7]. Shahraki et al. (2021) present a study zeroing in on 6G organizations' applications, center administrations, empowering advances, and future difficulties, adding to an exhaustive comprehension [8]. Garg et al. (2023) dig into convention security in 6G organizations, tending to key perspectives that are vital for guaranteeing secure correspondence [9]. Osorio et al. (2022) center around the intermingling of 6G with the Web of Vehicles, tending to security and protection worries for an associated vehicular environment [10–12]. Asghar et al. (2022) add to the talk by investigating the development of remote correspondence to 6G, featuring possible applications and exploration headings [13]. Ramezanpour et al. (2023) lead an overview on security and protection weaknesses of 5G/6G and WiFi 6, framing difficulties and examination bearings in the conjunction viewpoint [14]. Aslam et al. (2021) spin around the mental radio affiliation application in 6G, keeping an eye out for security issues and key difficulties in this space [15]. Akbar et al. (2022) offer a broad framework covering inconveniences, necessities, applications, empowering movements, use cases, man-made awareness coordination issues, and security perspectives regarding 6G affiliations [16]. Henrique and Prasad (2022) give a wide outline of 6G movements, offering experiences into the way to future distant improvements in 2030 [17]. Lu and Zheng (2020) lead a focus on movements, conditions, challenges, and related issues in 6G, adding to a general comprehension of the causing situation [18]. Dangi et al. (2023) center around key advancements, headings, and advances in 6G versatile organizations, giving a broad outline of the progressions in this area [19]. At long last, Gupta and Fernando (2021) investigate secure noticeable light correspondence in future (6G) IoT, framing potential security contemplations in this arising correspondence worldview [20]. On the whole, these examinations add to a rich group of writing that completely investigates the subtleties, difficulties, and expected arrangements inside the spaces of organization security and information protection in the 6G period.

## 4.3 NETWORK SECURITY

Network security is an intensive discipline inside the more prominent field of association prosperity that brilliantly highlights protecting the fairness, secret, and receptiveness of PC affiliations and the information they send. The essential objective of affiliation security is to defend data and improvement framework from unapproved access, assaults, and disrupting impacts. Figure 5.1 presents the component of network security.

Network security wraps a perplexing technique for overseeing protecting motorized correspondence and data frameworks against a substitute demonstration of electronic dangers. Access control shapes a fundamental layer, including attestation to check client or contraption characters and support to yield fitting consents thinking about position. Firewalls go most likely as a sincere obstruction, administering network traffic between confided in inner affiliations and outside ones, facilitated by predefined security rules [21]. Obstruction disclosure and avoidance frameworks (IDPS) carefully screen network works out, expediently making heads or making robotized moves fittingly aware of seen chances. Network division unequivocally distributes to contain and limit the effect of safety occasions, especially in impeding sidelong headway by aggressors. Distant affiliation prosperity tries, including encryption, certification shows, and run of the mill seeing, guarantee the strength of faraway correspondence. As a dynamic and making field, network security requires the mix of improvement, procedures, and client mindfulness in regards to fan out areas of strength for and against a reliably making mechanized risk scene. It stays as a foundation of online security, adding to the general adaptability and steadfast nature of motorized correspondence and data structures.

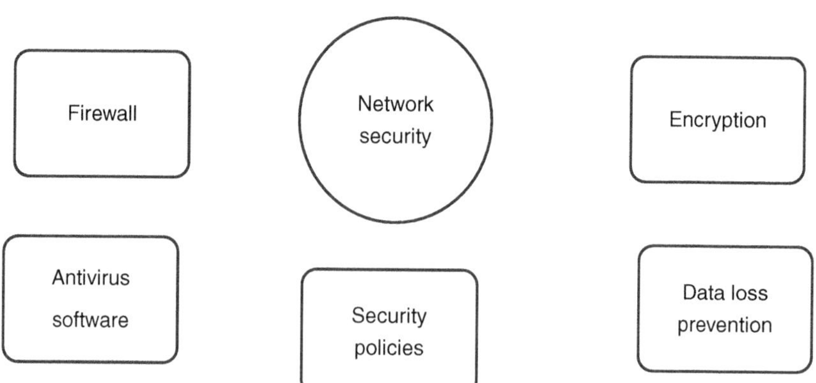


*Figure 5.1* Component of network security.

## 5.4 DATA PRIVACY

Anyway, this real impact in context raises issues about where and how data is taken care of. Ensuring data security transforms into a test when delicate information is taken care of at the association edge, away from ordinary concentrated server ranches.

*Artificial understanding and farsighted assessment* – The wire of man-made mental ability (man-made knowledge) in 6G familiarizes new angles with data taking care of and assessment. While man-made brainpower-driven applications offer redid and powerful organizations, they in a similar manner rely upon wide educational records. Changing the benefits of PC-based knowledge with the need to protect individuals from intruding profiling and perception transforms into a delicate endeavor. The fast progression of development regularly outflanks the improvement of managerial frameworks. The overview of 6G technology is presented in Figure 5.2.

*Ethical considerations in PC-based knowledge and data use* – The ethical usage of PC-based knowledge and data in 6G associations requires wary ideas of anticipated inclinations, isolation, and possibly adverse outcomes. Discovering some sort of congruity among improvement and moral norms transforms into a test as development moves at a fast speed.

## 5.5 METHODOLOGY

Keeping an eye out for the intricacies of connection security and data affirmation in the 6G time span requires an expansive, adaptable framework sorting out mechanical game-plans, methodology considerations, and supportive endeavors. This confusing strategy starts with a wide-forming

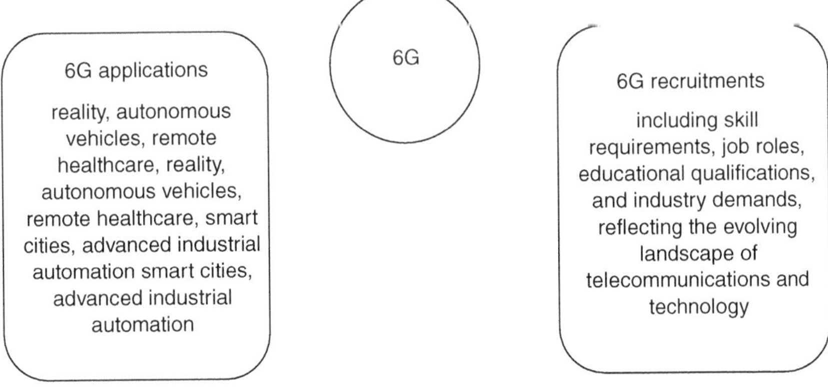


*Figure 5.2* 6G overview.

study to thoroughly figure out current difficulties, arising advances, and best practices. Security influence assessments review the repercussions for individual and authentic protection, guaranteeing consistence with information affirmation rules. Fanning out an administrative consistence system concurred with advancing legitimate necessities is focal, close by making moral rules and association structures for reenacted knowledge and information use. Client planning and care programs are basic to foster a security-cognizant outlook, kept up with by supportive endeavors with industry frill, government working environments, and association prosperity informed authorities. Persistent checking and change to the pushing risk scene, got together with occasion reaction and recuperation coordinating, guarantee adaptability. At last, ordinary testing and age practices are fundamental to see and direct potential lacks proactively, guaranteeing the power of the security planning.

## 5.6 CASE STUDY

In the quickly causing situation of faraway correspondence, the oncoming of the sixth period, 6G, guarantees amazing openness and mechanical levels of progress. In any case, with the tremendous potential for progress comes the essential need to address the complexities of affiliation security and information affirmation in this extraordinary period. This consistent assessment dives into a broad evaluation of affiliation security difficulties and information confirmation thoughts in the 6G climate, using a dataset named "Data_of_Attack_Back.csv." With 968 passages and 41 segments, this dataset gives a granular perspective on network traffic, working with a huge assessment of assault plans and the complexities of protecting sensitive data in the unprecedented space of 6G affiliations. Through a baffling system enveloping mechanical evaluations, risk evaluations, and administrative contemplations, this setting-focused assessment desires to uncover information into sensible methodology for setting the security act and saving information protection in the beginning of the 6G time span.

Detailed description of the dataset columns:

1. duration (float64):
    - Represents the duration of the network connection.
2. protocol_type (int64):
    - Categorizes the protocol used in the network connection.
3. service (int64):
    - Specifies the type of service associated with the network connection.
4. flag (float64):
    - Indicates the status or state of the network connection.

5. src_bytes (float64):
    - Denotes the number of source bytes transmitted in the connection.
6. dst_bytes (float64):
    - Represents the number of destination bytes transmitted in the connection.
7. land (int64):
    - Indicates whether the connection is from/to the same host/port.
8. wrong_fragment (int64):
    - Represents the number of wrong fragments in the network connection.
9. urgent (int64):
    - Indicates the urgency of the network connection.
10. dst_host_srv_diff_host_rate (int64):
    - Represents the rate of connections to different hosts through the same service at the destination host.
11. dst_host_serror_rate (float64):
    - Represents the error rate for connections to the destination host.
12. dst_host_srv_serror_rate (float64):
    - Represents the service error rate for connections to the destination host.
13. dst_host_rerror_rate (float64):
    - Represents the rate of connections that have a "REJ" error at the destination host.
14. dst_host_srv_rerror_rate (float64):
    - Represents the rate of service connections with a "REJ" error at the destination host.

The information kinds of the segments incorporate float64 for mathematical elements and int64 for all out highlights. The memory utilization for the DataFrame is 310.2 KB. This dataset fills in as the establishment for examining network assault designs and executing compelling safety efforts with regards to the 6G climate.

An exactness score of 0.9990781273705034 for a KNeighborsClassifier shows an elevated degree of execution in the order task. The exactness score is a metric that actions the proportion of accurately anticipated examples to the all-out number of cases in the dataset. For this situation, the score is extremely near 1, and that implies that the model has made exact expectations for by far most of the examples in the dataset, as shown in Table 5.1.

Breakdown of what this exactness score implies:

- **Accuracy score: 0.9990781273705034 (99.91%):**
    - This implies that the KNeighborsClassifier correctly predicted the target variable (or class labels) for approximately 99.91% of the instances in the dataset.

*Table 5.1* Confusion matrix

| | *Predicted negative* | *Predicted positive* |
|---|---|---|
| Actual negative | 450 (TN) | 5 (FP) |
| Actual positive | 2 (FN) | 511 (TP) |

Precision (Positive Predictive Value) = TP / (TP + FP) ≈ 0.9903

Recall (Sensitivity, True Positive Rate) = TP / (TP + FN) ≈ 0.9961

F1 Score = 2 * (Precision * Recall) / (Precision + Recall) ≈ 0.9932

Explanations for high accuracy:

1. **Well-separated classes:**
   - The classes in the dataset might be well-separated, making it easier for the KNeighborsClassifier to accurately distinguish between them. When classes are distinct and have clear boundaries, the model can make precise predictions.
2. **Appropriate model complexity:**
   - The choice of the KNeighborsClassifier and its parameter settings (such as the number of neighbors) might be appropriate for the characteristics of the dataset. The model's ability to capture the underlying patterns in the data contributes to its high accuracy.
3. **Feature relevance:**
   - The features used in the classification task may be highly relevant and informative. When features provide clear distinctions between classes, the model can leverage this information to make accurate predictions.
4. **Sufficient training data:**
   - Having a sufficiently large and diverse training dataset is crucial for building a robust model. If the dataset used for training is representative of the overall data distribution, the model is more likely to generalize well to unseen instances.

## 5.7 CONCLUSION

For this situation review, we applied the KNeighborsClassifier to address network security challenges utilizing the "Data_of_Attack_Back.csv" dataset with regards to the 6G climate. The model showed an uncommonly high exactness score of 99.91%, demonstrating its viability in precisely grouping cases inside the dataset. The amazing exhibition of the KNeighborsClassifier proposes that the model effectively caught and utilized the basic examples present in the information. A few elements add to this high exactness, including very much isolated classes, fitting model intricacy, and the importance of elements utilized for characterization. Moreover, taking into account the powerful idea of organization security dangers, persistent observing and variation of the model will be significant to keep up with its viability in true situations. This contextual analysis exhibits the capability of AI, explicitly the KNeighborsClassifier, in tending to organize security challenges. As innovation keeps on propelling, utilizing refined models becomes basic for guaranteeing the honesty and secrecy of information in the developing scene of 6G organizations

## 5.8 FUTURE SCOPE

In the always developing domain of organization security inside the 6G time, what is in store holds promising open doors for additional progressions. The combination of versatile AI models, for example, those fit for dynamic acclimations to arising assault designs, is a basic road for investigation. Consolidating logical man-made intelligence will improve model interpretability, giving straightforwardness and responsibility in direction. Troupe learning draws near and the combination of ongoing danger knowledge feeds will add to a heartier guard against different digital dangers. Protection safeguarding AI strategies, conduct examination, and peculiarity identification are foremost in guaranteeing the security of touchy information.

## REFERENCES

[1] Wang, M., Zhu, T., Zhang, T., Zhang, J., Yu, S., & Zhou, W. (2020). Security and privacy in 6G networks: New areas and new challenges. *Digital Communications and Networks*, 6(3), 281–291.

[2] Porambage, P., & Liyanage, M. (2023). *Security and Privacy Vision in 6G: A Comprehensive Guide*. John Wiley & Sons.

[3] Abdel Hakeem, S. A., Hussein, H. H., & Kim, H. (2022). Security requirements and challenges of 6G technologies and applications. *Sensors*, 22(5), 1969.

[4] Sandeepa, C., Siniarski, B., Kourtellis, N., Wang, S., & Liyanage, M. (2022). A survey on privacy for B5G/6G: New privacy challenges, and research directions. *Journal of Industrial Information Integration*, 55, 100405.

[5] Zaman, S., Tariq, F., Khandaker, M., & Khan, R. T. (2023). A Comprehensive overview of security and privacy in the 6G era. *6G Wireless: The Communication Paradigm Beyond*, *2030*, 203.

[6] Bhat, J. R., & Alqahtani, S. A. (2021). 6G ecosystem: Current status and future perspective. *IEEE Access*, 9, 43134–43167.

[7] Lopez, M. A., Barbosa, G. N., & Mattos, D. M. (2022, July). New barriers on 6G networking: An exploratory study on the security, privacy and opportunities for aerial networks. In *2022 1st International Conference on 6G Networking (6GNet)* (pp. 1–6). IEEE.

[8] Shahraki, A., Abbasi, M., Piran, M. J., & Taherkordi, A. (2021). A comprehensive survey on 6G networks: Applications, core services, enabling technologies, and future challenges. *arXiv preprint arXiv:2101.12475*.

[9] Garg, T., Kaur, G., & Rana, P. S. (2023, October). Protocol security in 6th generation (6G) networks. In *International Conference on Computation Intelligence and Network Systems* (pp. 47–62). Springer Nature Switzerland.

[10] Osorio, D. P. M., Ahmad, I., Sánchez, J. D. V., Gurtov, A., Scholliers, J., Kutila, M., & Porambage, P. (2022). Towards 6G-enabled internet of vehicles: Security and privacy. *IEEE Open Journal of the Communications Society*, *3*, 82–105.

[11] Kumar, R., Jain, V., Leong, W. Y., Teyarachakul, S. (2023). *Convergence of IoT, Blockchain and Computational Intelligence in Smart Cities*. Taylor & Francis, CRC Press, ISBN: 9781032404240.

[12] Sun, Y., Liu, J., Wang, J., Cao, Y., & Kato, N. (2020). When machine learning meets privacy in 6G: A survey. *IEEE Communications Surveys & Tutorials*, *22*(4), 2694–2724.

[13] Asghar, U.S., Kanani, R., Roylance, R., & Mittnacht, S. (2022). Systematic Review of Molecular Biomarkers Predictive of Resistance to CDK4/6 Inhibition in Metastatic Breast Cancer. JCO Precis Oncol. 2022 Jan;6:e2100002. doi: 10.1200/PO.21.00002. PMID: 35005994; PMCID: PMC8769124.

[14] Ramezanpour, K., Jagannath, J., & Jagannath, A. (2023). Security and privacy vulnerabilities of 5G/6G and WiFi 6: Survey and research directions from a coexistence perspective. *Computer Networks*, *221*, 109515.

[15] Aslam, M. M., Du, L., Zhang, X., Chen, Y., Ahmed, Z., & Qureshi, B. (2021). Sixth generation (6G) cognitive radio network (CRN) application, requirements, security issues, and key challenges. *Wireless Communications and Mobile Computing*, *2021*, 1–18.

[16] Akbar, M. S., Hussain, Z., Sheng, Q. Z., & Mukhopadhyay, S. (2022). 6G survey on challenges, requirements, applications, key enabling technologies, use cases, AI integration issues and security aspects. *arXiv preprint arXiv:2206.00868*.

[17] Henrique, P. S. R., & Prasad, R. (2022). *6G: The Road to the Future Wireless Technologies 2030*. CRC Press.

[18] Lu, Y., & Zheng, X. (2020). 6G: A survey on technologies, scenarios, challenges, and the related issues. *Journal of Industrial Information Integration*, *19*, 100158.

[19] Dangi, A., Husain, I., Jordan, C. Z., Yu, S., & Luo, X. (2023). Conversion of CD73hiFR4hi anergic T cells to IFN-γ-producing effector cells disrupts established immune tolerance. *J Clin Invest*. 2023 Mar 1; *133*(5), e163872. doi: 10.1172/JCI163872. PMID: 36649085; PMCID: PMC9974094..

[20] Gupta, A., & Fernando, X. (2021, June). Exploring secure visible light communication in next-generation (6G) internet-of-things. In *2021 International Wireless Communications and Mobile Computing (IWCMC)* (pp. 2090–2097). IEEE.
[21] Rana, R., & Kumar, R. (2019). Performance analysis of AODV in presence of malicious node. *Acta Electronica Malaysia (AEM)*, *3*(1), 1–5.

Chapter 6

# 6G network security and artificial intelligence

*Dina Darwish and Utku Köse*

## 6.1 INTRODUCTION

The advancement of 6G application domains necessitates a novel network architecture that surpasses existing network designs [1]. The Nokia Bell Labs architecture for 6G architectural building blocks consists of four main interworking components [2]. These encompass platform, functional, specialized, and orchestration aspects, spanning from the physical layer to the service layer, each having its own unique characteristics. The "het-cloud" refers to a cloud environment that is heterogeneous in nature, meaning it allows for the creation, placement, and scaling of dynamic cloud services in a simplified manner. The service-based architecture of the 5G core network will be expanded to include the radio access network (RAN), a concept known as "RAN-core convergence". The objective is to integrate the RAN and core activities in order to streamline network operations [2]. Smaller networks, including in-body networks, will develop and function independently while still taking advantage of the wide area network. The implementation of 6G will involve the transformation of network software and cloud infrastructure into intelligent networks, leading to a significant advancement in wireless networks, shifting from just connecting devices to enabling intelligent connectivity. Therefore, AI is an essential component of the network and plays a pivotal role. In order to achieve the objectives of 6G, it is necessary to have AI services that are available everywhere in the scattered and diverse networks. The utilization of artificial intelligence is emphasized in various elements of 6G networks, such as intelligent wireless communications, closed-loop optimization of networks, and big data analytics.

The network softwarization technologies employed in 5G, namely software-defined networking (SDN), network function virtualization (NFV), multi-access edge computing (MEC), and network slicing, will remain significant in 6G systems. Therefore, the security risks linked to these technologies would continue to exist in 6G as well. Notable security issues in SDN involve deliberate assaults on the SDN controller, the northbound and



DOI: 10.1201/9781003583127-6

southbound interfaces, and the inherent vulnerabilities of the platforms utilized for installing SDN controllers and applications [3]. The security vulnerabilities associated with NFV include intentional attacks on virtual machines (VMs), virtual network functions (VNFs), hypervisors, VNF administrators, and NFV orchestrators [4]. 6G networks are susceptible to physical security concerns, distributed denial-of-service (DDoS) attacks, and man-in-the-middle assaults due to their extensive deployment [5]. The network slicing framework is vulnerable to possible security risks, including denial-of-service (DoS) attacks and unauthorized access to information through compromised slices [6]. The dynamicity and full automation of the 6G network are impeded by attacks targeting network softwarization technologies. The goal of 6G is to realize the deployment of the Internet of Everything (IoE), which encompasses a wide array of heterogeneous devices. Implementing IoE in 6G, especially for small-sized devices such as in-body sensors, is not viable using the SIM card-based device security paradigm. The major distribution and management processes demonstrate considerable inefficiency within a network of this size [7]. Due to the constrained resources of Internet of Things (IoT) devices, they are unable to incorporate intricate cryptographic methods to guarantee strong security [8], rendering them very vulnerable to attacks. These devices are susceptible to compromise and can potentially be exploited to initiate attacks. The collection of data via the hyper-connected IoE to support 6G applications gives rise to concerns over privacy. Exploiting resource-constrained IoT devices to steal data will have a negative impact on data privacy, location privacy, and identity privacy.

Hence, this chapter has examined a wide range of publications, encompassing both specialized pieces and review articles that provide a comprehensive analysis of the 6G to illustrate the interconnections among the topics that served as the foundation for the creation of the development trajectory. Also, this chapter includes several topics discussing the security of 6G using artificial intelligence along with other disruptive technologies, and finally, the conclusion section is provided.

## 6.2 BACKGROUND

The 1G network was introduced in the 1980s primarily to offer telephone services. The dissemination of information relies on analogue transmissions and lacks a globally recognized wireless protocol. As a result, there are other disadvantages that occur, including difficult transfers of responsibility, insufficient guarantees of confidentiality and privacy, and inadequate efficiency of the gearbox. Phone services lack encryption, rendering data transmissions and phone conversations devoid of security and confidentiality. Consequently, the whole network and its users encounter substantial security and privacy obstacles, such as cloning, eavesdropping, and unauthorized access [9, 10].

The 2G network utilizes digital modulation methods, namely time division multiple access (TDMA) and code division multiple access (CDMA), to facilitate voice and short messaging services. The GSM protocol, also known as the global system for mobile communications, is the dominant and widely utilized standard for mobile communication on the 2G network [11]. The objective of the GSM is to ensure a security level that is on a par with that of a public switched telephone network (PSTN). The system provides comprehensive security and privacy services across four key domains: (1) ensuring anonymity, (2) verifying authentication, (3) safeguarding signals, and (4) preserving user data [12]. Anonymity is attained by the utilization of ephemeral identifiers that impede the tracing of the user's genuine identity. Nevertheless, the authentic identity must be utilized during the first activation of the device, subsequent to which a temporary identifier is provided. Authentication is mostly employed by network providers to authenticate the identities of users. The authentication process utilizes an encryption-based technique known as "challenge and response". Encryption is utilized to provide the security of signals and user data, with the encryption keys being securely managed by the subscriber identity module (SIM). Two primary techniques for safeguarding user privacy include radio route encryption and temporary mobile subscriber identity (TMSI) [13]. Nevertheless, although 2G exhibits significant enhancements in terms of security and privacy compared to 1G, it still possesses some vulnerabilities. A significant security problem arises due to the unidirectional nature of the authentication process. Although the network is capable of verifying the user's identity, the user does not have the capability to verify the network's identity, which exposes security risks. Illicit equipment, such as base stations, can impersonate permitted network participants, deceiving users and stealing their data [14]. Moreover, the encryption fails to encompass the entirety of the communication process. Only a fraction of the wireless channel is secured with encryption, while the fixed network lacks any encryption, hence creating vulnerabilities for possible assaults from adversaries [15]. Both radio path encryption and TMSI have inherent limitations in terms of privacy and are vulnerable to many types of attacks, such as eavesdropping. In the year 2000, the 3G network was implemented to provide rapid data transmission and internet connectivity, with a minimum speed of 2 Mbps. However, this speed has the ability to provide advanced features that were not possible in the 1G and 2G networks, such as browsing the internet, streaming television, and video services [16].

The security of the 3G system relies on the underlying 2G technology. It was vital to include GSM and other components in 2G, but it was also crucial to integrate additional strong security measures. The 3G network addresses the security vulnerabilities seen in the 2G network.

The network is established by implementing the Authentication and Key Agreement (AKA) protocol, which enables mutual authentication. The 3rd

Generation Partnership Project (3GPP) provides a comprehensive security framework comprising two primary components: air interface security, primarily utilized to protect users and the signaling data transmitted via wireless connections, and user network authentication, ensuring the trustworthiness of users and both ends of the network. The dangers of wireless interface attacks can be classified into the following categories: unauthorized data access, threats to data integrity, DOS attacks, and unauthorized service access [17].

The LTE system integrates both conventional and state-of-the-art technology, such as coordinated multi-point transmission and reception (CoMP), multiple input multiple output (MIMO), and orthogonal frequency division multiplexing (OFDM) [18]. 4G systems consist of several components, including IP backbone networks, wireless core networks, wireless access networks, and intelligent mobile terminals. Therefore, the main sources of significant security risks include vulnerabilities in wireless connections, interception of data, unauthorized modification, insertion, or deletion of data, and concerns about authentication for both wireless and wired network entities, among other problems [19].

LTE networks are prone to several security vulnerabilities, such as DoS attacks, data integrity attacks, unauthorized exploitation of users and mobile devices, and location tracking at the medium access control (MAC) layer [20]. It is expected that the upcoming deployment of the 5G network will lead to increased speed, enhanced system integration, and strengthened security measures [10]. The main breakthrough of 5G networks is in their ability to provide uninterrupted connectivity to an increasing number of devices while concurrently delivering improved services to all devices.

In addition, the variety of devices that can be utilized is not restricted solely to smartphones, but also include other devices such as IoT equipment, which are capable of establishing a connection with the network [21, 22]. The security and privacy issues in 5G networks can be classified according to the network structure.

The existence of a network can be divided into three levels: the access networks, the backhaul networks, and the core network.

The presence of several nodes and access mechanisms in access networks gives rise to novel security considerations. The growing frequency of handovers between various access technologies increases the susceptibility to attacks. Backhaul communication refers to the transmission of data between the base station and the core network in the backhaul network. Communication can be established using several methods, including wireless channels, microwaves, satellite links, or wired connections [23].

Devices that are not connected to networks have reduced security and privacy vulnerabilities in comparison to devices that are connected to access networks. The vulnerabilities often stem from modifications made to access network components, such as the EUTRAN Node B (eNB) or the Mobility

Management Entity (MME) in the core network. However, the GPRS Tunnel Protocol (GTP) has the capability to provide additional guarantees of security. Furthermore, the integration of the backhaul network into the data plane is achieved by employing techniques such as NFV. This strategy guarantees the reduction of security concerns by transferring them to the defined networking (SDN) system [24]. The core network in 5G networks has a wide range of distinct functions [25].

Another method involves utilizing protocols that categorize devices into groups utilizing different group-based AKA protocols. The introduction of innovative methods designed to improve the effectiveness of the 5G network unintentionally creates vulnerabilities in terms of security [26]. MIMO systems efficiently address both passive and active threats. The use of OpenFlow in SDN magnifies the risks associated with intentional eavesdropping efforts caused by malicious apps or actions.

Furthermore, the relocation of a service or function from one resource to another creates security weaknesses within the context of NFV [27]. The lack of restrictions on the platform can result in the introduction of additional privacy concerns in 5G networks. Facilitate the seamless and regular movement of a user's confidential data from a restricted condition to an open one. As a result, the contact's status changes from being offline to being online, resulting in a substantial increase. Therefore, the imminent arrival of 5G technology would undoubtedly lead to a heightened susceptibility to data breaches.

Massive MIMO systems help to prevent both passive and intentional eavesdropping efforts. The use of OpenFlow in SDN exacerbates the susceptibilities presented by malevolent apps or behaviors. Furthermore, the movement of a service or function from one resource to another raises security risks in the context of NFV [27]. The multitude of firm types and application situations in 5G networks gives rise to additional privacy problems. The platform's lack of restrictions can lead to the frequent and easy transfer of a user's sensitive information from a limited state to an available state. As a result, the contact's status changes from being offline to being online, which greatly increases the susceptibility to leaks. Therefore, the imminent arrival of 5G technology will undoubtedly lead to privacy concerns that require immediate and efficient resolution in the future. Advancements in data mining and machine-learning technology have led to the development of robust privacy protection solutions, which are anticipated to improve their effectiveness in the future.

## 6.3 MAIN FOCUS OF THE CHAPTER

Each iteration of network technology exhibits imperfections. While there are some mechanisms in place to mitigate exploitation, the challenge of enhancing fundamental protocols still leaves significant risk. 6G security architecture and applications are vulnerable to several types of attacks, such as signaling DoS and DDoS. Authentication servers, energy depletion attacks, and user tracking and resource limits have a universal impact on all network generations and provide significant challenges. The following are the key insights gained from the obstacles encountered in legacy network security:

- The security of newly developed applications is frequently compromised. Contemporary network protocols exceed the performance of earlier network standards in novel applications. Nevertheless, they may present additional hazards. Multiple studies have predicted the susceptibility of these developing applications to impersonation and DoS assaults [28–30]. Enhancing technology security prior to implementation is of utmost importance. Compatibility between an outdated protocol and a modern protocol may expose vulnerabilities. The primary reason is the lack of compatibility.
- Compatibility is often bypassed by requesting authentication using an obsolete architecture. This access control mechanism has the potential to expose past problems. Decreases in performance, namely in the range of 52–54, result in the utilization of outdated networks for 4G-LTE devices. The process of mutual verification between the user equipment (UE) and authentication servers in 2G/3G standards is crucial.

  The assailant can subsequently obtain the UE's IMSI. It is important to mention the availability of dual network access. 6G has security challenges in authentication and identity management. When implementing protocols, the number of new vulnerabilities decreases compared to when designing protocols.
- Numerous operators and consumers are likely to experience financial repercussions. Prior to deploying a new architectural or software system, it is imperative to do thorough security testing.
- A permanent modification to the current architecture is still required to address the existing problems.
- The problems of mutual authentication and end-to-end encryption have not yet been resolved.

Due to the significant processing and communication requirements, it seems improbable that 5G would be able to fulfil these demands. The implementation of encryption and mutual authentication in 6G networks has the potential to negatively impact services that are highly sensitive to latency. Artificial intelligence (AI) integrated with 6G technology will be employed to enhance the security of networks.

Academic institutions and private companies alike have begun doing research on the sixth generation of wireless network technology, often known as 6G, following the introduction of the fifth generation of wireless technology, or 5G. When the year 2030 arrives, it is projected that the introduction of 6G will take place. There is a need to provide insightful information on the significant difficulties and complexity that are related to the problems of security, privacy, and trust that are associated with 6G networks. In addition, details on the particular technology and the accompanying security problems need more investigation. Comparing the security framework of the 5G architecture to that of the 6G architecture is important. Figure 6.1 illustrates the potential technologies that are expected to build the foundation of 6G networks.

6G employs artificial intelligence to facilitate the operation of completely autonomous networks. Hence, any assaults on AI systems, particularly machine-learning (ML) systems, would have an impact on the development of 6G technology. Poisoning attacks, data injection, data manipulation, logic corruption, model evasion, model inversion, model extraction, and membership inference attacks pose significant security risks to machine-learning systems. Increased feature diversity enhances the performance of AI systems. Privacy concerns arise from both data breaches and the unauthorized use of personal data, as the process of data manipulation is typically concealed from users. Blockchain is a crucial technology for unlocking the full potential of 6G systems. Blockchain is well-suited for managing resources in a decentralized manner, sharing spectrum, and managing services in highly

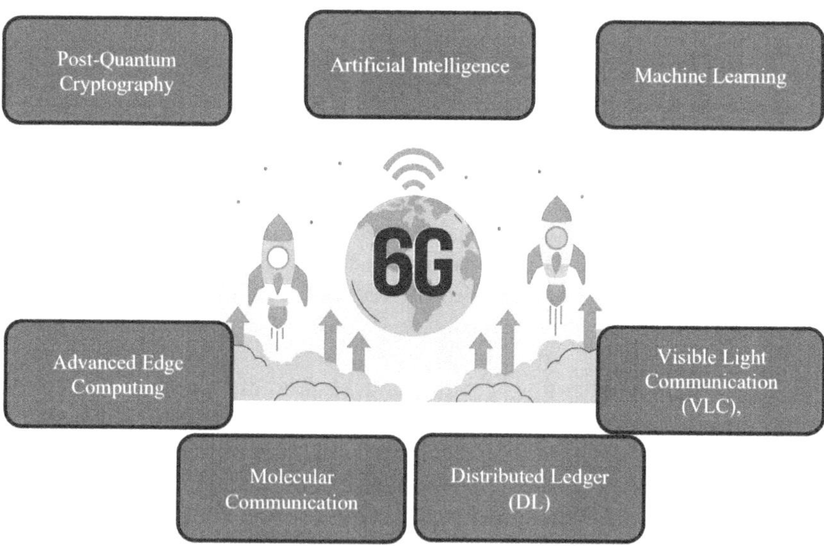


*Figure 6.1* The potential technologies that are expected to build the foundation of 6G networks.

extensive and scattered 6G networks. Quantum computers have the potential to carry out 51% attacks, which can disrupt the stability of the blockchain. Due to the public nature of blockchain networks, maintaining anonymity is a difficult task. The existing 5G standard does not address the security challenges posed by quantum computing. Instead, it relies on conventional cryptographic methods such as elliptic curve cryptography (ECC). The current security measures that rely on asymmetric key cryptography are susceptible to assaults from quantum computers, which will become prevalent in the 6G era. Consequently, the effectiveness of secure 5G communications, which rely on asymmetric key cryptography, may become obsolete in terms of post-quantum security unless quantum-safe cryptographic algorithms are developed. Visible light communication (VLC) is a technology that can be used for both indoor and outdoor systems, including positioning systems and vehicle-to-vehicle communication. The presence of eavesdropping, jamming, and node compromise systems poses significant threats to the secure use of VLC systems.

Implementing a system for detecting and preventing intrusions in SDN/NFV-enabled networks using deep reinforcement learning and deep neural networks (DNN) is feasible [31]. Compared to many traditional methods, they successfully protect against IP spoofing attacks, flow table overloading attacks, DDoS attacks, control plane saturation attacks, and host location hijacking attacks. Machine-learning techniques, including decision trees and random forests, have been demonstrated to be effective in identifying DDoS assaults in SDN systems. This is owing to their quick processing time and high accuracy, as mentioned in reference [32]. ML-based adaptive security techniques are successful in countering assaults on SDN and NFV due to the dynamic deployment of virtual functionalities as required by 6G networks. The assaults also employ AI approaches to acquire knowledge about weaknesses in a highly dispersed network. Therefore, detection methods that rely on rules are not effective. The current authentication and authorization solutions are inadequate for providing acceptable security in large-scale IoT due to on-device resource limits, challenges in key management in heterogeneous networks, and the extensive creation of device data. Anomaly-based intrusion detection systems in Industrial IoT (IIoT) [33] identify malicious packets by analyzing their activity. These detection systems that rely on learning utilize different characteristics of the data as the input, making them well-suited for identifying zero-day assaults. Utilizing machine learning to analyze communication connection properties and user behaviors for authentication and authorization, is a more effective strategy for future networks that have limited resources. By doing so, gadgets avoid allocating their finite resources towards implementing further intricate security measures. The sub-networks in 6G, which can be seen as an extension of local 5G networks outside specific areas of expertise, can gain advantages from employing security solutions based on machine learning within each sub-network and between multiple sub-networks. Machine-learning algorithms implemented at the network boundary can analyze the activities of different sub-networks and identify any suspicious network traffic originating from those sub-networks. Transferring large amounts of data between sub-networks may be futile, as these networks typically function independently. For the purpose of improving communication efficiency, a sub-network can exclusively transmit its acquired security intelligence to another network. Another sub-network can utilize collective intelligence, incorporate it into its machine-learning models, identify the malevolent network traffic of other networks, and implement adaptable regulations.

AI's capacity for extensive data analysis, coupled with the rapid processing capabilities of future computers and the increasing demand for automation in networks, makes it very susceptible to privacy breaches. The implementation of 6G necessitates the collection of an extensive volume of user data from billions of devices. However, users are no longer able to anticipate how external systems manage their data. For instance, the proposed intelligent authentication methods rely on physical features [34] and may utilize private user data. Vulnerable IoT devices, such as low-powered sensors, that transmit personal data to AI systems, are susceptible to data theft. Model inversion attacks on ML might potentially expose the training data, leading to privacy breaches [35]. AI-driven 6G networks require fewer human interventions in network operations. Machines possess a distinct approach to learning that diverges from human perception, and they lack the capacity to engage with ethical problems as people do. AI systems exhibit behavior based on the instructions and training they have received. However, they lack the ability to deviate from reasoning in extraordinary circumstances like humans may.

The 6G network should possess significantly enhanced AI capabilities. According to numerous academics, the 6G network is expected to be an "AI-empowered" network, where AI serves as both its driving force and most conspicuous characteristic. It should not just incorporate AI into its architecture like the 5G network does. The 6G network should entail a comprehensive fusion of presently evolving artificial intelligence tools and networking functionalities. Furthermore, given the growing significance of network security and privacy concerns in recent times, it is imperative to incorporate risk mitigation as a fundamental element of the design. Hence, exploring the potential enhancements to the existing 5G network that could form the foundation of the future 6G network is important. Additionally, it will address the security and privacy concerns related to each crucial technology and examine the potential applications of 6G networks in the future. Real-time intelligent edge, distributed artificial intelligence, intelligent radio, and three-dimensional intercoms are the four components that are necessary for a 6G network. Because of the enormous contributions that these four areas have made to the current research on 6G, it is essential to focus attention on investigating them. In addition to this, they are vulnerable to the most severe levels of privacy and security issues. The present study makes use of a variety of technologies, including software that is based on artificial intelligence, molecular communications, quantum communications, blockchain, technology that operates at speeds calculated with terahertz (THz), and technology that operates at VLC.

There is a significant possibility that these technologies might be utilized in a wide range of applications that are associated with 6G networks. These applications include, but are not limited to, multi-sensory X-Reality (multi-sensory XR) applications, networked robotics and autonomous systems, wireless brain–computer interfaces, blockchain and distributed ledger technologies, and a variety of other applications. The first three domains that include technologies that overlap with one another are the distributed artificial intelligence domain, the real-time intelligent edge domain, and the intelligent radio domain. Specifically, these three domains are the first three domains. Additionally, artificial intelligence is present at the intersection of all three domains as a result of our assumption that 6G networks would contain capabilities related to cognitive computing. Nevertheless, a number of technologies demonstrate a heightened awareness of particular risks. The video-link controller (VLC), in conjunction with both the real-time intelligent edge and the intelligent radio, is particularly susceptible to malicious activity and the processes of data transmission. The development of intelligent radio is made possible by both molecular communication and THz technology. The use of molecular communication technology poses problems regarding security and privacy, particularly in connection to authentication, encryption, and communication. These concerns are particularly relevant in the context of technology. The same may be said about THz technology, which is especially susceptible to authentication security issues and malicious activity. In the realms of distributed artificial intelligence and intelligent radio, the areas of blockchain technology and quantum communication connect with one another. In the context of this discussion, the most important aspects of security and privacy are the verification of identity, the restriction of access, the transfer of data, and the use of encryption. Different vulnerabilities are present in 6G applications. It is common for connected robots and autonomous systems to rely on AI and video-link communication (VLC) technologies. This may present difficulties in terms of malicious conduct, encryption, and the transfer of data. Multi-sensory augmented reality applications employ molecular communication technology, THz technology, and quantum communication technology. Consequently, they are prone to vulnerabilities associated with access control, malicious activity, and vulnerability to data transmission. Wireless brain–computer interfaces pose substantial security and privacy problems, even if they employ similar methods as the multi-sensory XR application. Conversely, the primary vulnerabilities lie in malevolent conduct and encryption. While blockchain and distributed ledger technologies, which heavily rely on blockchain, are generally considered secure in the context of the 6G network, they are nevertheless vulnerable to malicious actions. Typically, these recently established regions are susceptible to five primary classifications of security and privacy issues. The categories encompassed are authentication, access control, criminal behavior, encryption, and data transit. Some research has been effective in identifying and presenting the problems regarding privacy and security in the essential areas of the 6G network. The special privacy and security challenges that exist in 6G networks are presented and investigated in this research. Future 6G applications will have demanding prerequisites and necessitate enhanced network capabilities in contrast to the now established 5G networks.

These entities were created to facilitate a diverse range of essential 6G applications and can be classified accordingly. Furthermore, they have significant ramifications for the implementation of 6G security. The latency impact of security workflows will be taken into account for enhanced ultra-reliable, low latency communication (ERLLC/eURLLC) to ensure high service quality. Similarly, strict reliability requirements require highly effective security solutions that protect the availability of services and resources. Introducing further enhanced mobile broadband (FeMBB) will result in substantial enhancements in data transmission speeds. Nevertheless, the substantial data rates will present challenges in processing traffic for security objectives, including activities like attack identification, AI/ML workflows, traffic examination, and comprehensive encryption. To address the issue, one can employ distributed security solutions that entail processing traffic locally and in real-time across different segments of the network, ranging from the network edge to the central service cloud. Distributed ledger technology (DLT) will have a pivotal role in ensuring transparency, security, and redundancy. UMMTC, also known as ultra massive machine type communication, is specifically developed to cater to critical use cases that demand significantly enhanced security measures compared to 5G. The deployment and management of security solutions, such as distributed artificial intelligence and machine learning and privacy concerns, will face significant challenges due to the wide range of capabilities offered by the IoE. A crucial factor is the integration of new security features in a large number of devices that have limited resources. However, the implementation of security measures will become more intricate due to the increased mobility of network entities, as they regularly switch across edge networks and access services in various administrative domains. The new architecture incorporates advanced technologies such as state-of-the-art circuits, antennas, meta-material-based structures, and highly improved AI processors. This represents a significant shift in hardware design.

6G transceivers are designed with a modular architecture that enables the independent functioning of hardware components and transceiver algorithms. As a result, the transceiver algorithms can adjust and change their settings based on environmental and hardware data. Intelligent radio will utilize advanced artificial intelligence and machine-learning techniques to effectively tackle specific tasks in the wireless domain, including precise channel modeling, optimized physical layer design, adaptable spectrum access, complex network deployment, automation, optimization, and management challenges. Therefore, it is crucial to anticipate and prevent any suspicious activities carried out by hostile nodes during communication protocols in order to ensure the security of the radios. Edge intelligence, or EI, involves the application of AI/ML algorithms to gather, store, and analyze data at the network edge [36]. Within the domain of edge intelligence (EI), an edge server consolidates data produced by numerous networked devices. The data is distributed across numerous edge servers for training the model and subsequently used for analysis and prediction. As a result, devices can gain advantages such as faster feedback, less delay, and decreased expenses, all while improving their performance. Nevertheless, due to the aggregation of data from many sources and the heavy reliance of AI/ML algorithms on data, EI is exceedingly susceptible to numerous security vulnerabilities. Opponents can take advantage of this interconnectedness to carry out various forms of attacks, such data manipulation, data circumvention, or privacy breaches, which eventually undermine the effectiveness of AI/ML applications and reduce the advantages of EI. The diverse range of requirements for 6G and the planned full automation of network and service management using AI necessitate a significant restructuring of network service orchestration and management within the 6G architecture.

The ETSI ZSM (zero touch network and service management) [37] architecture for 5G is a highly promising concept aimed at optimizing the deployment of intelligent network management. The objective of 6G is to create autonomous networks that can attain self-X capabilities, such as self-configuration, self-monitoring, self-healing, and self-optimization, with minimal human involvement [38]. The ongoing efforts to integrate AI/ML into future networks, like the ETSI ZSM design, entail the implementation of closed-loop operation and the use of AI/ML methods to automate network management operations, including security, on a large scale. These efforts are vital measures in order to accomplish this goal. Due to the extensive integration of AI/ML, it is anticipated that these technologies would be utilized in a distributed and large-scale system to tackle various use cases, including network management. To effectively manage the large amount of data produced in 6G networks, it is imperative to employ distributed AI/ML methods, which facilitate swift control and analysis. In the realm of 6G, artificial intelligence and machine learning will be carefully placed in close proximity to the relevant data sources to achieve exceptionally low latency. The widespread adoption of AI/ML also presents significant challenges in terms of cybersecurity. This includes both the potential for AI/ML to enhance cybersecurity measures, as well as the possibility of AI/ML itself becoming a source of security vulnerabilities in certain situations. One key concern is the trustworthiness of AI/ML components in future networks. This issue becomes more significant when crucial network operations, such as security, are under the control of artificial intelligence. Trusted computing enablers, formal verification approaches, and integrity checks are crucial instruments for achieving this objective. In order to have control and be able to hold others accountable, it is essential to have visibility. Security specialists and monitoring personnel necessitate transparent and comprehensible understanding of AI-based systems, surpassing mere black box functionality. The research inquiry pertains to the effective and prompt surveillance of AI occurrences that breach security protocols. AI ethics and liability arise when integrating AI/ML into 6G security. A crucial concern is the fairness and ethical implications of AI-based optimization. More precisely, the debate arises as to whether AI-powered security solutions provide equal protection to all users. Another ambiguous aspect pertains to the party responsible in the event of a malfunction in AI-controlled security services. Managing liability is a complex undertaking involving independent companies functioning in an ICT context, which includes 6G security activities. Ensuring scalability and feasibility in distributed machine learning setups, such as federated learning, necessitates the implementation of secure data transports that respect privacy. Scalability poses a challenge for AI/ML controlled security functions due to the need for significant compute, communication, and storage resources. For example, the use of FeMMB results in significant volumes of data. When combined with AI/ML security controls, these processes can result in substantial additional workload. Model and data resilience refers to the ability of models to be protected and strong during both the learning and inference phases, specifically against poisoning attempts. Blockchain has the ability to serve as a solution for a framework that allows for the distribution, transparency, and security of data sharing [39]. Various machine-learning approaches, such as neural networks, deep learning, and supervised learning, can be utilized to safeguard privacy in relation to data, images, location, and communication. This can be applied to platforms such as Android, intelligent vehicles, and the IoT.

Quantum computing is anticipated to be employed in 6G communication networks for the purpose of identifying, mitigating, and maybe preventing security problems. Quantum computing-assisted communication is an emerging field of study that studies the idea of replacing quantum channels with noiseless conventional communication channels in order to achieve outstanding dependability in 6G mobile networks. With the progress that has been made in quantum computing, it is anticipated that the post-quantum age would require the implementation of encryption that is secure against quantum computing. The discrete logarithmic issue, which serves as the foundation for contemporary asymmetric cryptography, has the potential to be solved in polynomial time with the advent of quantum algorithms, such as Shor's algorithm [40]. Quantum computing is a technique that makes use of the inherent quantum properties of information. This technique has the potential to offer unrivalled randomness and increased security, hence enhancing the quality of transmission [41]. In order to guarantee the safety of 6G communication networks, it is possible to combine post-quantum cryptography systems with physical layer security measures. In the context of 6G networks, the implementation of machine learning-based cyber-security and quantum encryption in communication lines has the potential to bring in new eras of research. By leveraging quantum advancements in supervised and unsupervised learning to carry out tasks such as clustering and classification, quantum machine-learning techniques have the potential to improve the level of security and privacy in communication networks. There is the possibility of apps developed for 6G that can make use of quantum security approaches. Quantum communication protocols, such as quantum key distribution (QKD), have the potential to be utilized in some applications of 6G technology. For instance, ocean communication, satellite communication, terrestrial wireless networks, and THz communication systems are all examples of applications that might possibly exploit this technology [41]. QKD is a methodology that may be utilized in conventional key distribution methods to create a secret key between two authorized entities. This is accomplished by utilizing concepts that are derived from quantum physics.

It has already been demonstrated that some technologies are effective in key areas through the deployment of 6G networks. Within the context of 6G networks, they offer robust security measures, minimum latency, and effective communication services. For the most part, however, the majority of the new 6G technologies imply greater issues over privacy and security. This section examines the primary technologies that are utilized in 6G, as well as the security and privacy concerns that are linked with them. The random physical features and external interference that are present in wireless networks are taken into consideration while thinking about the strategies that are offered to safeguard the physical layer. On the other hand, the adaptability of PLS mechanisms, particularly in circumstances when resources are few, in conjunction with the potential of ground-breaking 6G technologies, may herald in a new age of PLS in the era of 6G. Terahertz communications, which are often referred to as THz. Terahertz communications operate throughout the whole frequency range, from 0.1 to 10 THz. In order to provide extremely high transmission rates, substantial immunity to interference, and seamless integration of sensing and communications, these frequency bands mix optical waves with a broad range and microwaves.

One of the most common applications for THz communications is to satisfy the requirements of systems that require transmission speeds in the region of Tbps. THz communications will significantly contribute to the extension of the transmission technology that is now in use. These devices will mostly be utilized for the purposes of communicating through holographic communications with a delay, communicating on a tiny scale, conveying vast volumes of data at extraordinarily high capacities, and transmitting across short distances at extremely high speeds. These examples are just a small sample of the many other applications that may be pursued. The use of THz communication signals for precise localization and high-fidelity detection are two additional demanding applications, which are currently being developed. A collection of important technologies and challenges associated with THz communication is presented in the following paragraph. The direct modulation architecture, the solid-state frequency mixing modulation architecture, and the optoelectronics modulation architecture are the three different types of transceiver architecture that are often used. Optimal compatibility, enhanced energy efficiency, and cost-effectiveness are the primary considerations that should be taken into account while designing architectural structures. THz signal source, mixer, multiplier, detector, and amplifier [42] are the key components that make up the RF end of a THz system. Other components include a multiplier and a detector. In the first place, the technology of THz communication has the capability to permit the transfer of data at speeds of one hundred gigabits per second or greater. The use of transmitters that have a narrow beam and a short pulse duration would also reduce the chance of eavesdropping, which would result in an improvement in the security of communication. It is also limited in its ability to dampen THz oscillations caused by particular compounds, which is another limitation.

VLC, which stands for visible light communications, is a technology that is not only practical but also capable of meeting the needs of the 6G wireless network. Furthermore, VLC has been extensively researched in a variety of sectors for a considerable amount of time. This research has included its application in indoor localization systems as well as vehicle-ad-hoc-network (VANET2) [43]. VLC technology is more resistant to interferences than RF technology, which is susceptible to large interference and considerable delay. VLC technology has broader bandwidths, which allow it to be more resistant to interferences. Specifically, the user is interested in learning about the various approaches that may be utilized to protect VLC communications from eavesdropping, DoS or jamming attacks, as well as node compromise attempts. Research [44] outlines the conditions that must be met in order to comply with the confidentiality, integrity, authenticity, and availability (CIAA) standards. The confidentiality of data ensures that it can only be accessed by those who are entitled to do so and prevents it from being shared with those who are not authorized to receive it. The term "integrity" refers to the process of guaranteeing that the information that is sent is accurate, whereas "authenticity" refers to the confirmation of the identity of network nodes. The verification of one's identity and the validation of the information that is supplied are the two components that are necessary for authentication. The main objective is to verify the identity of the individual who is utilizing the system while simultaneously assuring the integrity of the data that is being transferred. Both of the authentication components are completely necessary in order to guarantee the safety of the resources and the information. Within the context of wireless networks, the term "availability" refers to the capacity of users to connect to the network without any restrictions imposed on them in terms of the time or location at which they can do so.

Within the realm of 6G technology, molecular communication has a great deal of potential. The strategy, on the other hand, has not yet reached its full maturity and is still in its first phases. The use of biological signals as a means of information transmission is the fundamental premise upon which the technology of molecular communication is structured. In the field of molecular communication, the transmission and verification protocols have been shown to raise a number of privacy and security problems. Molecular communication is vulnerable to a variety of assaults, which aim to compromise its security by either rejecting the data that is being conveyed or modifying the information that is being passed back and forth between the transmitter and the receiver. According to the classification system [43], the attacks may be broken down into four distinct categories: attacks on the transport layer, attacks on the link layer, attacks on the network layer, and signaling attacks at the physical layer. The management of sessions and the maintaining of security are both the responsibilities of the molecular transport layer. There are a number of common security vulnerabilities that can affect both earlier and more recent communication networks. Some examples of these vulnerabilities include desynchronization and flooding. When it comes to the connection layer, collision and inequality are two of the most typical types of problems. Through the use of error-correction strategies, collision attacks may be avoided at the link layer. If, on the other hand, the diffusion approach is employed to convey data, then attackers have the ability to commit acts of injustice and conduct collision assaults. The objective of the molecular network layer is to identify the entities that are engaged and to offer a description of the significant skills that they possess. These functions are accountable for the formation of molecular networks as well as their migration across the network.

The importance of AI and ML technology in the network design of all 6G network technologies has been lately acknowledged. Hence, artificial intelligence has garnered significant interest within the realm of 6G networking. The integration of artificial intelligence and machine learning into 5G networks occurs in fields that are distinguished by ample training data and high-performance processing units. AI and ML are becoming important components of the 6G networks. AI and ML are employed to safeguard several aspects of the security framework of 6G and ensure its security. Implementing AI and ML in security systems improves their independence and accuracy, allowing them to have predictive capacities for security analysis. This research examines the challenges associated with implementing artificial intelligence and machine learning in the 6G system [43]. It is essential to consider the following:

(1) The trustworthiness of machine-learning models and components is of utmost importance when AI is tasked with network security.
(2) Surveillance: Continuously monitoring security operations using artificial intelligence and machine learning in real-time to guarantee oversight and trustworthiness.
(3) Ethical and legal considerations arise when employing AI-driven optimization approaches that may exclude some customers or applications. AI-driven security solutions may or may not provide consistent protection for all users. The responsibility for any failure in security services handled by AI lies with the entity or organization overseeing them.
(4) Extensibility and viability: The privacy of federated learners relies on the necessity of secure data transfers. The issue for AI/ML lies in the scalability of the necessary computational, networking, and storage resources.
(5) Controlled security duties can lead to excessive overhead when AI/ML security systems are linked to extensive data operations.
(6) The flexibility of models should be both reliable and adaptable during the learning and inference processes. The anticipated 6G system aims to incorporate advanced AI methods and approaches to meet the demanding service needs, necessary capabilities, and emerging use cases. Figure 6.2 shows important considerations associated with implementing artificial intelligence and machine learning in the 6G system.

Risks associated with artificial intelligence and machine learning can be mitigated in a variety of ways. The inclusion of changed cases in the training dataset that are similar to potential dangers is one way that adversarial training helps to better increase resilience. Defensive distillation is an additional method of protection that relies on the conveyance of information across neural networks through the utilization of software labels. To do this, it is necessary to take into account the outcomes of a network that has been trained in the past and to identify the various categories. Rather than being complicated by labels that allocate all of the data to a single category, these software labels are also utilized for training purposes. The

*Figure 6.2* Important considerations associated with implementing artificial intelligence and machine learning in the 6G system.

effectiveness of both of these methods lies on their ability to defend against both evasive and aggressive strikes.

On the other hand, there are inherent dangers involved in the process of adding security measures during the training phase in order to prevent data corruption and authenticate data. The distributed, transparent, and secure platform for data exchange that blockchain technology provides is a significant advantage. In addition to that, approaches for input validation and target protection are utilized. Furthermore, the latter is advantageous in terms of protecting against attacks from hostile environments. In addition, the security mechanism that was implemented to properly restrict access to machine-learning APIs is able to successfully avoid inversion attacks. Research [44] presented a wide range of artificial intelligence approaches that are utilized on a variety of physical levels. These techniques include unsupervised learning, *K*-means, and artificial neural networks. The optimization of interoperability is one of the ways in which these technologies have the potential to improve the efficiency of physical layers and to expand the range of prediction and safety. An unsupervised machine-learning approach was presented by Sattiraju et al. [45] with the intention of enhancing the authentication process and enhancing the security of the physical layer. A unique antenna design was presented by Hong et al. [46] with the intention of enhancing the classification of physical layer communication jobs in order to lessen the chance of data loss. In addition, Nawaz et al. [47] highlighted the fact that the protection of 6G connections is achieved by the use of encryption systems and machine-learning techniques. An investigation of artificial intelligence technologies was carried out by Zhou et al. [48], who asserted that these technologies have the capacity to properly identify advanced computer threats. On the other hand, they suggest carrying out extra study procedures. It was said by Dang et al. [49] that artificial intelligence may be of assistance in discovering network difficulties in 6G security and that it can also suggest strategies to prevent and protect against these problems. As stated by Tomkos et al. [50], network edge devices have the potential to enhance network security in 6G networks by exchanging information through the use of federated distributed artificial intelligence. It was brought to the attention of Zhou et al. [48] that data transfers have the potential to result in an increase in the number of privacy breaches because they have an effect on particular machine-learning algorithms.

A communication system that possesses extraordinary capabilities may be achieved through the use of quantum communications, which is a technology that is extremely adaptable and has a significant potential in 6G networks. Increasing the capabilities of quantum communication has the potential to greatly improve the essential aspects of security and dependability. If any part of the quantum communication is altered or copied by an unauthorized party, the quantum state will be altered as a consequence of this action. Quantum communication, in principle, offers unrivalled dependability and is highly appropriate for communication over large distances that requires accurate progress. It provides a wide range of creative solutions that improve communication to a sufficient degree compared to other options. In accordance with the information provided in references [51, 52], the adversaries are able to carry out assaults by utilizing quantum technologies. The integration of post-quantum encryption solutions that are resistant to assaults from quantum computers into Internet of Things devices is an ongoing challenge. The development of future 6G standards that are immune to quantum assaults is complicated by the presence of devices that use quantum cryptography. In traditional information transmission, a technique known as oblivious transfer (OT) is utilized. This approach enables a sender to communicate a single piece of information to a destination without the sender having awareness of the specific data that was conveyed. Due to the fact that any leakage would result in the destruction of the link, quantum information is unable to sustain this feature. Quantum cloning attacks include the procedure of replicating a randomly selected information state, which ultimately results in the creation of an identical clone that preserves the information in its initial state. Despite the fact that it is not possible to produce precise clones of quantum states, there are a number of sophisticated cloning techniques that may be employed to copy a quantum state with an impressive level of accuracy. Attacks that include quantum cloning express themselves as instances of quantum hacking in secure quantum channels, notably in high-dimensional quantum key distribution systems.

The expected integration of distributed ledger technology (DLT) with 6G networks is expected to have a significant impact on fixing the security weaknesses in blockchain and smart contracts within the 6G network infrastructure. The occurrence of these attacks might be attributed to issues in software development, constraints in programming languages, and weaknesses in network connectivity [53–57]. In addition, both public and private blockchain systems are vulnerable to similar types of attacks. As a result, these factors lead to reduced accuracy, financial losses in Bitcoin, and a higher frequency of system failure.

Ling et al. [58] developed a novel architecture for mobile service authorization that makes use of blockchain technology. This design improves the radio access network architecture in order to guarantee the safety of the network and improve the efficiency with which network access is managed among the many network entities. The use of blockchain technology to get access to the unconstrained licensed radio spectrum was proposed by Kotobi et al. [59] as a means of improving the safety and media access protocol of cognitive radio. Decentralizing the design of the 6G network has been accomplished, which is a significant success. Therefore, it is plausible to conclude that the data was altered by the attackers if a sizeable portion of the nodes, which is greater than a particular threshold, accounts for 51% of the total number of nodes that the hacker controls. In the aftermath, security flaws were discovered. In addition, the absence of a trustworthy intermediary in the process of safeguarding the storage of the network and monitoring the data might lead to vulnerabilities in the security system [60]. The ability of the hash chain to authenticate transactions across blockchain networks was postulated by Ferraro et al. [61], who suggested that this capability might have a substantial impact on the security of blockchains. There is brevity in the user's input. Figure 6.3 shows technologies to enhance 6G in the future.

*Figure 6.3* Technologies to enhance 6G in the future.

## 6.4 CONCLUSION

This chapter presents a comprehensive study of the requirements and issues that are associated with the security of the 6G network. The purpose of this chapter is to illustrate the development of security protocols in prior wireless networks, beginning with the 1G network and progressing all the way up to the next 6G network. Also, to provide the vision of the 6G network and to outline the research methodologies that are being followed in both the academic and industry sectors. An investigation of the potential assaults that may be launched against the various physical layer technologies utilized in 6G networks was carried out, and some remedies were discussed. The forthcoming sixth-generation wireless network (also known as 6G) will use artificial intelligence technologies in order to improve network protection and security. In this chapter, the security architecture of the 6G network, which is relying on artificial intelligence and machine-learning technologies, is investigated. A number of security concerns that were associated with the physical layer have been discussed. These concerns include molecular communication, THz communication, and VLC communication. When it comes to privacy and security, the majority of the forthcoming 6G technologies present considerable challenges. The state-of-the-art technologies that were discussed earlier have been emphasized, along with an explanation of their security flaws, their potential vulnerabilities, and the preventative security measures that are associated with them.

With each new generation of network technology comes the introduction of applications that are both inventive and imaginative. The sixth-generation wireless network is able to make use of applications that were developed for prior generations of radio technology. There are a multitude of future technologies that will dramatically transform human civilization in the 2030s and beyond, and the 6G network is quickly becoming the motivating force behind all of these technologies. In order to meet the extensive communication needs of 6G applications, high-performance capabilities and stringent security measures are required. In addition to this, the chapter contains a discussion of the potential advancements and challenges associated with the different implementations of 6G. As part of future scopes, it is important to carry out a comprehensive investigation into the wide variety of dangers that are intended to compromise the safety of the 6G network. Investigating different methods to safeguard 6G networks is an important matter that has to be addressed in the future. Artificial intelligence plays a crucial role in enabling the development of the upcoming 6G mobile networks. Ensuring security is of utmost importance in order to achieve the envisioned goals of 6G. AI-powered 6G security offers advanced, resilient security solutions. This chapter presents a comprehensive analysis of the various potential and problems associated with incorporating intelligent security and privacy measures into the role of artificial intelligence in 6G networks. Furthermore, it also identifies potential areas for future research by examining the difficulties in providing AI-based security and privacy and proposing feasible solutions.

## KEY TERMS AND DEFINITIONS

*Time division multiple access (TDMA):* is a channel access method for shared-medium networks. It allows several users to share the same frequency channel by dividing the signal into different time slots.

*Code division multiple access (CDMA):* is a method of encoding several sources of data so they may all be transmitted over a single RF carrier by one transmitter, or by using a single RF carrier frequency with multiple transmitters.

*Internet of Everything (IoE):* is the intelligent connection of people, process, data, and things.

*Internet of Things (IoT):* describes the network of physical objects—"things"—that are embedded with sensors, software, and other technologies for the purpose of connecting and exchanging data with other devices and systems over the internet.

*Virtualization:* is technology that you can use to create virtual representations of servers, storage, networks, and other physical machines.

*Industrial IoT (IIoT):* is the use of smart sensors, actuators and other devices, such as radio frequency identification tags, to enhance manufacturing and industrial processes.

*Denial-of-service (DoS) attack:* is an attack meant to shut down a machine or network, making it inaccessible to its intended users. DoS attacks accomplish this by flooding the target with traffic, or sending it information that triggers a crash.

## REFERENCES

[1] M. Giordani, M. Polese, M. Mezzavilla, S. Rangan, and M. Zorzi, "Toward 6G networks: Use cases and technologies," *IEEE Commun. Mag.*, Vol. 58, No. 3, 2020, pp. 55–61.

[2] V. Ziegler, H. Viswanathan, H. Flinck, M. Hoffmann, V. Räisänen, and K. Hätönen, "6G architecture to connect the worlds," *IEEE Access*, Vol. 8, 2020, p. 173.

[3] Ahmad, T. Kumar, M. Liyanage, J. Okwuibe, M. Ylianttila, and A. Gurtov, "Overview of 5G security challenges and solutions," *IEEE Commun. Stand. Mag.*, Vol. 2, No. 1, 2018, pp. 36–43.

[4] R. Khan, P. Kumar, D.N.K. Jayakody, and M. Liyanage, "A survey on security and privacy of 5g technologies: Potential solutions, recent advancements, and future directions," *IEEE Commun. Surv. Tutor.*, Vol. 22, No. 1, 2019, pp. 196–248.

[5] P. Ranaweera, A.D. Jurcut, and M. Liyanage, "Survey on multi-access edge computing security and privacy," *IEEE Commun. Surv. Tutor.*, Vol. 23, No. 2, 2021, pp. 1078–1124.

[6] S. Wijethilaka and M. Liyanage, "Survey on network slicing for internet of things realization in 5g networks," *IEEE Commun. Surv. Tutor.*, Vol. 22, No. 17, 2022, 6623.

[7] N. Wang, P. Wang, A. Alipour-Fanid, L. Jiao, and K. Zeng, "Physical layer security of 5G wireless networks for IoT: Challenges and opportunities," *IEEE Internet Things J.*, Vol. 6, No. 5, 2019, pp. 8169–8181.

[8] Y. Yang, L. Wu, G. Yin, L. Li, and H. Zhao, "A survey on security and privacy issues in internet-of-things," *IEEE Internet Things J.*, Vol. 4, No. 5, 2017, pp. 1250–1258.

[9] P. Chandra, D. Bensky, T. Bradley, C. Hurley, S.A. Rackley, J. Rittinghouse, J.F. Ransome, C. Cism, T. Stapko, G.L. Stefanek, et al., *Wireless Security: Know it All*, Newnes, 2011.

[10] L. Gupta, R. Jain, and G. Vaszkun, "Survey of important issues in UAV communication networks," *IEEE Commun. Surv. Tutorials*, Vol. 18, No. 2, 2015, pp. 1123–1152.

[11] G. Arunabha, J. Zhang, J. G. Andrews, and R. Muhamed, *Fundamentals of LTE*, The Prentice Hall Communications Engineering and Emerging Technologies Series, Pearson, September 2010.

[12] C. Brookson, "GSM security: A description of the reasons for security and the techniques," in: *IEEE Colloquium on Security and Cryptography Applications to Radio Systems*, GSM MoU Security Rapporteur, British Telecommun. plc, London, 06 August 1994, pp. 2/1–2/4.

[13] S. Gindraux, "From 2G to 3G: A guide to mobile security," in: *Third International Conference on 3G Mobile Communication Technologies*, IET, London, UK, 2002, pp. 308–311.

[14] G. Cattaneo, G. De Maio, and U.F. Petrillo, "Security issues and attacks on the GSM standard: A review," *J. UCS*, Vol. 19, No. 16, 2013, pp. 2437–2452.

[15] M. Toorani and A. Beheshti, "Solutions to the GSM security weaknesses," in: *2008 The Second International Conference on Next Generation Mobile Applications, Services, and Technologies*, IEEE, Cardiff, UK, 2008, pp. 576–581.

[16] P. Sharma, "Evolution of mobile wireless communication networks-1G to 5G as well as future prospective of next generation communication network," *Int. J. Comput. Sci. Mobile Comput.*, Vol. 2, No. 8, 2013, pp. 47–53.

[17] M.C. Chuah, "Universal mobile telecommunications system (UMTS) quality of service (QOS) supporting variable QOS negotiation," *US Patent*: 7668176, 2010.

[18] U. Varshney, "4G wireless networks," *IT Prof.*, Vol. 14, No. 5, 2012, pp. 34–39.

[19] J. Cao, M. Ma, H. Li, Y. Zhang, and Z. Luo, "A survey on security aspects for LTE and LTE-A networks," *IEEE Commun. Surv. Tutorials*, Vol. 16, No. 1, 2013, pp. 283–302.

[20] N. Seddigh, B. Nandy, R. Makkar, and J.-F. Beaumont, "Security advances and challenges in 4G wireless networks," in: *2010 Eighth International Conference on Privacy, Security and Trust*, IEEE, Ottawa, ON, Canada, 2010, pp. 62–71.
[21] R.N. Mitra and D.P. Agrawal, "5G mobile technology: A survey," *ICT Express*, Vol. 1, No. 3, 2015, pp. 132–137.
[22] N. Panwar, S. Sharma, and A.K. Singh, "A survey on 5G: The next generation of mobile communication," *Phys. Commun.*, Vol. 18, 2016, pp. 64–84.
[23] M. Jaber, M.A. Imran, R. Tafazolli, and A. Tukmanov, "5G backhaul challenges and emerging research directions: A survey," *IEEE Access*, Vol. 4, 2016, pp. 1743–1766.
[24] J. Prados-Garzon, O. Adamuz-Hinojosa, P. Ameigeiras, J.J. Ramos-Munoz, P. Andres-Maldonado, and J.M. Lopez-Soler, "Handover implementation in a 5g sdnbased mobile network architecture," in: *2016 IEEE 27th Annual International Symposium on Personal, Indoor, and Mobile Radio Communications (PIMRC)*, IEEE, Valencia, Spain, 2016, pp. 1–6.
[25] J. Kim, D. Kim, and S. Choi, "3GPP SA2 architecture and functions for 5G mobile communication system," *ICT Express*, Vol. 3, No. 1, 2017, pp.1–8.
[26] Ahmad, S. Shahabuddin, T. Kumar, J. Okwuibe, A. Gurtov, and M. Ylianttila, "Security for 5G and beyond," *IEEE Commun. Surv. Tutorials*, Vol. 21, No. 4, 2019, pp. 3682–3722.
[27] M. Pattaranantakul, R. He, Q. Song, Z. Zhang, and A. Meddahi, "NFV security survey: From use case driven threat analysis to state-of-the-art countermeasures," *IEEE Commun. Surv. Tutorials*, Vol. 20, No. 4, 2018, pp. 3330–3368.
[28] R. Rana and R. Kumar, "Performance analysis of AODV in presence of malicious node," *Acta Electron. Malays.*, Vol. 3, No. 1, 2019, pp. 1–5.
[29] H. Hussein, H. Elsayed, and S. Abd El-kader, "Intensive benchmarking of D2D communication over 5G cellular networks: Prototype, integrated features, challenges, and main applications". *Wirel. Netw.* Vol. 26, 2019, pp. 3183–3202.
[30] S.R. Hussain, M. Echeverria, O. Chowdhury, N. Li, and E. Bertino, "Privacy attacks to the 4G and 5G cellular paging protocols using side channel information," in: *Proceedings of the 26th Network and Distributed Systems Security (NDSS) Symposium*, San Diego, CA, USA, 24–27 February 2019, pp. 1–15.
[31] I. H. Abdulqadder, S. Zhou, D. Zou, I. T. Aziz, and S. M. A. Akber, "Multi-layered intrusion detection and prevention in the SDN/NFV enabled cloud of 5G networks using AI-based defense mechanisms," *Comput. Netw.*, Vol. 179, 2020, p. 107364.
[32] R. Santos, D. Souza, W. Santo, A. Ribeiro, and E. Moreno, "Machine learning based detection of DDoS attacks in software defined network," *Indonesian Journal of Electrical Engineering and Computer Science*, Vol. 32, No. 3, December 2020, p. 1503. 10.11591/ijeecs.v32.i3.pp1503-1511
[33] M. Zolanvari, M. A. Teixeira, L. Gupta, K. M. Khan, and R. Jain, "Machine learning-based network vulnerability analysis of industrial internet of things," *IEEE Internet Things J.*, Vol. 6, No. 4, 2019, pp. 6822–6834.

[34] H. Fang, X. Wang, and S. Tomasin, "Machine learning for intelligent authentication in 5G and beyond wireless networks," *IEEE Wirel. Commun.*, Vol. 26, No. 5, 2019, pp. 55–61.

[35] Y. Sun, J. Liu, J. Wang, Y. Cao, and N. Kato, "When machine learning meets privacy in 6G: A survey," *IEEE Commun. Surv. Tutor.*, Vol. 22, No. 4, 2020, pp. 2694–2724.

[36] G. Plastiras, M. Terzi, C. Kyrkou, and T. Theocharidcs, "Edge intelligence: Challenges and opportunities of near-sensor machine learning applications," in: *2018 IEEE 29th International Conference on Application-Specific Systems, Architectures and Processors (ASAP)*, Milan, Italy, 2018, pp. 1–7.

[37] ETSI ISG ZSM, "ETSI GS ZSM 002: ZSM reference architecture," 2019. Available: www.etsi.org/deliver/etsi gs/ZSM/ 001 099/002/01.01.01 60/gs ZSM002v010101p.pdf

[38] S. Zhang and D. Zhu, "Towards artificial intelligence enabled 6G: State of the art, challenges, and opportunities," *Comput. Netw.*, Vol. 183, 2020, p. 107556.

[39] W. Li, Z. Su, R. Li, K. Zhang, and Y. Wang, "Blockchain-based data security for artificial intelligence applications in 6G networks," *IEEE Netw.*, Vol. 34, No. 6, 2020, pp. 31–37.

[40] S.J. Nawaz, S.K. Sharma, S. Wyne, M.N. Patwary, and M. Asaduzzaman, "Quantum machine learning for 6G communication networks: State-of-the-art and vision for the future," *IEEE Access*, Vol. 7, April 2019, pp. 46317–46350.

[41] S. Tarantino, B. Da Lio, D. Cozzolino, and D. Bacco, "Feasibility of quantum communications in aquatic scenarios," *Optik*, 2020, p. 164639.

[42] A.U. Gawas, "An overview on evolution of mobile wireless communication networks: 1G–6G". *Int. J. Recent Innov. Trends Comput. Commun.*, Vol. 3, 2015, pp. 3130–3133.

[43] J. Luo, L. Fan, and H. Li, "Indoor positioning systems based on visible light communication: State of the art". *IEEE Commun. Surv. Tutor.*, Vol. 19, 2017, pp. 2871–2893.

[44] G. Blinowski, " Security of visible light communication systems—A survey," *Phys. Commun.*, Vol. 34, 2019, pp. 246–260.

[45] R. Sattiraju, A. Weinand, and H.D. Schotten, "Ai-assisted Phy Technologies for 6G and beyond wireless networks," 1st 6G WIRELESS SUMMIT, Levi, Finland, 24-25 2019.

[46] T. Hong, C. Liu, and M. Kadoch, "Machine learning based antenna design for physical layer security in ambient backscatter communications." *Wirel. Commun. Mob. Comput.*, 2019, pp.1–10.

[47] S.J. Nawaz, S.K. Sharma, S. Wyne, M.N. Patwary, and M. Asaduzzaman, "Quantum machine learning for 6G communication networks:State-of-the-artandvisionforthefuture," *IEEEAccess*,Vol.7,2019, pp. 46317–46350.

[48] Z. Zhou, H. Liao, B. Gu, K.M. Huq, S. Mumtaz, and J. Rodriguez, "Robust mobile crowd sensing: When deep learning meets edge computing," *IEEE Netw.*, Vol. 32, 2018, pp. 54–60.

[49] S. Dang, O. Amin, B. Shihada, and M.-S. Alouini, "What should 6G be?" *Nat. Electron.*, Vol. 3, 2020, pp. 20–29.
[50] I. Tomkos, D. Klonidis, E. Pikasis, and S. Theodoridis, "Toward the 6G network era: Opportunities and challenges," *IT Prof.*, Vol. 22, 2020, pp. 34–38.
[51] S. Tarantino, B. da Lio, D. Cozzolino, and D. Bacco, "Feasibility of quantum communications in aquatic scenarios," *Optik*, Vol. 216, 2020, p. 164639.
[52] L. Gyongyosi, S. Imre, and H.V. Nguyen, "A survey on quantum channel capacities," *IEEE Commun. Surv. Tutor.*, Vol. 20, 2018, pp. 1149–1205.
[53] J. Partala, "Post-quantum cryptography in 6G," Computer Communications and Networks 6G Mobile Wireless Networks, 2021, pp. 431–448
[54] L.U. Khan, I. Yaqoob, M. Imran, Z. Han, and C.S. Hong, "6G wireless systems: A vision, architectural elements, and future directions," *IEEE Access*, Vol. 8, 9163104, pp. 147029–147044.
[55] W. Li, Z. Su, R. Li, K. Zhang, and Y. Wang, "Blockchain-based data security for artificial intelligence applications in 6G networks," *IEEE Netw.*, Vol. 34, 2020, pp. 31–37.
[56] T. Maksymyuk, J. Gazda, M. Volosin, G. Bugar, D. Horvath, M. Klymash, and M. Dohler, "Blockchain-empowered framework for decentralized network management in 6G," *IEEE Commun. Mag.*, Vol. 58, 2020, pp. 86–92.
[57] S. Velliangiri, R. Manoharn, S. Ramachandran, and V.R. Rajasekar, "Blockchain based privacy preserving framework for emerging 6G wireless communications," in: *IEEE Transactions on Industrial Informatics*; IEEE, Piscataway, NJ, USA, 2021.
[58] X. Ling, J. Wang, T. Bouchoucha, B.C. Levy, and Z. Ding, "Blockchain radio access network (B-ran): Towards decentralized secure radio access paradigm," *IEEE Access*, Vol. 7, 2019, pp. 9714–9723.
[59] K. Kotobi and S.G. Bilen, "Secure blockchains for dynamic spectrum access: A decentralized database in moving cognitive radio networks enhances security and user access," *IEEE Veh. Technol. Mag.*, Vol. 13, 2018, pp. 32–39.
[60] L. Qiao, S. Dang, B. Shihada. M.-S. Alouini, R. Nowak, and Z. Lv, "Can blockchain link the future?" Digital Communications and Networks, Vol. 6, No. 5, 2021, pp. 687–694.
[61] P. Ferraro, C. King, and R. Shorten, "Distributed ledger technology for smart cities, the sharing economy, and social compliance," *IEEE Access*, Vol. 6, 2018, pp. 62728–62746.

Chapter 7

# The future of 6G in marketing

*Mohammed Majeed, Jonas Yomboi,*
*Seidu Alhassan, and Sulemana Ibrahim*

## 7.1 INTRODUCTION

The emergence of technologies, like 6G has an impact on individuals who will use them, leading to both positive and negative outcomes (Alsharif et al., 2020; Jiang & Schotten, 2021). The main goal of 6G is to ensure connectivity for not only humans but also a wide range of devices and machines. This will create a network environment (Jiang et al., 2021). Additionally, the integration of technology such as AI, edge computing, and advanced antenna technologies is expected to optimize resource distribution on the network strengthen security measures and improve the user experience. The exponential growth of artificial intelligence (AI), machine learning (ML), deep learning (DL), augmented reality (AR), virtual reality (VR), 3D media or 360-degree video, the Internet of Things (IoT), Enterprise Internet of Things (E IoT), biological and nano Internet of Things (IoBNT), and various other applications has led to an increase in global communication traffic (Giordani et al., 2020). According to Nayak and Patgiri (2021), there is projected to be a significant rise of approximately 55% in global mobile traffic. The expansion of the 6G market is propelled by the imperative for enhanced data rates and network capacity, which are essential for facilitating the implementation of novel technologies and applications. The proliferation of IoT technology in many sectors has necessitated a demand for dependable and high-speed connectivity to accommodate the substantial volume of interconnected devices (Giordani et al., 2021). The aforementioned rise exemplifies the direct influence of sophisticated communication systems on traffic conditions. The ecosystem industry, the robotics industry, and the field of machine interaction are increasingly gaining prominence in contemporary society (Nguyen et al., 2022). To enhance the quality of intelligent existence and the widespread implementation of automation, numerous sensors will be deployed across urban areas, automobiles, residences, industrial settings, food production, playthings, and various other domains (Akyildiz et al., 2020; Imoize et al., 2021). Hence, it is plausible that forthcoming advancements would

 DOI: 10.1201/9781003583127-7

necessitate the provision of high-speed data transmission and exceptionally robust reliability to effectively sustain these aforementioned applications. The collection and sharing of environmental information, equipment status information, sensor data, user behavior information, and workflow in a cloud computing platform are facilitated by a 6G connection (Nguyen et al., 2022). These data are processed to enable the realization of various processes, such as environment monitoring, equipment management, transaction analysis, action prediction, insight generation, decision-making, and optimization. The outcomes of these processes are then made available on terminal devices (Vaigandla et al., 2021). This type of service is extensively utilized within the service industry for the purposes of user assistance, strategic decision-making, process enhancement, and managerial oversight. Furthermore, these technologies find practical implementation in diverse sectors, such as education, healthcare, and electricity, among others. Hence, this chapter primarily addresses the issue of enhancing the total accessibility of creative digital marketing for organizations by integrating sixth-generation technology into marketing and advertising information operations.

## 7.2 LITERATURE

### 7.2.1 6G

The goal of 6G is to build a network infrastructure that allows connections, among people, devices, and machines (Ahokangas et al., 2020). 6G aims to offer data speeds reaching terabits per second, which will enable uninterrupted streaming and support applications like augmented reality virtual reality, mixed reality, and the IoT (Ahokangas et al., 2023). The development of the 6G will bring four changes in the way we communicate. To achieve connectivity 6G technology will expand beyond terrestrial networks by incorporating non-terrestrial connections, such as satellite and unmanned aerial vehicle (UAV) communication networks. This will establish a communication network that includes space, air, ground, and sea. Additionally, thorough research will be conducted on all spectrums to improve data rates and connection density. This research will cover bands, like sub 6 GHz frequencies, millimeter wave (mmWave) terahertz (THz) and optical frequencies as shown in Figure 7.1. Moreover, considering the amount of data generated by networks various communication scenarios, advanced antenna arrays, wide bandwidths, and evolving service demands the introduction of 6G networks will enable the emergence of innovative intelligent applications by integrating AI and big data methodologies. Additionally ensuring network security is crucial, in the development of 6G networks. Implementing 6G with advanced technical requirements, than its predecessor, 5G will significantly enhance and expand communication capabilities to the extent

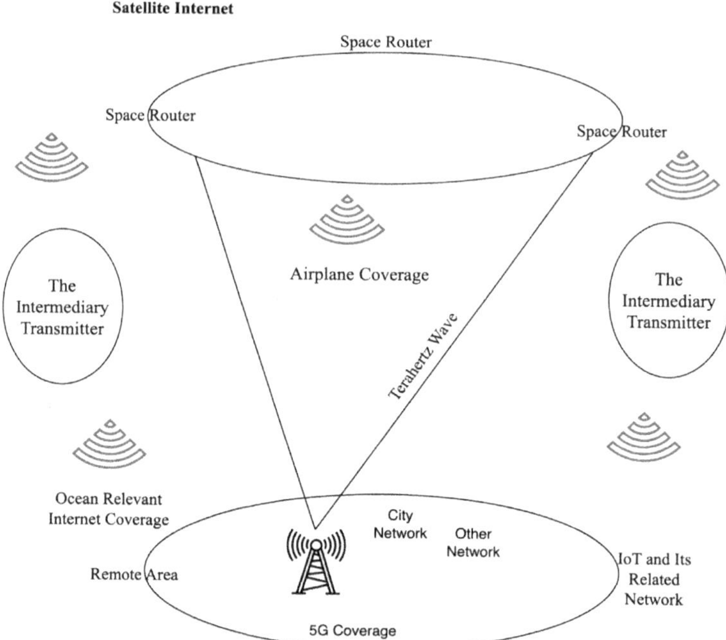


*Figure 7.1* 6G comprehensive dimensional framework. (Authors' construction.)

where it becomes challenging to distinguish between the physical and cyber realms (Bertin et al., 2021; Matinmikko-Blue et al., 2023).

### 7.2.2 Aim of 6G

The objective of 6G is to establish a network infrastructure that enables seamless connectivity between individuals, devices, and machines. After a long and winding road from the original 1G, the most recent generation of wireless communication technology is 6G. With 6G, the 6G wireless communication system can finally come to fruition. Releases of new generations of wireless devices typically occur every 10 years. 6G is anticipated to be a part of our lives around 2030 if this trend persists. To make the progression between generations since the 1G is easy to understand, we will go over the general traits of each generation (Alwis et al., 2021).

Initially, the 1G made use of 2 Kbps frequency division multiple access (FDMA) multiplexing (Al-Ansi et al., 2021). There was a poor connection and extremely limited capacity with this mobile phone service. We later observed that 2G speeds had increased to 14–64 Kbps (Ahokangas & Aagaard, 2023). Even more of a change from an analogue to a digital network occurred with the upgrade from 1G to 2G. 2G networks use 30–200 kHz bandwidth for technologies like multimedia messaging service (MMS) and short message service (SMS) (Katz et al., 2019). The bandwidth in 3G ranged from 15 to 20 MHz, and the speed was further increased to 2 Mbps. Significant advancements in mobile technology resulted from this (Wen, 2023). Among these developments is the technology that has allowed the "smart" phone to become an integral part of our everyday lives. Then came 4G, which could only manage download speeds of 100–300 Mbps; 5G, on the other hand, can handle much faster speeds. Up to 1 Gbps is the planned throughput speed (Nayak & Patgiri, 2021; Ziegler et al., 2020). To compare and evaluate the development of 6G, this information regarding the gradual evolution from 1G to 5G is provided.

### 7.2.3 6G vs 5G

The global rollout of generation (5G) wireless communication networks began in 2020 with efforts to standardize additional features, like widespread connectivity, high reliability, and low latency guarantees (Giordani

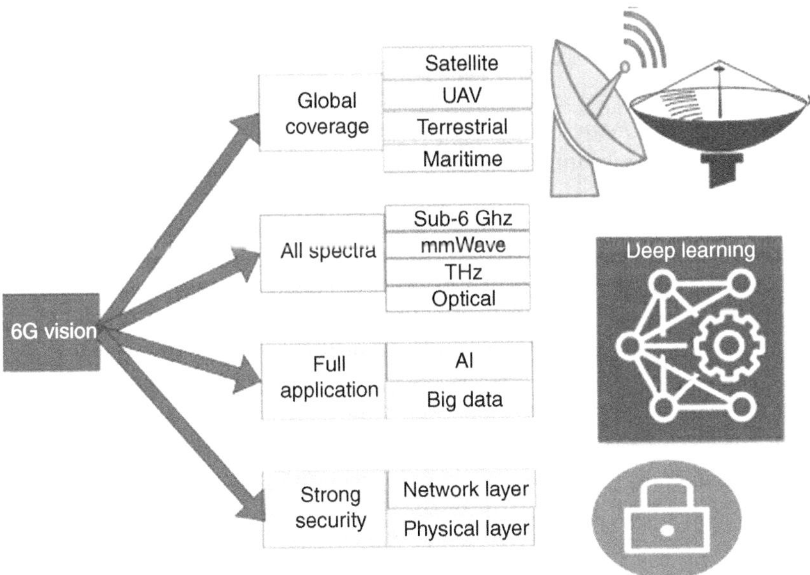


*Figure 7.2* Vision of 6G wireless communication. (Authors' own construction.)

et al., 2021). However, it is expected that 5G technology might not meet all requirements beyond 2030. As a result, experts predict that the next generation of communication networks, known as 6G will provide coverage along with improved efficiency in terms of spectrum utilization, energy consumption and cost effectiveness. Additionally, 6G networks will incorporate intelligence and security measures (Nguyen et al., 2022). To meet these specifications the operation of 6G networks will rely on emerging technologies, such as air connections and spreading methods. Furthermore, innovative network architecture elements like waveform design, multiple access techniques, channel coding schemes, multiantenna technologies, network slicing approaches cell-free building designs as cloud/fog/edge computer science will play crucial roles (Uusitalo et al., 2021). The development of the 6G network can be seen as a progression from its predecessor: the 5G network. Thus, establishing a connection, between the two (Chowdhury et al., 2020).

One key difference, between the 6G network and its previous iteration the 5G network lies in the capabilities of the former as emphasized by Zhang et al. (2019) as appeared in Figure 7.2. Hence, an examination of the practical implementations of 5G network technology can provide insights into the extensive potential uses of 6G network technology in forthcoming times.

## 7.3 POTENTIAL IMPACT OF 6G ON MARKETING AND CUSTOMERS

The proliferation of digital media and other new technology has caused a sea change in the marketing industry (Saad, 2021). Nevertheless, 6G technology is about to arrive, further altering the marketing landscape. Businesses' approaches to advertising and marketing are about to undergo a radical transformation, thanks to this state-of-the-art technology (Tataria et al., 2021). A major improvement to the electronic commerce scene that 6G will usher in is faster and more efficient transmission of information. 6G networks are anticipated to deliver data transfer rates of up to 1 terabyte per second, which is more than 100 times faster than the speeds provided by 5G networks at the moment. Marketers will gain unrivalled consumer insight by analyzing large volumes of data in real-time. 6G networks will revolutionize remote work and corporate collaboration. Although 6G is new, professionals and academics are interested in its potential (Hassnain et al., 2020).

Businesses may provide VR and AR experiences using 6G. This technology may make their events more interactive and immersive, making clients more engaged (Ji et al., 2021). Wide coverage is a significant feature of 6G networks. Covering the same area requires smaller towers. This is helpful if you wish to build towers in areas with frequent rainfall or abundant vegetation. According to Saad et al. (2020), 6G is also designed to accommodate more mobile connections than 5G. This bodes well for the future of service quality since fewer gadgets will interfere with one another.

Sixth-generation wireless capacity to massively support the IoT is another important aspect. With an anticipated 50 billion connected devices on Earth by 2030, 6G will play a crucial role in controlling and utilizing the data produced by these devices (Zadorozhnyi et al., 2023). Marketers will be able to learn more about customer habits and preferences, which will allow them to craft ads that are more tailored to each individual.

The potential for Internet advertising and marketing is always expanding due to technological advancements. Businesses will be able to launch more effective marketing with the advent of 6G technology. Remote workers are about to enter a golden age of efficiency, productivity, and adaptability because of improved connectivity, immersive collaboration, access to cutting-edge technology, and worldwide cooperation opportunities. According to Viswanathan and Mogensen (2020), the lines between virtual and physical workspaces will continue to melt, allowing businesses and individuals to fully utilize remote work and change the way work is done in the future. The lightning-fast speeds are the major selling point of 6G technology. According to Gustavsson et al. (2021), this will allow enterprises to deliver their information at a faster rate than before, which will give them a competitive advantage. With the advent of 6G, there are ways to enhance user experience by reducing latency and speeding up page loads. Downloading 100 hours of video will be a matter of seconds allowing lightning streaming capabilities. This technology may improve corporate operations and customer service. Correct claims help insurers. Nguyen et al. (2021) suggested that 6G will let marketers customize adverts. AI and ML algorithms improve using this technology. 6G networks should boost AR and VR. These networks reduce latency and speed up data transfer, increasing user experiences (Mahmood et al., 2020). Effective campaigns that promote brand awareness and client involvement will also be possible for marketers. Additionally, 6G simplifies advertisement distribution. Due to its rapid data transfer speeds, this technology lets organizations access large amounts of data. Tailoring messages focuses on marketing.

Alsharif et al. (2020) claimed this method will help them generate customer-focused advertisements. 6G lets workers control industrial equipment from home in real-time. Network data transport efficiency may enable haptic sensory suites. Workers may utilize a joystick and headset to keep safe while engaging with coworkers in these suits.

In addition, with 6G technology, businesses will have greater command over their campaigns thanks to cutting-edge capabilities like real-time optimization. What this implies is that they may optimize their campaigns by making real-time adjustments. According to Jiang and Schotten (2021), this will help businesses maximize their marketing expenditure and achieve the most ROI potential. The BUS has been discussing metaverse meetups and stores. A new era is dawning on humanity, one in which the line between the two worlds will dissolve entirely. Perhaps a consumer would prefer to shop from the convenience of his own home, engage in real-time conversation with a salesperson, and enjoy the store's ambience in person.

On top of that, 6G will allow marketers to provide customers with even more immersive experiences. Thanks to 6G's minimal latency, marketers will have an unprecedented opportunity to craft immersive and compelling VR and AR experiences (Mahmood et al., 2020). 6G technology and linked sensors will merge the digital and physical worlds by creating digital replicas of actual objects, people, and their environments (Jiang et al., 2021). With its speed and efficiency, 6G will improve clients' metaverse and mixed reality experiences (Kumar et al., 2023). 6G lets advertisers watch customers closely. Akhtar et al. (2020) suggests employing 6G technology to better identify customers' purchase behavior to improve marketing communications.

*Phygital smoothness.* Digital representations may create digital copies of actual objects, people, and their surroundings using 6G technology and sensors, connecting the digital and physical worlds. Customers' experiences in the metaverse and mixed reality will be enhanced with 6G's speed and efficiency (Giordani et al., 2020). With 6G, smart and interconnected devices will be a reality: 6G is expected to be 100 times more dependable and support 10 times as many devices, according to experts (Kumar et al., 2023). With this new tech, we can create an environment where every single device is in perfect harmony with every other. As an example, autonomous cars of the future may be able to detect the locations of other vehicles. Vehicles may be able to divert their attention from monitoring heavily populated areas to coordinating their movements to reduce riders' trip times. If the network connection is reliable and quick, all of this could be accomplished. The real-time experience for customers is brought about by 6G (Raddo et al., 2021). Metaverse meeting spaces and shops have been discussed. A new era is dawning on humanity, one in which the line between the two worlds will dissolve entirely (Dang et al., 2020). Perhaps a consumer would prefer to shop from the convenience of his own home, engage in real-time conversation with a salesperson, and enjoy the store's ambience in person. Lastly, 6G will allow companies to take advantage of AI and ML to sharpen their audience targeting and understanding. As a result, companies will be able to design more effective and tailored marketing campaigns, which should increase their website traffic and revenue.

## 7.4 IMPLICATIONS

The study's premise is that product/service and technology company managers will be able to use this evaluation to their advantage to generate new knowledge, building on the findings presented in this chapter. Top research organizations and mobile corporations are hard at work on the sixth generation of wireless communication, all while the 5G rollout continues worldwide. One goal of 6G networks is to improve accessibility to immersive technologies and speed mobile-to-mobile communication to bridge the gap between the real and virtual worlds. To be future-proof and make the most of their wireless infrastructure, organizations should be familiar with 6G networks and how they function. Personalized advertising and promotion of goods and services is a key component of marketing's future growth. If contextual promotion now centers on considering how people use the World Wide Web, then the foreseeable future marketing field is tied to keeping tabs on how people handle different aspects of their lives daily. Consumers' actions throughout the research, evaluation, and acquisition of products and services are the primary emphasis of behavioral marketing. One way that marketing research is becoming more personalized is through the use of information and communication technology. A substantial data resource for advertising goods and services is the monitoring of Internet searches made by cellular service users in conjunction with the tracking of people's movements. While both 5G and 6G networks present serious security risks, the latter may be significantly more dangerous. The reason is that when we compare 5G networks to 6G networks the latter could have a vulnerability area and potentially introduce security risks due, to more complex scenarios. There are also concerns regarding the reliance on open-source software in 6G networks, which could further impact security. To mitigate these threats businesses must enhance their security measures and stay up to date with the advancements in network technology. Protection against actors requires implementing security techniques, such as authentication, encryption, and access control. Moreover, businesses should be aware of the risks posed by components and infrastructure and take appropriate actions to safeguard national security.

## 7.5 THE FUTURE OF 6G IN MARKETING

The field of technology has seen advancements over time starting from the first generation (1G) that used analogue communication to the current fifth generation (5G), which enables seamless connectivity across different devices and systems. The rise of communication has greatly influenced how people live their lives leading to the digitalization and integration of information technology, into our society. Naturally, the upcoming 6G services are expected to offer speed and capabilities that will attract users. However, there is still uncertainty around the design and elements needed to establish a robust 6G network. The integration of connectivity and automated processes in advanced iterations has the potential to support the development and expansion of networks discussed earlier. It is also important to address the limitations of existing networks to enhance network architecture and optimize performance for results. By utilizing intelligence (AI) systems along with high-performance edge computing, we can leverage the enhanced speed capabilities offered by 6G technology to better coordinate complex systems and ensure uninterrupted internet access.

It is expected that the upcoming 6G technology will offer speeds 20 times faster, than what we have with 5G. Alongside this, it will have a much larger capacity to handle more devices and services. This progress will be advantageous, for businesses as it enables them to offer a range of applications and resources to their customers resulting in immersive experiences. Moreover, businesses will have the capability to effectively reach their desired audience by providing advertisements and offers. The increasing fascination with 6G technology is driven by the developments witnessed with 5G, which has sparked curiosity, about its achievements. The true potential of the 6G network lies in its capacity to provide speed, extensive capacity, and minimal latency to the Internet. This opens up possibilities for interactions that were previously unimaginable. If implemented 6G technology has the potential to revolutionize marketing and promotional efforts for companies and entrepreneurs. Marketers would be able to deliver advertisements and content, with speed, enabling participation and interaction with clients. This would enable enterprises to effectively contact and engage their customers through the use of extremely customized and personalized messaging, leading to improved conversion rates and return on investment (ROI). Industries such as business-based enterprises, medical care, educational institutions, recreation, and transport are poised to see substantial advantages through the implementation of 6G connectivity technology. It is imperative to prioritize the alignment of corporate operations with emerging technologies, prospects, and the integration of global network devices at present.

## 7.6 CONCLUSION

The influence of technology on the human lifestyle is significant. The advent of wireless technologies has brought about a significant transformation in various domains of human existence, including enterprises, living standards, and infrastructure. Humanity is perpetually engaged in a relentless pursuit of sophisticated resolutions to diverse challenges, while actively seeking novel pathways for advancement. The desire of humanity has facilitated the evolution of wireless communication from the first generation (1G) to the fifth generation (5G). Nevertheless, the progress in this area has not ceased. Researchers worldwide are diligently engaged in the advancement of the 6G wireless communications system, with an anticipated deployment timeline of 2030. This chapter provides a concise overview of 6G technology, including an introductory discussion. Furthermore, a brief analysis of its core technologies has been conducted to ascertain its benefits. Additionally, an examination of the future market prospects of the 6G network has been undertaken, drawing upon data pertaining to ten crucial elements of humanity. Through a comparative analysis of the three primary features along with the application industries of 5G network technologies, it becomes evident that 6G network technology possesses three distinct advantages. These advantages not only enable it to dominate the existing 5G market applications but also facilitate the development of novel and exclusive technologies. In addition to its superior speed compared to 5G systems, 6G has distinctive technical applications, including XR technology and virtual interaction. The advent of 6G connectivity is poised to introduce a plethora of novel functionalities to the realm of electronic commerce, hence empowering marketers with unparalleled opportunities to acquire a comprehensive understanding of customer behavior and preferences. This technology will complicate data management and privacy. The increasing e-business marketing industry must monitor these breakthroughs and strategy modifications to compete. This requires existing network capabilities including rapid data transfer, low latency, robust security, flexibility, and intelligent analysis. 6G network installation is resource-intensive. Automation requires AI, machine learning, cognitive networks, better encryption, and IoT. Network security, reliability, and privacy to improve industry apps needs to be adequately managed.

## REFERENCES

Ahokangas, P., & Aagaard, A. (2023). The Changing World of Mobile Communications: 5G, 6G and the Future of Digital Services. In *The Changing World of Mobile Communications: 5G, 6G and the Future of Digital Services*. Springer International Publishing. https://doi.org/10.1007/978-3-031-33191-6

Ahokangas, P., Matinmikko-Blue, M., & Yrjola, S. (2023). Envisioning a Future-Proof Global 6G from Business, Regulation, and Technology Perspectives. *IEEE Communications Magazine*, *61*(2), 72–78. https://doi.org/10.1109/MCOM.001.2200310

Ahokangas, P., Yrjola, S., Matinmikko-Blue, M., Seppanen, V., & Koivumaki, T. (2020). Antecedents of future 6G mobile ecosystems. *2nd 6G Wireless Summit 2020: Gain Edge for the 6G Era, 6G SUMMIT 2020*. https://doi.org/10.1109/6GSUMMIT49458.2020.9083756

Akhtar, M. W., Hassan, S. A., Ghaffar, R., Jung, H., Garg, S., & Hossain, M. S. (2020). The shift to 6G communications: Vision and requirements. In *Human-centric Computing and Information Sciences* (Vol. 10, Issue 1). Springer. https://doi.org/10.1186/s13673-020-00258-2

Akyildiz, I. F., Kak, A., & Nie, S. (2020). 6G and Beyond: The Future of Wireless Communications Systems. *IEEE Access*, *8*, 133995–134030. https://doi.org/10.1109/ACCESS.2020.3010896

Al-Ansi, A., Al-Ansi, A. M., Muthanna, A., Elgendy, I. A., & Koucheryavy, A. (2021). Survey on Intelligence Edge Computing in 6G: Characteristics, Challenges, Potential Use Cases, and Market Drivers. *Future Internet*, *13*(5). https://doi.org/10.3390/fi13050118

Alsharif, M. H., Kelechi, A. H., Albreem, M. A., Chaudhry, S. A., Sultan Zia, M., & Kim, S. (2020). Sixth Generation (6G) Wireless Networks: Vision, Research Activities, Challenges and Potential Solutions. *Symmetry*, *12*(4). https://doi.org/10.3390/SYM12040676

Alwis, C. De, Kalla, A., Pham, Q. V., Kumar, P., Dev, K., Hwang, W. J., & Liyanage, M. (2021). Survey on 6G Frontiers: Trends, Applications, Requirements, Technologies and Future Research. *IEEE Open Journal of the Communications Society*, 2, 836–886. https://doi.org/10.1109/OJCOMS.2021.3071496

Bertin, E., Crespi, N., & Magedanz, T. (2021). Shaping future 6G networks: Needs, impacts, and technologies. In *Shaping Future 6G Networks: Needs, Impacts, and Technologies*. books.google.com. https://doi.org/10.1002/9781119765554

Chowdhury, M. Z., Shahjalal, M., Ahmed, S., & Jang, Y. M. (2020). 6G Wireless Communication Systems: Applications, Requirements, Technologies, Challenges, and Research Directions. *IEEE Open Journal of the Communications Society*. vol. 1, pp. 957-975, 2020 https://doi.org/10.1109/OJCOMS.2020.3010270

Dang, S., Amin, O., Shihada, B., & Alouini, M. S. (2020). What Should 6G Be? *Nature Electronics*, *3*(1), 20–29. https://doi.org/10.1038/s41928-019-0355-6

Giordani, M., Polese, M., Laya, A., Bertin, E., & Zorzi, M. (2021). 6G Drivers for B2B Market: E2E Services and Use Cases. *Shaping Future 6G Networks: Needs, Impacts, and Technologies*, vol.5, 9–22. https://doi.org/10.1002/9781119765554.ch2

Giordani, M., Polese, M., Mezzavilla, M., Rangan, S., & Zorzi, M. (2020). Toward 6G Networks: Use Cases and Technologies. *IEEE Communications Magazine*, *58*(3), 55–61. https://doi.org/10.1109/MCOM.001.1900411

Gustavsson, U., Frenger, P., Fager, C., Eriksson, T., Zirath, H., Dielacher, F., Studer, C., Pärssinen, A., Correia, R., Matos, J. N., Belo, D., & Carvalho, N. B. (2021). Implementation Challenges and Opportunities in Beyond-5G and 6G Communication. *IEEE Journal of Microwaves*, *1*(1), 86–100. https://doi.org/10.1109/JMW.2020.3034648

Hassnain Mohsan, S. A., Mazinani, A., Malik, W., Younas, I., Hamood Othman, N. Q., Amjad, H., & Mahmood, A. (2020). 6G: Envisioning the Key Technologies, Applications and Challenges. *International Journal of Advanced Computer Science and Applications*, *11*(9), 14–23. researchgate.net. https://doi.org/10.14569/IJACSA.2020.0110903

Imoize, A. L., Adedeji, O., Tandiya, N., & Shetty, S. (2021). 6G Enabled Smart Infrastructure for Sustainable Society: Opportunities, Challenges, and Research Roadmap. *Sensors*, *21*(5), 1–57. https://doi.org/10.3390/s21051709

Ji, B., Wang, Y., Song, K., Li, C., Wen, H., Menon, V. G., & Mumtaz, S. (2021). A Survey of Computational Intelligence for 6G: Key Technologies, Applications and Trends. *IEEE Transactions on Industrial Informatics*, *17*(10), 7145–7154. https://doi.org/10.1109/TII.2021.3052531

Jiang, W., Han, B., Habibi, M. A., & Schotten, H. D. (2021). The Road Towards 6G: A Comprehensive Survey. *IEEE Open Journal of the Communications Society*, 2, 334–366. https://doi.org/10.1109/OJCOMS.2021.3057679

Jiang, W., & Schotten, H. D. (2021). The kick-off of 6G research worldwide: An overview. *2021 7th International Conference on Computer and Communications, ICCC 2021*, 2274–2279. https://doi.org/10.1109/ICCC54389.2021.9674614

Katz, M., Pirinen, P., & Posti, H. (2019). Towards 6G: Getting ready for the next decade. *Proceedings of the International Symposium on Wireless Communication Systems, 2019 August*, 714–718. https://doi.org/10.1109/ISWCS.2019.8877155

Kumar, R., Jain, V., Leong, W. Y., Teyarachakul, S. (2023). *Convergence of IoT, Blockchain and Computational Intelligence in Smart Cities*. Taylor & Francis, CRC Press, ISBN: 9781032404240.

Kumar, R., Jain, V., Tan, G.W.H. & Touzene, A. (Eds.) (2023). *Immersive Virtual and Augmented Reality in Healthcare – An IoT and Blockchain Perspective*. Taylor & Francis, CRC Press. https://doi.org/10.1201/9781003340133

Mahmood, N. H., Alves, H., Lopez, O. A., Shehab, M., Osorio, D. P. M., & Latva-Aho, M. (2020). Six key features of machine type communication in 6G. *2nd 6G Wireless Summit 2020: Gain Edge for the 6G Era, 6G SUMMIT 2020*. https://doi.org/10.1109/6GSUMMIT49458.2020.9083794

Matinmikko-Blue, M., Yrjölä, S., & Ahokangas, P. (2023). Multi-perspective approach for developing sustainable 6G mobile communications. In *Telecommunications Policy*. Elsevier. https://doi.org/10.1016/j.telpol.2023.102640

Nayak, S., & Patgiri, R. (2021). 6G Communication Technology: A Vision on Intelligent Healthcare. *Studies in Computational Intelligence*, 932, 1–18. https://doi.org/10.1007/978-981-15-9735-0_1

Nguyen, D. C., Ding, M., Pathirana, P. N., Seneviratne, A., Li, J., Niyato, D., Dobre, O., & Poor, H. V. (2022). 6G Internet of Things: A Comprehensive Survey. *IEEE Internet of Things Journal*, 9(1), 359–383. https://doi.org/10.1109/JIOT.2021.3103320

Nguyen, V. L., Lin, P. C., Cheng, B. C., Hwang, R. H., & Lin, Y. D. (2021). Security and Privacy for 6G: A Survey on Prospective Technologies and Challenges. *IEEE Communications Surveys and Tutorials*, *23*(4), 2384–2428. https://doi.org/10.1109/COMST.2021.3108618

Raddo, T. R., Rommel, S., Cimoli, B., Vagionas, C., Perez-Galacho, D., Pikasis, E., Grivas, E., Ntontin, K., Katsikis, M., Kritharidis, D., Ruggeri, E., Spaleniak, I., Dubov, M., Klonidis, D., Kalfas, G., Sales, S., Pleros, N., & Tafur Monroy, I. (2021). Transition Technologies towards 6G Networks. *Eurasip Journal on Wireless Communications and Networking*, *2021*(1). jwcn-eurasipjournals.springeropen. https://doi.org/10.1186/s13638-021-01973-9

Saad, W. (2021). *6G Wireless Systems: Challenges and Opportunities BT - 5G and Beyond: Fundamentals and Standards* (X. Lin & N. Lee (eds.); pp. 201–229). Springer International Publishing. https://doi.org/10.1007/978-3-030-58197-8_7

Saad, W., Bennis, M., & Chen, M. (2020). A Vision of 6G Wireless Systems: Applications, Trends, Technologies, and Open Research Problems. *IEEE Network*, *34*(3), 134–142. https://doi.org/10.1109/MNET.001.1900287

Tataria, H., Shafi, M., Molisch, A. F., Dohler, M., Sjoland, H., & Tufvesson, F. (2021). 6G Wireless Systems: Vision, Requirements, Challenges, Insights, and Opportunities. *Proceedings of the IEEE*, *109*(7), 1166–1199. https://doi.org/10.1109/JPROC.2021.3061701

Uusitalo, M. A., Rugeland, P., Boldi, M. R., Strinati, E. C., Demestichas, P., Ericson, M., Fettweis, G. P., Filippou, M. C., Gati, A., Hamon, M. H., Hoffmann, M., Latva-Aho, M., Parssinen, A., Richerzhagen, B., Schotten, H., Svensson, T., Wikstrom, G., Wymeersch, H., Ziegler, V., & Zou, Y. (2021). 6G Vision, Value, Use Cases and Technologies from European 6G Flagship Project Hexa-X. *IEEE Access*, *9*, 160004–160020. https://doi.org/10.1109/ACCESS.2021.3130030

Vaigandla, K. K., Bolla, S., & Karne, R. (2021). A Survey on Future Generation Wireless Communications-6G: Requirements, Technologies, Challenges and Applications. *International Journal of Advanced Trends in Computer Science and Engineering*, *10*(5), 3067–3076. academia.edu. https://doi.org/10.30534/ijatcse/2021/211052021

Viswanathan, H., & Mogensen, P. E. (2020). Communications in the 6G Era. *IEEE Access*, *8*, 57063–57074. https://doi.org/10.1109/ACCESS.2020.2981745

Wen, H. (2023). Webcast Marketing Platform Optimization via 6G R&D and the Impact on Brand Content Creation. *PLoS ONE*, *18*(10 October). journals.plos.org. https://doi.org/10.1371/journal.pone.0292394

Zadorozhnyi, Z. M., Muravskyi, V., Pochynok, N., & Ivasechko, U. (2023). Application of the Internet of Things and 6G Cellular Communication to Optimize Accounting and International Marketing. *Virtual Economics*, *6*(1), 38–56. https://doi.org/10.34021/VE.2023.06.01(3)

Zhang, Z., Xiao, Y., Ma, Z., Xiao, M., Ding, Z., Lei, X., Karagiannidis, G. K., & Fan, P. (2019). 6G Wireless Networks: Vision, Requirements, Architecture, and Key Technologies. *IEEE Vehicular Technology Magazine*, *14*(3), 28–41. https://doi.org/10.1109/MVT.2019.2921208

Ziegler, V., Viswanathan, H., Flinck, H., Hoffmann, M., Räisänen, V., & Hätönen, K. (2020). 6G Architecture to Connect the Worlds. *IEEE Access*, *8*, 173508–173520. https://doi.org/10.1109/ACCESS.2020.3025032

Chapter 8

# Emerging technologies for security and privacy in 6G wireless communication networks

*Shikha Aggarwal, Deepti Mehrotra, and Anchal Garg*

## 8.1 INTRODUCTION

Over time, wireless communication networks have seen significant advancements, with new generations offering innovative capabilities that have transformed communication and interaction. The sixth generation (6G) of wireless networks promises never-before-seen speed, low latency, and ubiquitous access, but it also carries with it a host of security and privacy issues that need to be resolved. The way we plan, construct, and safeguard the integrity of 6G networks must change dramatically in response to the exponential growth of data traffic, the multitude of connected devices, and the emergence of new threat vectors. The advent of 6G wireless networks heralds a new era in which it is getting harder to distinguish between the physical and digital worlds and an intelligent, technologically driven, hyper-connected ecosystem is beginning to take shape. Encouraging confidence among users, organizations, and society at large will depend on ensuring the robust security and privacy of the underlying infrastructure as this vision takes shape.

In this context, new technologies are critical to protecting user privacy and enhancing 6G networks' security posture. The foundation for a discussion of the cutting-edge innovations influencing 6G security and privacy is laid forth in this introduction. The security challenges faced by 6G networks are multifaceted [1]. The advent of quantum computing poses a tangible threat to traditional cryptographic algorithms, necessitating the adoption of quantum-safe cryptography techniques to safeguard sensitive information. Meanwhile, the increasing sophistication of cyber attacks necessitates dynamic, adaptive security solutions that can foresee and mitigate hazards in real time [2].

User privacy concerns are becoming more important in an era of widespread data collection, which highlights the need for innovative privacy-preserving technology. Techniques like homomorphic encryption, safe multi-party computation, and differential privacy become essential tools for striking a balance between the demands of data-driven insights and privacy

DOI: 10.1201/9781003583127-8

protection. This study examines the diverse array of technologies equipped to address these concerns as the 6G landscape evolves [3]. The nuances of these technologies, from artificial intelligence-driven threat detection to blockchain-enabled secure authentication, and from biometric identity verification to the use of network slicing for customized security approaches, are examined in the following sections along with their combined effects on the security and privacy framework of 6G wireless communication networks [4]. As we follow this path of revolution, integrating state-of-the-art technologies not only shields us from ever-evolving threats but also sets the stage for a future in which 6G networks will be associated with reliability, durability, and a commitment to protecting the digital exchanges that define our international community [5].

## 8.2 RELATED WORK

1. Tang et al. (2023) present an overview of several machine learning (ML), reinforcement learning (RL), and deep learning (DL) approaches in connection to the advancement of 6G technology. The significance of these algorithms for creating new technologies that align with the objectives and ambitions of the 6G network is emphasized. Integration of new technologies and ML algorithms is necessary for meaningful advancements in communication.
2. Quantum computing poses a threat to traditional cryptographic methods. Wang et al. (2021) discuss the implementation of quantum-resistant secure communication in 6G networks.
3. Alhammadi et al. (2024) explore the opportunities, challenges, and solutions of integrating artificial intelligence into 6G networks.
4. Sarveshwaran et al. (2021) investigate the use of blockchain for secure and privacy-preserving communication in 6G systems.
5. Sivizaca Conde et al. (2024) review privacy-preserving data-sharing techniques in the context of 6G wireless networks.
6. Li et al. (2023) discuss opportunities and challenges associated with the integration of post-quantum cryptography in 6G networks.
7. Nguyen et al. (2021) analyze the opportunities and challenges of implementing biometric authentication in 6G mobile networks [6].
8. C. R. García et al. (2023) explore the challenges, opportunities, and solutions associated with secure edge computing in the context of 6G networks Li et al. (2021) investigate security and privacy issues in network slicing for 6G, providing insights into challenges and potential solutions.
9. Mucchi et al. (2021) discuss opportunities and challenges associated with implementing physical layer security in 6G wireless networks.
10. Ramezanpour et al. (2021) present principles and deployment considerations for zero-trust networks in 6G.

11. Alawadhi et al. (2023) covers the main features of 6G networks, such as 3D intercoms, distributed artificial intelligence, intelligent radio, and real-time intelligent edge computing. Along with pertinent security and privacy concerns, that also examines promising new technology in each field.
12. Siriwardhana et al. (2021) offer information on a range of 6G technologies, such as terahertz (THz) communication, intelligent surfaces, and artificial intelligence (AI). They also talk about possible privacy and security issues with 6G wireless networks.

## 8.3 6G VISION, REQUIREMENTS, AND CAPABILITIES

With 5G's ability to improve communication, there are high hopes for 6G. Those projections are strengthened even more by the assumption of a large user base and high connectivity in the 6G network. It was demonstrated by the previously issued 6G network KPIs, which were also covered during the ITU-R workshop on IMT. Various capabilities and requirements are discussed here are as follows:

1. **Network flexibility:** The network flexibility is enhanced in the 6G network. It targets enhancing the development speed, reducing installation cost as well energy usage. 6G ensures the deployment of a low-cost network with higher capacity. For more flexibility of the 6G network enhancing the transport layer in terms of reliability and scalable to handle 6G use cases [7].
   6G separates the control part from the data part of the network. This makes it easier and faster to change the network to fit what the user needs right now. 6G is made to use lots of different frequencies, like really high waves and light waves. This flexibility allows spectrum resources to be used as needed for things like fast data or quick reactions to changes in the environment. Future deployments will use the same platforms for the main networks and radio networks, which will make things less centralized. In addition, they use RAN and CN abilities together to stop unnecessary network functions [8]. To have control over network operations and development, it is important to have the right combination of high performance and deployment flexibility to enable new uses.
2. **End-to-end communication:** End-to-end communication in a 6G network refers to the transfer of data from a source to a destination, which must transit through several network components and technologies before reaching its final destination. For the high performance of the 6G services in the future the connections required extraordinary bandwidth and robustness. For better end-to-end connection, an excellent collaboration is required for all layers of the

system. Many services and applications require the lowest latency with maximum tolerance. Predictable latency enables unique use cases in both centralized and scattered deployments. Standard open interfaces between network components facilitate the integration of new technologies and services. This creates a more open environment for online innovation [9].

3. **High coverage:** New radio technology should be able to do many things well and allow new services to be offered at a fair price. First, we need data to travel faster and with less delay time. We need really good network connections to provide services to more people. 6G required a higher range to cover the area for the signal transmission, which is less in the previous networks. New radio technologies, such as terahertz frequencies, are being investigated for 6G. While these frequencies enable greater data speeds, their range is often limited. However, advances in antenna design and signal processing seek to increase coverage at high frequencies. Energy efficiency is also a major consideration for 6G [10]. Lower power consumption in network infrastructure and devices will enable the broader deployment, especially in remote locations where power supplies may be limited.
4. **Embedded devices:** Embedded devices played a vital role in the 6G network, enabling a wide range of applications across different sectors. IoT and sensor devices are one of important devices in the 6G network for better and faster collection of the information from the surrounding environment. These embedded devices collect as well as transmit the data and also process it [11].
5. **Trusted systems:** The key to making strong systems is being able to withstand, notice, respond to, and bounce back from attacks and unexpected problems. Four important things make systems trustworthy. They are using private computing, secure identifications and protocols, ensuring services are always available, and having security insurance and defense. AI will have a big effect on how technology grows and how we stay safe in the future in these four areas. Also, AI components must be reliable. Many people are putting a lot of importance on security support and certification nowadays. The latest security designs are good at keeping a specific product safe. However, we need to make things better. We need to think about virtualization, cloud computing, continuous integration, continuous delivery, and AI improvements. The ways we keep things secure today need to be improved in the future to check everything in the system. It is very important to make clear rules and steps that everyone agrees upon when making security plans. Confidential computing is a strong way to keep data safe when it is being used or stored. Encryption is a way to keep data safe from being changed in the cloud. The hardware takes care of making sure everything is certified and processed, so the

cloud provider cannot change the hardware. The root of trust (RoT) is really important for making sure that computers stay safe. We can keep our information safe by using methods to protect it, keeping it private on our computers and devices, and separating tasks on the network. We were discovering the different pieces of a computer and what the computer can do with RoT. We want to make a system that keeps information secure and protects the privacy of apps. Resources for important services are arranged to give different types of help. Automatic recovery systems can look at and put together information from all parts of the communication system. We need a plan that is organized in layers so we can see how well things are working as they happen. AI can be used to see if a service is working by looking at data and watching closely. Real-time AI analytics can also help make the traffic flow better and make adjustments to radio settings [12].

6. **Cognitive networks:** We need to use the intelligence of these networks more efficiently to create better networks at a lower cost. Cognitive networks will help make energy use more efficient and services more available. We think this could happen in two ways: using ML and AI to help with hard optimizations that are usually difficult to do, and creating control systems that can manage the system by themselves. When people tell systems what they want them to do, it will affect how the systems work. Automated management needs better ways for people to tell the machines what to do, and for the machines to understand and figure out the goals. People and computer programs will share what they know and what they have learned to help each other learn. A cognitive network [13] uses these parts to deal with different situations, figure out the best things to do, and make a plan for how to work well in the network.

   Cognitive systems need to naturally change and learn from what's around them, always keeping an eye on and learning from past events. Information about how earlier operations went is sent back to improve settings, procedures, and software immediately. We will keep getting better at defending against attacks that target both the physical and logical parts of a network. This constant improvement can make the whole system much more secure and dynamic than it is now. Intelligence will be available in different ways through a network.

7. **URLLC:** URLLC is a method of communication that is highly dependable and fast. This means that URLLC is good for things that need instant communication and where even a little bit of data loss or delay can be very important. As we work on 6G, we expect to use new technologies to meet the need for very reliable and fast communication. URLLC has many benefits for 6G networks. A URLLC

could make critical infrastructure and systems safer by making sure they can communicate quickly and without any errors. It will help users do things more easily and make better choices quickly.

In simple words, URLLC can use less power and internet space because it focuses on being dependable and quick. This means that URLLC networks are designed to send little bits of data fast and without errors, even when it is difficult. The advantages of URLLC make it good for many high-tech tools, like AI-powered optimization. This will improve the URLLC networks when it comes to predicting issues, deciding where to put resources, keeping things secure, and fixing network problems. The Internet of Everything (IoE) creates new security issues because of the various functions and challenges of using distributed artificial intelligence. Because these new devices move a lot, they often change between networks and use services from different networks. This can make the data less safe and private. In 6G, they believe it will take much less time for a signal to go from one place to another, maybe only one millisecond, or even less for certain types of communication. Lately, many studies have looked at how ML can be used to improve future communications because it has a lot of potential benefits. In [14], the writers talked about using different types of machine learning at each level of communication between devices. They said that the machine learning program used in applications and infrastructure can meet the needs of 6G [14–16]. Meanwhile, Tang et al. talked about one of the requirements for 6G networks, called URLLC. The authors showed how machine learning can be used to make channels work better, decide the best ways to send data through a network, control traffic when it gets busy, and adjust how videos are streamed [29]. Also, many studies talked about how ML algorithms can help improve m-MIMO communication by optimizing parameters. In [30], the authors studied how machine learning helps with transmitting multiple signals in 5G networks. They [15] worked on different problems like figuring out the best way to send signals through channels, directing beams of signals, improving signal quality, setting up networks without using individual cells, and using multiple antennas to send signals with multiple users on the same channel. Another study looked at how to improve communication for the 6G network. They focused on using a DL algorithm called a transformer, which helps with exchanging information more smartly. ML algorithms can be used to make sensing and communication better. Demirhan and Alkhateeb talked about ten important jobs for machine learning in sensing and communication [16]. They put these jobs into three groups: working together for sensing and communication, using senses to help with communication, and using communication to help with sensing. The writers of [17] explained how using a smart surface (RIS) can make wireless communication better by using a special computer program (ML algorithm) to change its settings. They said that using RIS as a new technology with algorithms based on data could make communication better.

Some previous studies have described how machine learning algorithms can be used in 6G networks. However, there are not many studies that talk about using ML in new technologies for 6G.

8. **Scalability:**

   Scalability means the network can simply handle the increasing number of users and their devices without affecting the network performance in terms of data rates, low latency, maximum throughput massive connectivity, etc. 6G provides better scalability for machine-to-machine (M2M refers to communication between two end nodes without human intervention). Scalable networks will become more and more necessary as M2M connections rise. The widespread deployment of 6G networks will enable a significantly greater number of M2M devices to be linked to the network, which will be necessary for applications like the Internet of Things. As a result, 6G networks are well-positioned to address this need. By guaranteeing that users have a dependable and constant connection to the network, scalability will also contribute to an enhanced user experience [18]. This would make it possible for brand-new, cutting-edge M2M uses in wireless networks, like smart cities, transportation, and industrial automation.

   Software-defined networking (SDN), network slicing, AI, heterogeneous networks, and hybrid clouds are some of the elements that affect 6G networks' scalability. With SDN, the network will be able to be managed and controlled by software, making it simpler for the network to adjust to variations in traffic demand. Heterogeneous networks and network slicing can grow networks for enormous applications and services. A hybrid cloud deployment paradigm integrates the advantages of both private and public clouds. By adding more resources, the public cloud can expand the network horizontally, but the private cloud can expand the network vertically by adding more potent resources.

9. **Personalized QoE:**

   It is anticipated that 6G wireless networks would offer a substantially better quality of experience than earlier models. Many of the reasons for this are covered in the preceding article, including increased data speeds, decreased latency, increased reliability, improved security, and greater coverage. However, QoE must be customized for each user or application on the network to maximize overall network efficiency toward a particular goal. This way, network operators may make

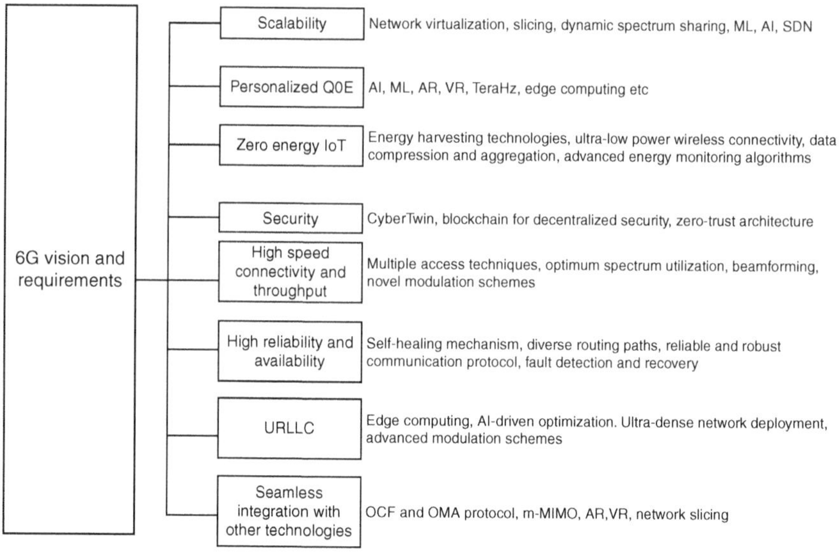

*Figure 8.1* 6G vision and capabilities.

sure that resources are used effectively. Aside from that, a variety of users and applications with varying QoE requirements are anticipated to be supported by 6G networks. Therefore, tailoring QoE can guarantee that customers can quickly and efficiently maximize their experience and attain the greatest potential outcome. Network slicing, edge computing, ML, and AI can increase QoE [19]. AI and ML can analyze user activity and preference data to personalize network experiences. Network slicing can personalize QoE by separating the network into many virtual networks that can be configured for users or applications. Radio access solutions like THz communication could support more demanding applications and improve user QoE by increasing data rates or lowering latency. VR, AR, self-driving cars, Industrial IoT, and smart cities benefit from personalizing QoE.

## 8.4 TECHNOLOGIES ENABLED FOR 6G SECURITY

### 8.4.1 Quantum-resistant secure communication in 6G networks

The aim of studying quantum-resistant secure communication in 6G networks is to create ways to keep information safe from quantum computers trying to hack it. Quantum computers could break codes that keep communication secret on the internet. To keep our messages safe as 6G networks get better, we need to create and use new ways to encrypt them that are resistant to quantum computers [11].

Post-quantum cryptography is using codes that are thought to be safe from attacks by regular and super-powerful computers. This group includes types of encryption that are hard for quantum computers to break, like hash-based cryptography, code-based encryption, and lattice theory-based algorithms [12]. Quantum physics is used in quantum key distribution, a quantum communication technique, to secure a communication channel. QKD, which is based on quantum mechanics, ensures security while permitting the spread of cryptographic keys. Any attempt to intercept the quantum key, even by an eavesdropper, must be discovered due to the laws of quantum physics. Hybrid cryptosystems combine classical and quantum-resistant cryptography techniques to provide a transitional solution [13]. This allows for the integration of quantum-resistant techniques without requiring a total overhaul of the existing cryptographic architecture. For upcoming communication networks like 6G, standardization committees are developing and approving quantum-resistant cryptographic standards. Leading the charge to identify and standardize post-quantum cryptography techniques is the National Institute of Standards and Technology (NIST). Researchers are hard at work developing new cryptographic primitives and protocols that are resistant to attacks by quantum computers. This integrates mathematical analysis, algorithm design, and testing for practical implementation. Secure communication resistant to quantum errors must be easily integrated into the entire architecture of 6G networks. This includes topics like key management, authentication, and secure data transport [14].

### 8.4.2 AI-powered security in 6G networks

AI algorithms continuously monitor network traffic and system logs to detect anomalies or suspicious patterns that may indicate a security threat. Deployed machine learning models to analyze historical data to identify normal behavior and detect deviations that may indicate potential security incidents [15]. AI-powered systems provide real-time monitoring of network activities, allowing for immediate detection and response to security threats. Automated response mechanisms, guided by AI algorithms, are triggered during a security incident. This may include isolating affected components, blocking malicious traffic, or alerting security personnel. AI algorithms perform behavioral analysis to understand patterns in user and device behavior, enabling the identification of abnormal or suspicious activities. AI-enhanced authentication mechanisms use biometric data, behavioral patterns, and contextual information to ensure secure user access. Machine learning models can predict potential security threats based on historical data and emerging patterns, allowing for proactive security measures [16].

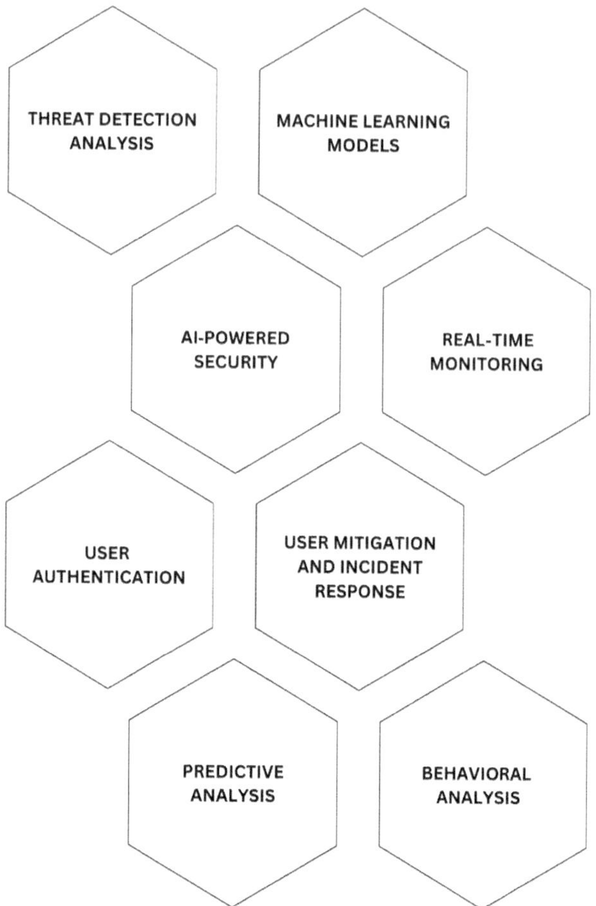


*Figure 8.2* Interconnected components of AI-powered security.

Figure 8.2 is a high-level diagram representing the interconnected components of an AI-powered security system in 6G networks [17]. The arrows indicate the flow of information and actions between the different components. Feel free to adapt and enhance this diagram based on the specific details and nuances of your AI-powered security implementation [18].

### 8.4.3 Zero trust is a security model

It assumes no entity, whether inside or outside the network, should be trusted by default. It requires strict verification of anyone trying to access resources in the network, regardless of their location or connection method.

The zero-trust model's core is verifying user or device identity. This involves robust authentication mechanisms, including multi-factor authentication (MFA) and continuous monitoring of user behavior. A central policy engine defines and enforces access policies based on user roles, device types, and other contextual factors. Policies specify who can access what resources under what conditions. Network segmentation is implemented on a micro-level, meaning that even within the network, access is restricted based on the principle of least privilege [19]. This limits lateral movement in case of a security breach. Continuous monitoring of user and device behavior helps in detecting anomalies or suspicious activities in real-time. This can include monitoring network traffic, device health, and user activity. Before granting access, the security model verifies the health and compliance of devices. This can include checks for updated security software, patched operating systems, and adherence to security policies. End-to-end encryption is enforced to secure data in transit, ensuring the data remains unreadable even if a network is compromised [20].

Zero trust extends beyond traditional network security to include API security. Access to APIs is also controlled and monitored to prevent unauthorized access. Figure 8.3 shows the conceptual diagram, outlining the main components of a zero-trust security model [21]. It illustrates how user/device identity is verified, policies are enforced, network segmentation is applied, continuous monitoring is conducted, device health is checked, encryption is implemented, and API security is maintained. The arrows represent the flow of information and access controls in a zero-trust environment.

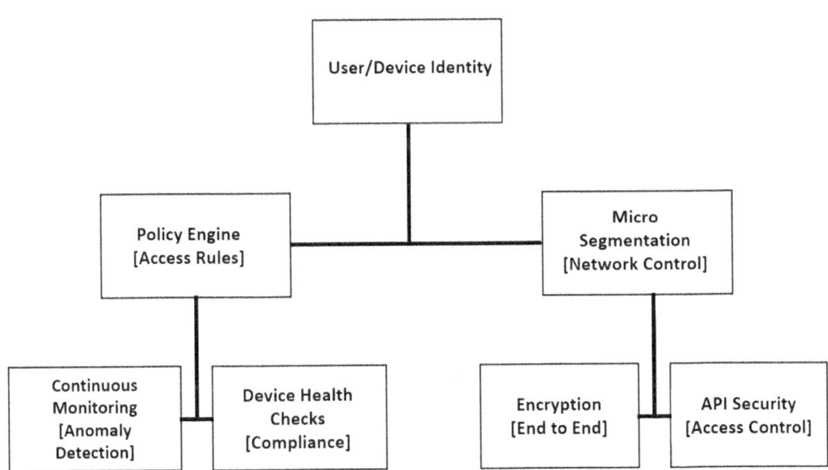


*Figure 8.3* Zero-trust security model.

### 8.4.4 Biometric Authentication in 6G

Biometric Authentication gives a guarantee of how the user verifies their details which revolves around the authenticity of the user. Figure 8.4 explains the data channel in the security paradigm and illustrates the implementation of authentication in 6G [22]. There are various ways to authenticate the biometric in a 6G environment as follows:

1. **Biometric data collection:** The process begins with the collection of biometric data from the user. This could include fingerprints, facial

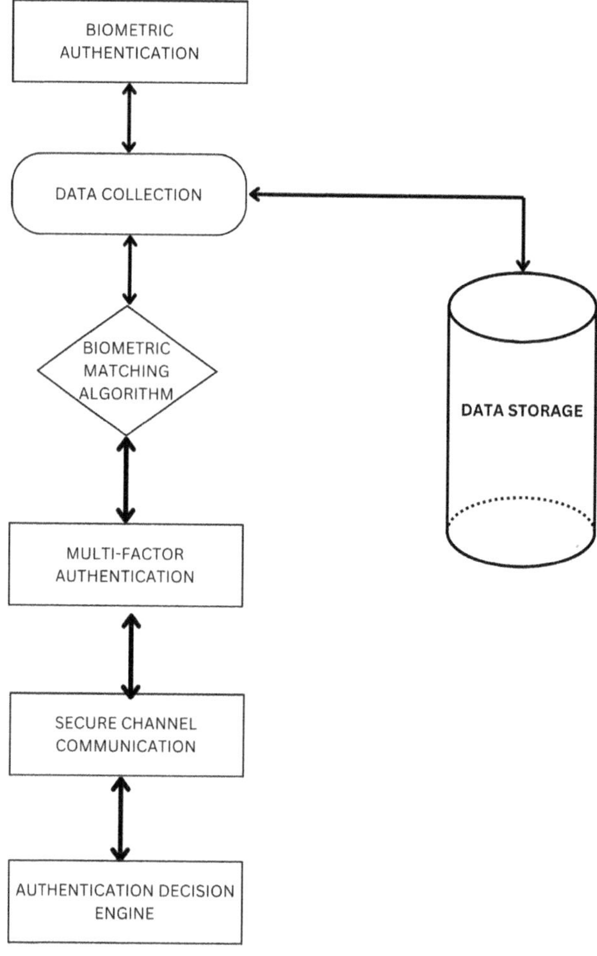


*Figure 8.4* Biometric authentication.

features, voice patterns, or other unique physiological or behavioral characteristics.

2. **Biometric data encryption:** The collected biometric data is encrypted to ensure its confidentiality and integrity during transmission and storage. End-to-end encryption is implemented to protect the data from unauthorized access.
3. **Biometric data storage:** Encrypted biometric data is stored securely in a central repository. Advanced security measures, such as secure enclaves or hardware-based security modules, may be employed to safeguard the stored data.
4. **Biometric matching algorithm:** A sophisticated biometric matching algorithm compares the live biometric data (captured during authentication attempts) with the stored encrypted biometric templates. This algorithm is designed to verify the identity of the user accurately.
5. **Multi-Factor Authentication (MFA):** Biometric authentication is often used as part of a MFA approach. Other factors, such as passwords or PINs, may be combined with biometrics to enhance security.
6. **Secure communication channel:** During the authentication process, a secure communication channel is established between the user's device and the authentication server to prevent eavesdropping or tampering.
7. **Authentication decision engine:** The authentication decision engine evaluates the results of the biometric matching and other authentication factors. It determines whether the user's identity is verified and grants or denies access accordingly.

## 8.5 NETWORK SLICING FOR SECURITY IN 6G

Network slicing is a key architectural concept in 6G networks, allowing the creation of multiple virtual networks (slices) within a single physical infrastructure. Each slice is customized to meet specific requirements, such as latency, bandwidth, and reliability [23]. Ensuring security within network slicing involves implementing tailored security measures for each slice. Below is a conceptual overview of network slicing for security in 6G, along with a simplified algorithm for security provisioning.

The network operator defines various slices based on specific service requirements. Slices can represent different services, industries, or user groups. A security policy is assigned to each slice, considering the unique security requirements of the services it supports. Policies may include access controls, encryption methods, and authentication mechanisms. Dynamic isolation mechanisms are implemented to ensure that the security boundaries of each slice are maintained [24]. This prevents unauthorized access or interference between slices. Dedicated resources, both hardware and virtual, are allocated to implement security functions within each slice. This may include intrusion detection, firewalling, and encryption/decryption processes. Users and devices connecting to a particular slice undergo authentication specific to that slice. This ensures that only authorized entities gain access to the services provided by that slice [25]. The security parameters of each slice are dynamically adjusted based on the current threat landscape, service demands, and the overall network condition. This adaptability enhances the network's resilience. Continuous monitoring of network slices is performed to detect anomalies or security breaches. Machine learning algorithms may be employed for advanced threat detection. In a security incident, an automated incident response mechanism is triggered to contain the threat. Remediation measures are applied to restore the security posture of the affected slice [26].

Simplified *n*etwork *s*licing for *s*ecurity *a*lgorithm*:*

```
For each Slice in the Network:
   Assign Security Policy to Slice
   Allocate Resources for Security Functions
   Implement Dynamic Isolation Mechanisms
While the Network is Operational:
   For each Connected Device in the Slice:
      Authenticate Device based on Slice-Specific Authentication
      Monitor Device Activity
      If Anomalies Detected:
         Trigger Incident Response Mechanism
         Apply Remediation Measures
   Adjust Security Parameters Dynamically
```

This simplified algorithm outlines the key steps involved in implementing network slicing for security in 6G [27]. The actual implementation would involve more granular details, protocols, and interfaces, but this provides a high-level view of the processes. Security in network slicing is dynamic and adaptive, responding to the unique demands of each slice and the evolving threat landscape [28].

## 8.6 PHYSICAL LAYER SECURITY (PLS)

It is a paradigm that leverages the characteristics of the physical transmission medium to enhance the security of communication systems. In the context of 6G networks, PLS becomes increasingly relevant as it addresses potential vulnerabilities and threats at the fundamental level of wireless communication. PLS exploits the unique features of the physical layer, such as channel characteristics and signal properties, to provide secure communication without relying solely on cryptographic techniques. This involves taking advantage of the inherent randomness and complexity of the wireless channel [29].

The wireless channel introduces noise, fading, and other impairments. PLS exploits these characteristics to ensure that an eavesdropper would have difficulty extracting meaningful information from the transmitted signals. PLS techniques involve intentionally designing signal transmissions to achieve secure communication. For instance, techniques like artificial noise injection, beamforming, and precoding can be employed to secure information transmission [30]. Artificial noise deliberately added to the transmission, can confuse eavesdroppers and make it challenging for them to distinguish between the actual signal and the noise. This technique enhances the confidentiality of the transmitted information. By shaping the transmitted signal in the spatial domain, beamforming and precoding can be used to direct the signal toward the intended receiver while minimizing its reception by unintended eavesdroppers, thus enhancing security. PLS may involve dynamically generating encryption keys based on channel state information. The varying characteristics of the wireless channel contribute to the randomness needed for secure key generation [31].

Integrating quantum principles for key distribution at the physical layer can enhance the security of communication by leveraging the principles of quantum mechanics to detect eavesdropping attempts. Multiple input multiple output (MIMO) and diversity techniques can be used to improve the reliability and security of wireless communication by exploiting the spatial and temporal characteristics of the wireless channel. PLS in 6G networks should be adaptable to changes in the communication environment. Algorithms that dynamically adjust security parameters based on real-time channel conditions contribute to the effectiveness of PLS. PLS is not a standalone security solution. It should be integrated with higher-layer security protocols and mechanisms to provide a comprehensive security framework for 6G networks. Physical layer security in 6G represents a forward-looking approach to addressing security challenges by exploiting the characteristics of the wireless channel itself [32]. The integration of PLS techniques enhances the confidentiality and integrity of communication, making it a crucial aspect of the security architecture in the next generation of wireless networks [33].

## 8.7 BLOCKCHAIN TECHNOLOGY

Blockchain technology has many uses in a 6G network, such as keeping records, spreading out the network, and sharing resources [34]. Dang and others investigate decentralizing [35] networks using blockchain technology and how it can make networks work better and be easier to manage. Using blockchain technology and distributed ledgers would greatly improve security when confirming identities. The blockchain is a very important technology for the internet. Additionally, by using blockchain technology and a system for sharing spectrum, we can solve the problem of not using the spectrum efficiently and one company controlling it. This will also protect the use of the spectrum.

The ways we communicate, prove who we are, and control who can access the information are all connected to blockchain security and privacy worries. Strinati, S. Barbarossa discuss a new way for companies to manage network access and security using blockchain technology in the radio access network architecture. McMahan and others [27] created a new way to approve mobile services using blockchain technology. McMahan suggested using blockchain technology to make media access protocols more secure and to enable cognitive radio. This would allow access to a licensed spectrum that is currently not being used.

## 8.8 MOLECULAR COMMUNICATION

Living organisms possess minuscule structures that they utilize for communication with one another. This is called molecular communication. Over the past 10 years, progress in tiny technology has been made in nanotechnology, bioengineering, and synthetic biology. This has led to the creation of very small technologies. Also, it takes only a small amount of energy to make and send a molecular communication signal. Communication researchers have just started to study this, even though biologists have been researching it for a long time. One very good technology for 6G communications is called molecular communication. It is a mix of different subjects and is still growing. Using chemical signals to transfer information is the main idea of molecular communication.

However, problems with keeping information safe and private when communicating, verifying identity, and encrypting data have been found. Zhang and his team [13] believed that enemies could disrupt the communication channels at a very small level in our bodies. Not many studies have looked into how to protect these channels. Lu and others [30] discuss how a new system was created to make sure data is safer when it is being sent. Also, Wang and others [19] suggest important ways to improve communication at a very small scale. They also talked about new ways to make sure data is secure and private. The writers talk about different ways to attack molecular communication, like flooding, jamming, and desynchronization. Even though it will take more time to make real-world molecular communication for the 6G network, this technology is expected to be better than what we have used before.

## 8.9 VISIBLE LIGHT COMMUNICATION (VLC)

VLC uses light that we can see to send messages. This light is between 400 and 800 THz. It has a wavelength of 780 to 375 nanometers. High-speed internet (THz) with almost no interruptions, lots of available space for communication, and using high frequencies repeatedly are all possible. Thus, VLC improves 6G short-range communications. In addition, 1G-5G wireless networks use sub-6 GHz microwave communications, which are practically exhausted. As more smart devices are added to the network and area traffic capacity explodes, 6G needs high capacity. VLC technology could meet 6G capacity and data rate needs.

D2D communication allows devices to communicate directly without base stations, using licensed (cellular) or unlicensed spectrum (WiFi). D2D enhances communication throughput, energy efficiency, latency, and fairness. With the explosion of edge devices, D2D is gaining popularity and adoption [32].

## 8.10 TERAHERTZ TECHNOLOGY

5G networks use mm-wave bands, but they are not fast enough for 6G because it requires even higher transmission rates. The radio frequency band is almost full and we cannot use it for new technology. These things have caused terahertz technology to be developed. Terahertz communication uses a range of frequencies between 0. 1 and 10 terahertz. This range has more available resources than the millimeter-wave range. Also, it uses energy from electricity and light. The THz communication technology can keep up data speeds of 100 gigabits per second or even faster [12]. Secondly, THz communication has a narrow beam and short pulses, which makes it harder for people to listen in to the communication, making it more secure. THz waves can travel through some materials without losing much energy, so they can be used for a lot of different things. Terahertz communication can be aimed very precisely, which reduces interference between cells.

## 8.11 MACHINE LEARNING AS A FRAME OF MIND

As 6G wireless networks are getting ready to launch, people are really excited about the extremely fast speeds and constant connectivity they will offer. However, while people are excited, it is important to make sure that there is strong security and protection of user privacy. In this scenario, machine learning comes out as a strong friend, providing new ways to solve the difficult security and privacy problems in 6G networks.

Machine learning can be used to find unusual things in 6G networks, which is really helpful. With more and more devices being connected and a lot of data being sent, old ways of keeping networks safe are not working well anymore. Machine learning programs can learn to find small changes in how a computer network usually works. This can help them notice when someone is trying to hack into the network or access it without permission. By watching network traffic all the time and adjusting to new dangers, machine learning systems can help protect 6G networks and make them more secure.

Besides finding unusual things, machine learning can also help make sure users are who they say they are in 6G networks. Old-fashioned ways of proving you are who you say you are, like using a password or PIN, are not very safe. They can be easily guessed or stolen by hackers. Computer programs that learn from fingerprint, voice, or face pictures can create strong and easy ways to confirm someone's identity. They are difficult to trick or fake. By using technology that can recognize things like how a person looks or acts, a computer system can make sure it is you before letting you use 6G networks and services.

Also, machine learning can help protect user privacy in 6G networks, especially when it comes to sharing data and working together on calculations. Differential privacy is a way to measure how private data analysis algorithms are. It can be made even better by using machine learning methods. By using machine learning models that protect privacy, differential privacy mechanisms can give useful information without exposing private data. Also, federated learning is a type of machine learning where models are trained on user devices. This allows for collaborative data analysis without sharing sensitive information with third parties or centralized servers. Using federated learning methods, 6G networks can use the knowledge of edge devices while still keeping user privacy and data control.

Additionally, machine learning can make 6G networks more secure by finding and stopping interference and attacks on wireless transmissions. Newer ways of sending signals and processing them can now change how they are sent to stay private and not get blocked or interrupted by others. Machine-learning-based beamforming systems can use past information and current observations to improve how efficiently they use the wireless network. This can make communication faster and more reliable while using less energy.

At the network level, machine learning can help manage internet traffic and allocate resources efficiently to make the network safer and work better. Reinforcement learning is a type of machine learning where a computer tries different things to figure out the best way to control something. It can be used to make 6G networks work better by organizing and controlling how they are used. By changing with the network and what users need, reinforcement learning algorithms can use resources wisely, reduce delays, and increase data transfer while still keeping things safe and private.

Additionally, machine learning can help create strong systems to detect and respond to intrusions in 6G networks. Machine learning software can detect security breaches or bad activities as they happen by looking at how data moves around a network and keeping track of system activity. Adversarial machine learning can help make intrusion detection systems stronger against tricky cyber attacks and ways of avoiding detection. Machine learning systems can help protect 6G networks and important buildings from cyber attacks. They can change and get better at tackling new threats.

In summary, machine learning has a lot of potential to help solve the security and privacy problems in 6G wireless networks. Machine learning can help make 6G networks safer, keep things private, and work better. It can find unusual things, use body scans to check who you are, and keep data private. As 6G networks are being introduced quickly, it is important to use machine learning to keep them safe and protect people's privacy. This will help create strong, reliable, and user-friendly communication systems for the new wireless technology.

## 8.12 RESULTS AND CONCLUSION

In our exploration of emerging technologies for security and privacy in 6G wireless communication networks, we conducted a comprehensive review and analysis of various advancements. The integration of these technologies contributes to establishing a robust and secure foundation for 6G networks. Table 8.1 summarizes the key findings.

*Table 8.1* Summarizes the security aspects of various 6G technologies

| *S. No.* | *Technology* | *Security aspect addressed* | *Key findings* |
|---|---|---|---|
| 1 | Quantum-safe cryptography | Encryption security | Resistance against quantum computing threats |
| 2 | AI-powered security | Threat detection and response | Real-time threat identification and mitigation |
| 3 | Blockchain for Security | Authentication and data integrity | Decentralized and tamper-resistant mechanism |
| 4 | Privacy-preserving technologies | Data privacy | Homomorphic encryption, differential privacy |
| 5 | Post-quantum cryptography | Cryptographic resistance to quantum threats | Adoption of quantum-resistant algorithms |
| 6 | Biometric authentication | Identity verification | Enhanced user identification beyond passwords |
| 7 | Secure edge computing | Decentralized security functions | Reduced latency and enhanced network resilience |
| 8 | Network slicing for security | Customized security approaches | Tailored security measures for diverse services |
| 9 | Physical layer security | Wireless transmission security | Exploiting channel characteristics for security |
| 10 | Zero-trust security models | Continuous authentication and authorization | Elimination of trust assumptions in network |

In conclusion, a proactive approach to security and privacy is important given the quick development of 6G wireless communication networks. The integration of developing technologies provides a holistic approach to addressing a range of risks and difficulties. Long-term encryption resilience is guaranteed by quantum-safe cryptography, while real-time attack response is improved by AI-powered security. Protecting data and ensuring secure authentication are made possible by blockchain technology.

Networks are getting ready for the coming era of quantum computing with the deployment of post-quantum cryptography. A personalized layer of identity verification is added via biometric authentication, and security operations are decentralized for greater efficiency using secure edge computing. Network slicing makes it possible to implement specialized security measures, and physical layer security uses wireless channel properties to ensure secure transmission.

Adopting a zero-trust security approach means rejecting conventional wisdom and supporting ongoing authorization and authentication. The combination of these technologies creates a reliable and private-focused communication ecosystem in addition to strengthening the security posture of 6G networks. The combination of these technologies is essential as we approach the deployment of 6G to protect wireless communication networks' availability, integrity, and secrecy against ever-evolving cyber threats.

## REFERENCES


1. C. Wang, and A. Rahman, "Quantum-Enabled 6G Wireless Networks: Opportunities and Challenges". doi: 10.36227/techrxiv.14785737.v1.
2. A. Alhammadi et al., "Artificial Intelligence in 6G Wireless Networks: Opportunities, Applications, and Challenges," *International Journal of Intelligent Systems*, 2024. doi: 10.1155/2024/8845070
3. V. Sarveshwaran, R. Manoharan, R. Sitharthan and V. Rajasekar, "Blockchain Based Privacy Preserving Framework for Emerging 6G Wireless Communications," *IEEE Transactions on Industrial Informatics*, 2021. doi: 10.1109/TII.2021.3107556
4. D. Sivizaca Conde, N. L. Kämpf, D. Rößler-von Saß, T. Schurig and N. Kliewer, "Privacy-Preserving Data Sharing: A systematic review and Future Research Directions," European Conference on Information Systems (ECIS), at Paphos, Cyprus, 2024, pp. 1–16.
5. S. Li, Yuxiang Chen, Lin Chen, Jing Liao, Chanchan Kuang, Kuanching Li, Wei Liang, and Naixue Xiong, "Post-Quantum Security: Opportunities and Challenges," *Sensors* vol. 23, no. 21, 8744, 2023. doi: 10.3390/s23218744.
6. V.-L. Nguyen, P.-C. Lin, B.-C. Cheng, R.-H. Hwang and Y. R. Lin, "Security and Privacy for 6G: A Survey on Prospective Technologies

and Challenges," *IEEE Communications Surveys & Tutorials*, 2021. doi: 10.1109/COMST.2021.3108618

7. A. Alawadhi, A. Almogahed and E. Azrag, "Towards Edge Computing for 6G Internet of Everything: Challenges and Opportunities," *1st International Conference on Advanced Innovations in Smart Cities (ICAISC)*, Jeddah, Saudi Arabia, 2023, pp. 1–6, doi: 10.1109/ICAISC56366.2023.10085007.
8. A. M. Alwakeel and A. Abdulrahman K Alnaim. "Network Slicing in 6G: A Strategic Framework for IoT in Smart Cities," *Sensors* vol. 24, no. 13, 4254, 2024. doi: 10.3390/s24134254
9. L. Mucchi, S. Jayousi, S. Caputo, E. Panayirci, S. Shahabuddin, J. Bechtold, I. Morales, R.-A. Stoica, G. Abreu and H. Haas, "Physical-Layer Security in 6G Networks," *IEEE Open Journal of the Communications Society*, 2021. doi: 10.1109/OJCOMS.2021.3103735
10. K. Ramezanpour and J. Jagannath, "Intelligent Zero Trust Architecture for 5G/6G Tactical Networks: Principles, Challenges, and the Role of Machine Learning", *Computer Networks*, vol. 217, 109358, ISSN 1389-1286, 2021.
11. C. R. García et al., "Secure and Agile 6G Networking – Quantum and AI Enabling Technologies." In *23rd International Conference on Transparent Optical Networks (ICTON)*, Bucharest, Romania, 2023, pp. 1–4.
12. Y. Siriwardhana, P. Porambage, M. Liyanage and M. Ylianttila. "AI and 6G Security: Opportunities and Challenges," *2021 Joint European Conference on Networks and Communications & 6G Summit (EuCNC/6G Summit)*, 2021, pp. 616–621,. doi: 10.1109/EuCNC/6GSummit51104.2021.9482503
13. L. Zhang, Y. C. Liang, and D. Niyato, "6G Visions: Mobile Ultra-Broadband, Super Internet-of-Things, and Artificial Intelligence," *China Communications*, vol. 16, no. 8, pp. 1–14, 2019.
14. Z. Zhang, Y. Xiao, Z. Ma, M. Xiao, Z. Ding, X. Lei, and P. Fan, "6G Wireless Networks: Vision, Requirements, Architecture, and Key Technologies," *IEEE Vehicular Technology Magazine*, vol. 14, no. 3, pp. 28–41, 2019.
15. H. H. Hussein, H. A. Elsayed, and S. M. Abd El-kader, "Intensive Benchmarking of D2D Communication over 5G Cellular Networks: Prototype, Integrated Features, Challenges, and Main Applications," *Wireless Networks*, pp. 1–20, 2019.
16. J. Kaur, M. A. Khan, M. Iftikhar, M. Imran, and Q. Emad Ul Haq, "Machine Learning Techniques for 5G and Beyond," *IEEE Access*, vol. 9, pp. 23472–23488, 2021.
17. F. Tang, B. Mao, Y. Kawamoto, and N. Kato, "Survey on Machine Learning for Intelligent End-to-End Communication Toward 6G: From Network Access, Routing to Traffic Control and Streaming Adaption," *IEEE Communications Surveys and Tutorials*, vol. 23, no. 3, pp. 1578–1598, 2021. Article 9403380. https://doi.org/10.1109/C.
18. P. K. Gkonis, "A Survey on Machine Learning Techniques for Massive MIMO Configurations: Application Areas, Performance Limitations and Future Challenges," *IEEE Access*, vol. 11, pp. 67–88, 2023.

19. Y. Wang, Z. Gao, D. Zheng, S. Chen, D. Gunduz, and H. V. Poor, "Transformer-Empowered 6G Intelligent Networks: From Massive MIMO Processing to Semantic Communication," *IEEE Wireless Communications*, pp. 1–9, 2022.
20. U. Demirhan and A. Alkhateeb, "Integrated Sensing and Communication for 6G: Ten Key Machine Learning Roles," *IEEE Communications Magazine*, vol. 61, pp. 113–119, 2023.
21. A. A. Puspitasari and B. M. Lee, "A Survey on Reinforcement Learning for Reconfigurable Intelligent Surfaces in Wireless Communications," *Sensors*, vol. 23, p. 2554, 2023.
22. K. Latva-aho, Leppanen, Key Drivers and Research Challenges for 6G Ubiquitous Wireless Intelligence (White Paper), 6G Flagship Research Program, University of Oulu, Finland, 2019.
23. SM. G. Kibria, K. Nguyen, G. P. Villardi, O. Zhao, K. Ishizu, and F. Kojima, "Big Data Analytics, Machine Learning, and Artificial Intelligence in Next-Generation Wireless Networks," *IEEE Access*, vol. 6, pp. 32328–32338, 2018.
24. F. Tariq, M. Khandaker, K.-K. Wong, M. Imran, M. Bennis, and M. Debbah, "A Speculative Study on 6G," *arXiv Preprint arXiv:1902.06700.*
25. F. Tang, Y. Kawamoto, N. Kato, and J. Liu, "Future Intelligent and Secure Vehicular Network Toward 6G: Machine-Learning Approaches," *Proceedings of the IEEE.*
26. I. Tomkos, D. Klonidis, E. Pikasis, and S. Theodoridis, "Toward the 6G Network Era: Opportunities and Challenges," *IT Professional*, vol. 22, no. 1, pp. 34–38, 2020.
27. H. B. McMahan, E. Moore, D. Ramage, and S. Hampson, et al., "Communication Efficient Learning of Deep Networks from Decentralized Data," *arXiv Preprint arXiv: 1602.05629*, 2017.
28. L. Loven, T. Leppanen, E. Peltonen, J. Partala, E. Harjula, P. Porambage, M. Ylianttila, and J. Riekki, "Edge AI: A Vision for Distributed, Edge-Native Artificial Intelligence in Future 6G Networks," *The 1st 6G Wireless Summit*, pp. 1–2, 2019.
29. E. C. Strinati, S. Barbarossa, J. L. Gonzalez-Jimenez, D. Ktenas, N. Cassiau, and C. Dehos, "6G: The Next Frontier," *arXivPreprint*, 2019, *arXiv:1901.03239.*
30. G. Gui, M. Liu, F. Tang, N. Kato, and F. Adachi, "6G: Opening New Horizons for Integration of Comfort, Security and Intelligence," *IEEE Wireless Communications*, vol. 27, no. 5, pp. 126–132, 2020.
31. C. Jiang, H. Zhang, Y. Ren, Z. Han, K.-C. Chen, and L. Hanzo, "Machine Learning Paradigms for Next-Generation Wireless Networks," *IEEE Wireless Communications*, vol. 24, no. 2, pp. 98–105, 2016.
32. S. Dang, O. Amin, B. Shihada, and M.-S. Alouini, "What Should 6G Be?," *Nature Electronics*, vol. 3, no. 1, pp. 20–29, 2020.
33. W. Saad, M. Bennis, and M. Chen, "A Vision of 6G Wireless Systems: Applications, Trends, Technologies, and Open Research Problems, *IEEE Network*, vol. 34, no. 3, pp. 134–142, 2020.

34. M. Katz, P. Pirinen, and H. Posti, "Towards 6G: Getting Ready for the Next Decade," In *2019 16th International Symposium on Wireless Communication Systems (ISWCS)*, IEEE, pp. 714–718, 2019.
35. S. Zhang, J. Liu, H. Guo, M. Qi, and N. Kato, "Envisioning Device-to-Device Communications in 6G," *IEEE Network*, vol. 34, no. 3, pp. 86–91, 2020.

Chapter 9

# Applications of 6G in the healthcare sector and the importance of network security and data privacy-encryption of medical images using the SHA 256 blockchain algorithm

*S.N. Kumar, Jomin Joy, I. Christina Jane, Alan James, Siju John, and Eduard Babulak*

## 9.1 INTRODUCTION

6G is probably going to have several uses in the healthcare industry that have the potential to completely transform the provision of healthcare services. A 6G architecture network's foundation is being projected by 5G technologies including cloud core networks and virtual radio access networks (vRANs) [1, 2]. The healthcare industry will face significant challenges related to user data ethics and privacy protection. The artificial intelligence (AI) model should now adhere to strict ethical guidelines for data gathering and the usage of user data for training models, as the AI will currently solely focus on performance. Furthermore, the AI model must adhere to the privacy norms and regulations enacted by the regulatory organizations. Strong network security and data privacy safeguards must be prioritized to protect sensitive health information, maintain patient and healthcare provider confidence, and assure the well-being of 6G users. The 6G technology has the potential to revolutionize the healthcare industry [1]. To provide future healthcare, 6G communication technology will need to provide high data rates, high operation frequencies, minimal end-to-end delays, high reliability, and high mobility. Thus, via device-to-device, terrestrial, and satellite connectivity, 6G technology will offer deeper and wider coverage. Thus, 6G technology needs the highest level of security. Federated AI, quantum machine learning, quantum computing, and quantum communication will all be used to thwart attacks. A patient could be killed by any modification to health data. Thus, protecting the data from hackers is essential [3].

Attacks on network automation technologies reveal the highly dynamic nature and extensive network automation capabilities of the 6G network.


DOI: 10.1201/9781003583127-9

Existing network security tools like firewalls and IPsec will not be able to prevent other parties from stealing data [4]. To safeguard healthcare systems from ransomware, malware, and other cyberattacks, 6G networks need to implement sophisticated security mechanisms. Healthcare is becoming more and more dependent on Internet of Things (IoT) devices, therefore protecting sensitive patient data from unwanted access requires making sure these devices are authenticated and permitted. Healthcare companies that use 6G technology need to be open and honest about their data practices, have patients' informed consent, and make sure that their use of data complies with moral guidelines [5]. 6G mobile networks have to deal with a variety of risks in the areas of space, air, and land as well as new technologies, an accessible user information explosion, and an integrated network environment. Data security is ensured in the encryption process when data is transferred from one node to another node [4]. AI algorithms plays vital role in the 6G technology for the processing and interpretation of real-time data especially in the healthcare sector [6].

## 9.2 RELATED WORKS

The applications of 6G communication in various industrial sectors were discussed in [7]. The merits and challenges are also discussed and 6G technology relies on blockchain and AI algorithms for efficient processing of data [7]. A novel e-healthcare system comprising fog computing and AI algorithms was highlighted in [8], the system comprises of fog layer, AI layer, and cloud layer. The European research project Hexa-X features are highlighted in [9], mobile communication system, 6G technology, and its challenges are addressed in this study. The healthcare monitoring system in hospitals and the deployment of IPv6 Routing Protocol for data communication in 6G technology healthcare sector were proposed in [10]. The cooja simulator was used for validation in terms of the performance metrics [10]. The non-orthogonal multiple access (NOMA) technology in 6G networks for AR/VR applications was proposed in [11]. The features of coupled system comprising of AI and self-learning 6G networks for smart cities was highlighted in [12]. The AI and 6G applications in smart cities with respect to safety, environment, health, and transportation were proposed in [12].

Distributed ledger technology (DLT) and physically unclonable functions (PUFs) are combined in PUFchain 3.0 to enhance authentication in healthcare cyber-physical systems (H-CPS). It aims to improve the security and access management of H-CPS settings for connected medical devices and the data they hold. Improved device authentication, decentralized security, and improved data integrity are some of the system's advantages. By making use of the IOTA Tangle, improving PUF accuracy, and cutting latency, PUFchain 3.0 may improve upon previous iterations. It is feasible to use applications like secure medical device communication, patient data access control, and counterfeit device identification. Among the difficulties include ensuring performance and scalability, integrating the system with the existing H-CPS infrastructure, and adhering to legal standards for data security and privacy in healthcare settings. Future study should primarily focus on improving DLT and examining state-of-the-art PUF technologies [13]. This study investigates ways to improve cybersecurity and patient-centric data retrieval in the healthcare sector, with a focus on privacy-preserving strategies. It tackles the challenges of enhanced cybersecurity and efficient data retrieval, all the while concentrating on patient-centric data processing. Federated learning, personalized information access, and ontology-based retrieval are important ideas. Intrusion detection and prevention systems, blockchain technology, and privacy-preserving encryption are examples of cybersecurity solutions [14].

To improve clinical research and patient care, the article suggests a service platform that would make it easier for real-world health data (RWD) to be shared. The platform intends to streamline RWD access for multiple stakeholders by addressing issues with data fragmentation and a lack of uniform access methods. This system gathers RWD from many sources, unifies and harmonizes data, and incorporates tools for statistical analysis, data visualization, and machine-learning applications. It provides secure access control as well. Among the benefits are enhanced research efficiency, more individualized care, and improved public health surveillance. Among the challenges are regulatory frameworks, data security and privacy, scalability, and sustainability. Future goals will include harmonized data standards promotion, user authorization, anonymization, a sustainable financial model, seamless integration with existing healthcare IT systems, and the deployment of robust security measures [15]. Real-time patient monitoring using a wireless body area network was proposed in [16] with patient data privacy as the highest priority. The healthcare sector involves big data and the article [17] describes the features of big data computing and its challenges. The various algorithms and models involved in ensuring the security of healthcare data were highlighted in [18–20]. Blockchain technology, machine learning, artificial intelligence, and decentralized data processing and storage are some of the future directions. The article [21] also discusses possible future paths and cutting-edge technology that could improve the security and privacy of health data even more [21]. Solutions include data anonymization, safer multi-party computing, federal education, standardization, stricter access control, user education, and advancements in data governance and regulatory frameworks in addition to technologies that safeguard privacy.

The Dobbs *v.* Jackson Women's Health Organization ruling, which reversed Roe *v.* Wade, sparked worries about possible limits on access to abortion and the implications for the privacy of health data. The authors in [22] examine the features of the Dobbs ruling for HIPAA, possible legal liabilities, and obstacles faced by patients seeking abortion care outside their state. The role of blockchain technology in health care was put forward in [23], a patient-controlled blockchain-enabled electronic health record (PcBEHR) system addressing the issues of classical electronic healthcare systems is the notable feature of this research work. The article [24] also explores the role of blockchain, explainable AI, and metaverse technologies in the healthcare sector. VR concepts are enabled through metaverse technologies.

## 9.3 BLOCKCHAIN-BASED MEDICAL IMAGE ENCRYPTION USING THE SHA 256 ALGORITHM

A cryptographic hash function called SHA-256, or 256-bit, transforms text into an almost unique 256-bit alphanumeric string. Because of its unique and irreversible output, it is regarded as being quite secure when it comes to cryptographic security [25, 26]. The National Security Agency (NSA) of the United States Government created SHA-256, which is a member of the SHA-2 family of hash functions. The National Security Agency (NSA) found flaws in the hash function and advised US federal agencies to move away from SHA-1. The United States government and other encryption systems employ SHA-256 to safeguard confidential data. To guarantee that a hash cannot be reversed back to its original content, the method employs a predefined algorithm. The text must be smaller than 264 bits to get a 256-bit hash, which permits random hash results.

Hashing is the process of changing a character string into a fixed-length, limited key that corresponds to the original string that was entered. Because of digests and hash values, the procedure is used in databases to safely store and retrieve data while facilitating quick access to the data. The hash values are preimage-resistant because only the confidential user can decrypt them, making all other users unable to do so. Because the hash cannot ever be decrypted back to the original text, it is not considered "encryption." It has a set length, regardless of how long the original text is. The following procedures, which include padding, appending length, dividing the input into 512-bit blocks, adding a chaining variable, and adding a process block, are used to generate SHA-256. A unique 256-bit, or 32-byte, hash code is produced by SHA-256. It goes through 64 cycles to determine the final hash value. Additionally, the text length up to 264 bits, or 2.3 exabytes, will be converted to a digest size of 256 bits, or 32 bytes. x86 architecture, 64-bit processor, 16.80 cycles per byte. 64 iterations make up a single cycle, which increases its security and safety [27].

The SHA256 algorithm is a hashing method that employs 16 32-bit words to represent a block size of 512 bits. A message scheduler expands this block into 64 32-bit words. The approach uses eight 32-bit working variables (A, B, C, D, E, F, G, and H) to perform operations on 32-bit words. The first message block of the SHA256 hashing algorithm uses a 256-bit initialization vector (IV), and the second message block uses an intermediate message digest acquired after the first 64 rounds as the IV. The IV is added to the hash function's output, which is created using the Davies-Meyer design, after 64 rounds. After adding the IV and executing the message compression algorithm 64 times, the method produces an intermediate message digest of 256 bits. A block cypher that employs a 512-bit message block key (message block) and a 256-bit message block size (IV) that is extended into 64 32-bit round keys for each of the cipher's 64 rounds using the message scheduler is comparable to the SHA256 hashing algorithm.

Standard safe hash algorithms (SHA) are often used in combination with other cryptographic approaches for keyed-hash message authentication. Several secure algorithms are available, depending on the length of the message. One-way hash algorithms are capable of converting a message into a message digest, which is a condensed representation, (1) to locate a message that matches a message digest that has been provided, or (2) to locate two distinct messages that provide the same message digest. Any modification to a message will almost certainly produce a distinct message digest. Preprocessing and hash computation are the two stages of a secure hash algorithm. Preprocessing involves setting initial values to be utilized in the hash calculation, padding a message, and parsing the padded message into *m*-bit blocks. The message block size (*m* bits) for secure hash methods varies depending on the algorithm. SHA-256 uses 512 bits per message block, which are encoded as 16 32-bit genesis blocks.

$$ch(x,y,z) = (x^{\wedge}y) \oplus (\sim x \wedge z)$$

$$maj(x,y,z) = (x^{\wedge}y) \oplus (x \wedge z) \oplus (y \wedge z)$$

$$\sum_{0}^{(256)} x = ROTR^{2}(x)\, ROTR^{13}(x)\, ROTR^{22}(x)$$

The first hash value, $H(0)$, for SHA-256, is made up of the eight 32-bit words that follow. The first 32 bits of the fractional parts of the square roots of the first eight prime numbers were taken to create these words. From the padded message, the hash computation creates a message schedule, which is then used with functions and constant operations to iteratively produce a sequence of hash values. The message digest is calculated using the final hash value that is produced by the hash computation. The suggested algorithm will have good security if SHA is used. Using secure hash function techniques, color images can be encrypted and compressed for quick and safe image transmission. Only the R, G, and B color values are used to compress images. The compressed image is then used to encrypt an image using the secure transmission function technique, which uses a 256-bit digest length cryptographic hash function. Figure 9.1 represents the hashing algorithm for cryptography.

Step 1: Insert the image as I.
Step 2: To prepare the image for pre-processing, convert it to RGB.
Step 3: Randomly store RGB values and compress them.
Step 4: Convert ($m$x$n$) rows and columns to encrypt each pixel with a represented bit by taking a compressed image of 512×512 size for the encryption process. Generate a block value of $m$ bits to calculate the key digest.
Step 5: Next, create a temporary $w$-bit word and an $m$-by-$n$ matrix of zeros for the hash computation using the image size. Compare the results.
Step 6: Use a 32-bit block to conduct ⊕ bitwise XOR, or "exclusive-OR," to create a digest.
Step 7: Use the bit block and reverse bit XOR procedures to compute the hash.
Step 8: $Z = (X + Y) \bmod 2w$ is the addition modulo $2w$. Rotate right (circular right shift): SHR $n$ $(x) = x >> n$ is a right shift operation. $(x >> n)\ U\ (x << w - n)$ = ROTR $n$ $(x)$. The message was padded by parsing the values into 512 block values, or 16 times the 32-bit values. The message's final 256-bit message digest. And reverse those procedures to begin the decryption process.
Step 9: The decompression procedure uses the only color values that are accessible to take each pixel of color as an adjacent average.

The blockchain-related SHA-256 cryptographic hash algorithm generates hashes for safe access and is primarily used for data identification, password verification, session time, and data integrity verification during transactions. Blockchain architecture, also known as blockchain technology, is the assembly of blocks, and hash references to produce a linked list. Every block,

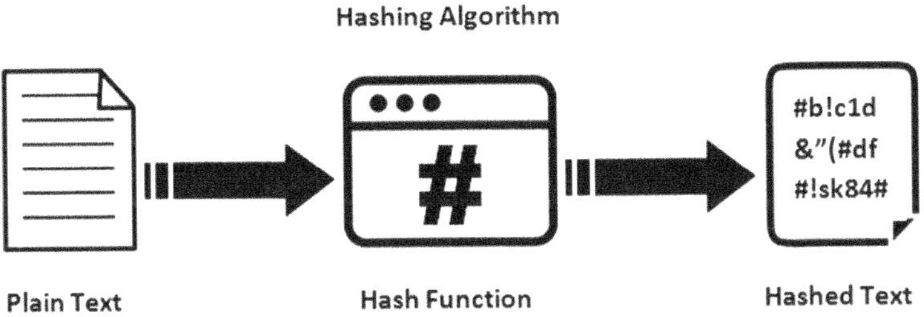


*Figure 9.1* Hashing algorithm for cryptography.

except for the first block, which has no previous hashing, contains data plus a unique hash value from the preceding block. The methods carried out by an authorized investigator include identifying, gathering, analyzing, and producing reports using the logical graph of evidence (LGoE). Every user registers with AS at first, and AS provides them with their secret key based on harmony search optimization (HSO).

Step 1: Digital data: Enter data from up to 100 IoT device nodes. Thus, the plain text's length is 1024 bits and 128 multiples.
Step 2: Add a 128-bit representation of the original plain text in step 2, making sure that length equals multiples of 1024 bits.
Step 3: Set the 64-bit initialization of the buffers ($a, s, d, f, g, h, k, l$).
Step 4: Perform 100 steps for every block and plain text.
Step 5: A hash code (SHA-256 cryptographic hash) is output in the buffer.

The plain text block size is 1024 bits, and there are 100 steps. Every step QWORD is 64 bits, which is generated from plain text. Each step is assigned a 64-bit buffer size. Intermediate results are stored in a buffer. The result is saved as a hash code. One hundred mobile nodes (IoT devices), open-flow switches, blockchain-based controllers, cloud servers, authentication servers (AS), and the researchers make up the cloud-based software defined network (SDN). The user possesses a digital signature, user_id, resource_id, public key, and URL address. In the hash block, digital signatures were employed to verify user identity and, finally, to receive verification IDs from smart contracts. Hash blocks are used to verify user identities and, finally, to send verification IDs from smart contracts.

## 9.4 RESULTS AND DISCUSSION

The MATLAB software was used for blockchain algorithm development and executed on the i5 processor with 1.19 GHz. Figures 9.4 and 9.5 depict the medical image encryption algorithm results. The sample inputs D1 to D5 and its results are depicted here. The blockchain-based medical image encryption algorithm was validated by the following performance metrics [28–30]; peak to signal noise ratio (PSNR), correlation values, structural content (SC), normalized cross-correlation (NCC), mean square error (MSE), average difference (AD), Laplacian mean square error (LMSE), and normalized absolute error (NAE). PSNR gauges the fidelity of the encrypted image compared to the original, calculated as the ratio between the maximum signal power and the noise affecting its representation. Correlation values evaluate the similarity between the original and encrypted images, indicating how well the encryption maintains its structural and statistical traits. SC assesses the retention of structural elements like edges and textures in the encrypted image, crucial for preserving diagnostic information in medical images

NCC measures the likeness between two signals, such as the original and encrypted images, by normalizing the correlation coefficient within a range of –1 to 1. MSE computes the average squared difference between pixel values of the original and encrypted images, with lower values indicating higher similarity and better encryption. AD determines the average absolute difference between corresponding pixel values in the original and encrypted images, offering insight into the overall distortion introduced by encryption. LMSE quantifies the mean squared error between the Laplacian-transformed versions of the original and encrypted images, useful for evaluating the preservation of high-frequency details. NAE calculates the normalized average absolute difference between corresponding pixel values in the original and encrypted images, offering a standardized measure of distortion in the encrypted image. Table 9.1 represents the correlation values of the encrypted images.

Figure 9.2 represents the histogram of the encrypted image corresponding to the input image (D1).

Tables 9.2 and 9.3 depict the performance metrics measures of the encryption model. A high value of PSNR > 35 dB favors the efficiency of the encryption model. The closer the value of SC and NK to 1 and the low value of MSE also indicates proficiency.

The lower value of AD, LMSE, and NAE indicates the proficiency of the medical image encryption model.

Figure 9.3 represents the correlation plot of the input and encrypted images. The correlation plot reveals the efficiency of the blockchain-based encryption model since the pixels are highly uncorrelated in the encrypted image.

The entropy value also indicates the proficiency of the encryption model and for all the tested images, the value of entropy for encrypted images was 8.

*Table 9.1* Correlation values of the encrypted images

| *Image details* | *Horizontal correlation* | *Vertical correlation* | *Diagonal correlation* |
|---|---|---|---|
| D1 | -5.246135e-03 | 4.105709e-03 | 1.499436e-03 |
| D2 | 2.665068e-03 | 2.657343e-03 | 5.403793e-03 |
| D3 | -1.495746e-03 | -1.072601e-03 | -9.481232e-04 |
| D4 | -1.349576e-03 | -1.042601e-03 | -6.481232e-04 |
| D5 | 1.829607e-03 | -7.230304e-03 | -6.297878e-03 |
| D6 | 8.945013e-04 | 1.156107e-03 | 1.300040e-03 |

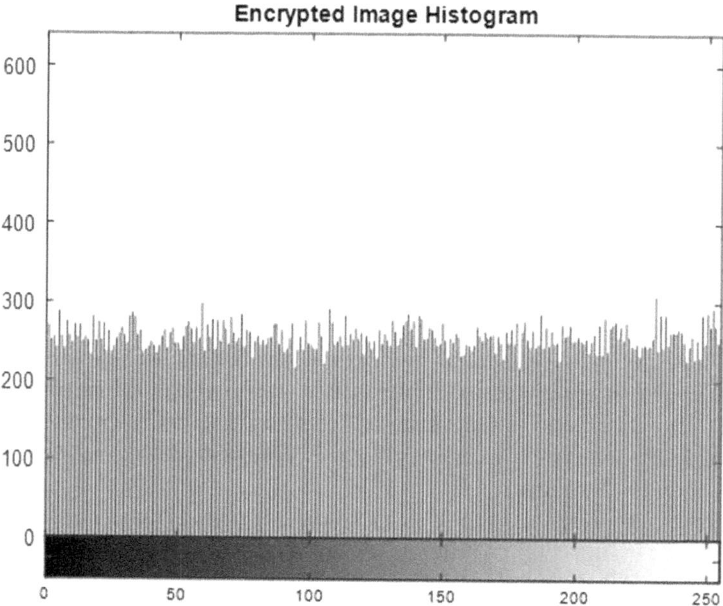


*Figure 9.2* Histogram of the encrypted image corresponding to the input image (D1).

*Table 9.2* PSNR, MSE, SC, and NK measures of the encryption model

| *Image details* | *PSNR* | *MSE* | *SC* | *NK* |
|---|---|---|---|---|
| D1 | 39.117 | 7.966 | 0.996 | 1.000 |
| D2 | 38.533 | 9.114 | 0.997 | 0.999 |
| D3 | 37.528 | 11.488 | 0.995 | 1.000 |
| D4 | 33.562 | 7.987 | 0.994 | 0.994 |
| D5 | 38.632 | 8.909 | 0.997 | 1.000 |
| D6 | 38.293 | 9.632 | 0.997 | 0.999 |

*Table 9.3* AD, LMSE and NAE measures of the encryption model

| *Image Details* | *AD* | *LMSE* | *NAE* |
|---|---|---|---|
| D1 | -0.0518 | 0.0783 | 0.0034 |
| D2 | -0.0435 | 0.1080 | 0.0028 |
| D3 | -0.0654 | 0.1764 | 0.0034 |
| D4 | 0.0354 | 0.1634 | 0.0029 |
| D5 | -0.0510 | 0.1494 | 0.0026 |
| D6 | -0.0469 | 0.1616 | 0.0023 |

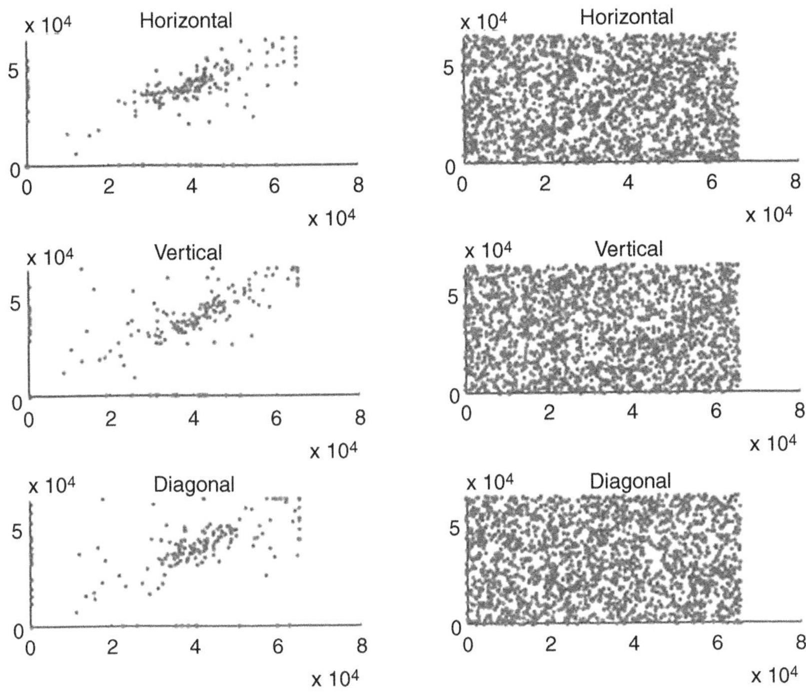


*Figure 9.3* Correlation plot of the input and the encrypted image (D1).

## 9.5 CONCLUSION

The blockchain technology is also a good choice for secure and transparent patient record management. The SHA 256 blockchain-based encryption model was found to be proficient in ensuring the security in healthcare sector. The following are essential considerations in the security aspect; robust encryption for data at rest and in transit is crucial to protect sensitive patient information, multi-factor authentication (MFA) to add an extra layer of security for accessing healthcare systems and patient data, design the network with security as a priority, employing segmentation, firewalls, and intrusion detection systems. A zero-trust approach is adopted that assumes no entity is inherently trusted and enforces strict access controls and continuous monitoring. Edge computing is utilized to process data closer to its source, reducing data transmission and ensuring the security of edge devices. 6G's swift response times and extensive bandwidth will facilitate real-time, high-definition telemedicine and remote surgical procedures, allowing specialists to treat patients from afar. 6G-fueled robots can undertake

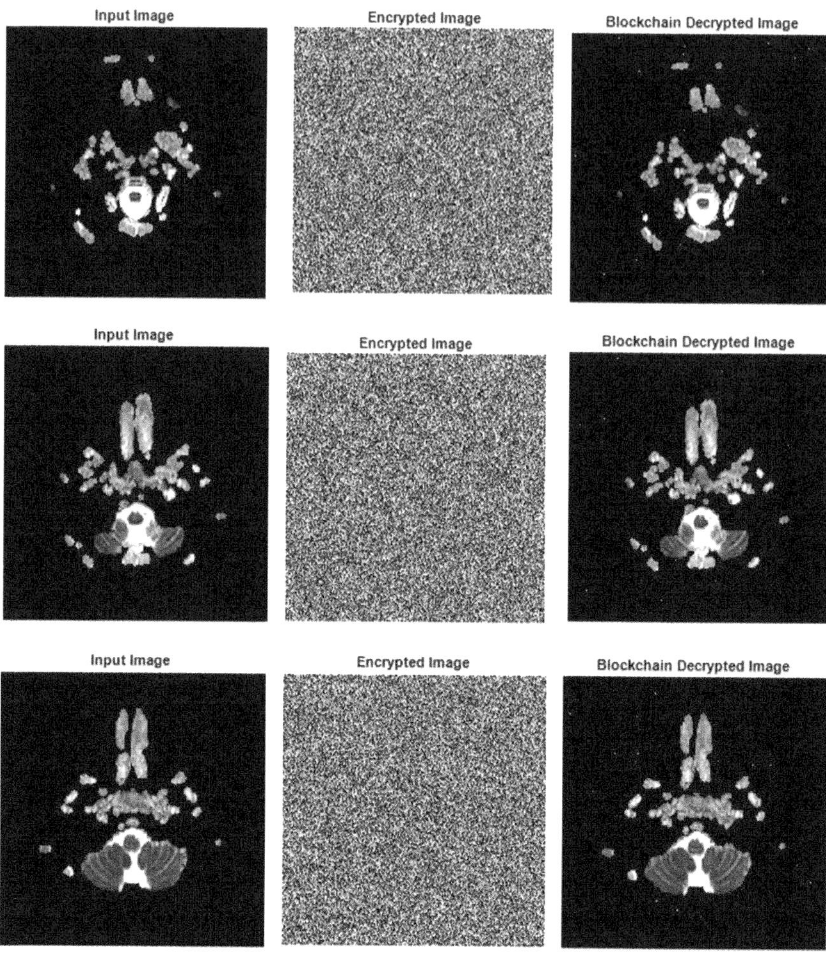


*Figure 9.4* The first column depicts the input images, the second column depicts the encrypted images, and the third column depicts the decrypted images (D1, D2, and D3).

various healthcare roles, from drug dispensing to surgical assistance, elevating patient care and reducing the workload on healthcare professionals. While 6G advances healthcare, robust security measures are vital to protect patient data, possibly through blockchain and advanced encryption methods. 6G extends healthcare services to remote and underserved regions, enabling previously inaccessible populations to access medical expertise and resources and streamlines remote clinical trials, making participation easier for patients and reducing drug development time and costs.

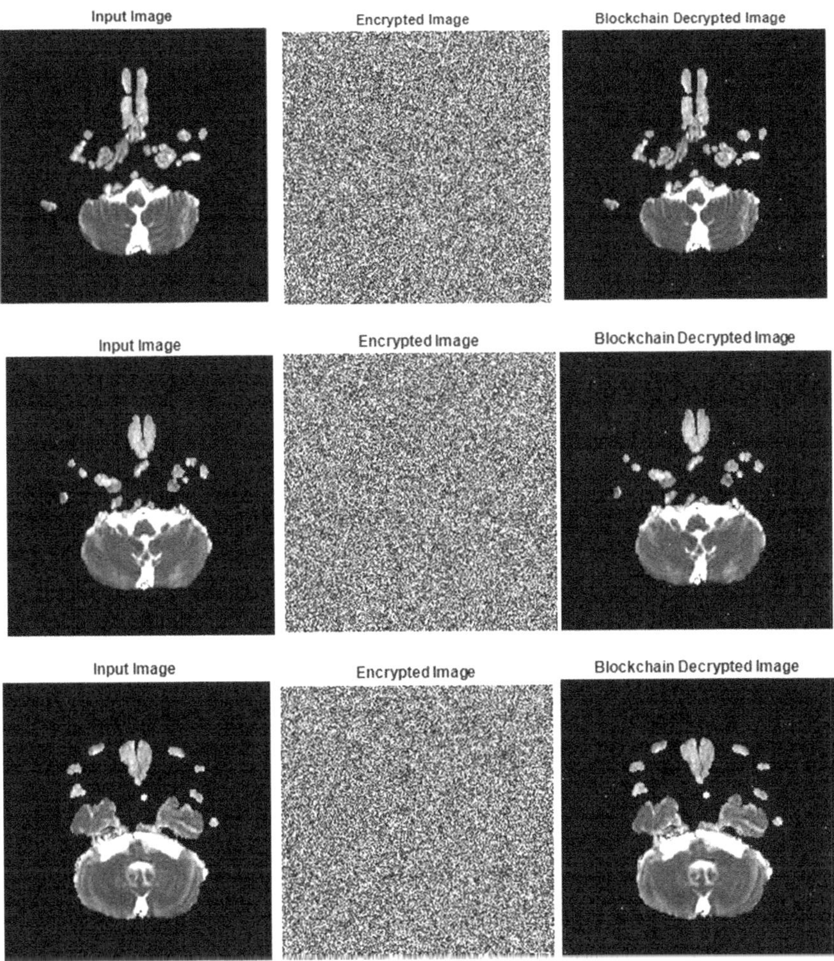


*Figure 9.5* The first column depicts the input images, the second column depicts the encrypted images, and the third column depicts the decrypted images (D1, D2, and D3).

## REFERENCES

1. Porambage P, Gür G, Osorio DP, Liyanage M, Gurtov A, Ylianttila M. The roadmap to 6G security and privacy. *IEEE Open Journal of the Communications Society*. 2021 May 10; 2:1094–1122. https://doi.org/10.1109/OJCOMS.2021.3073032
2. Kumar R, Jain V, Leong, WY, Teyarachakul, S. *Convergence of IoT, Blockchain and Computational Intelligence in Smart Cities*. Taylor &

Francis, CRC Press, 2023. ISBN: 9781032404240. www.crcpress.com/Convergence-of-IoT-Blockchain-and-Computational-Intelligence-in-sSmart-Cities/Kumar-Jain-Leong-Teyarachakul/p/book/9781032404240

3. Nayak S, Patgiri R. 6G communication technology: A vision on intelligent healthcare. *Health Informatics: A Computational Perspective in Healthcare.* Ripon Patgiri, Anupam Biswas, and Pinki Roy (Eds), 2021:1–8. Springer Nature. https://doi.org/10.1007/978-981-16-4911-5_2
4. Abdel Hakeem SA, Hussein HH, Kim H. Security requirements and challenges of 6G technologies and applications. *Sensors.* 2022 March 2; 22(5):1969. https://doi.org/10.3390/s22051969
5. Gundu SR, Charanarur P, Chandelkar KK, Samanta D, Poonia RC, Chakraborty P. Sixth-generation (6G) mobile cloud security and privacy risks for AI system using high-performance computing implementation. *Wireless Communications and Mobile Computing.* 2022 May 5; 2022:1–4. https://doi.org/10.1155/2022/4561024
6. Porambage P, Liyanage M. *Security and Privacy Vision in 6G: A Comprehensive Guide.* John Wiley & Sons, 2023 July 27. www.wiley.com/en-us/Security+and+Privacy+Vision+in+6G-p-9781119875248
7. Chowdhury MZ, et al. 6G wireless communication systems: Applications, requirements, technologies, challenges, and research directions. *IEEE Open Journal of the Communications Society.* 2020; 1:957–975. https://doi.org/10.1109/OJCOMS.2020.3010270
8. Sodhro AH, Noman Z. AI-enabled framework for fog computing driven e-healthcare applications. *Sensors.* 2021; 21(23):8039. https://doi.org/10.3390/s21238039
9. Uusitalo Mikko A, et al. 6G vision, value, use cases and technologies from european 6G flagship project Hexa-X. *IEEE Access.* 2021; 9:160004–160020. https://doi.org/10.1109/ACCESS.2021.3130385
10. Mardini W, Aljawarneh S, Al-Abdi A. Using multiple RPL instances to enhance the performance of new 6G and Internet of Everything (6G/IoE)-based healthcare monitoring systems. *Mobile Networks and Applications.* 2021; 26:952–968. https://doi.org/10.1007/s11036-020-01704-w
11. Liu Y, et al. Application of NOMA in 6G networks: Future vision and research opportunities for next generation multiple access. 2021; *arXiv preprint* arXiv:2103.02334. https://arxiv.org/abs/2103.02334
12. Ismail L, Buyya R. Artificial intelligence applications and self-learning 6G networks for smart cities digital ecosystems: Taxonomy, challenges, and future directions. *Sensors.* 2022; 22(15):5750. https://doi.org/10.3390/s22155750
13. Bathalapalli Venkata KVV, et al. PUFchain 3.0: Hardware-assisted distributed ledger for robust authentication in healthcare cyber–physical systems. *Sensors.* 2024; 24(3):938. https://doi.org/10.3390/s24030938
14. Arunprasath S, Annamalai S. Improving patient centric data retrieval and cyber security in healthcare: Privacy preserving solutions for a secure future. *Multimedia Tools and Applications.* 2024; 83(27):1–31. https://doi.org/10.1007/s11042-022-12900-2

15. Hoffmann K, et al. Streamlining intersectoral provision of real-world health data: A service platform for improved clinical research and patient care. *medRxiv.* [Preprint]. 2024; 2024–01. Available at: www.medrxiv.org https://doi.org/10.1101/2024.01.15.23284657
16. Khan FA, et al. A cloud-based healthcare framework for security and patients' data privacy using wireless body area networks. *Procedia Computer Science* 2014; 34:511–517. https://doi.org/10.1016/j.procs.2014.07.072
17. Abouelmehdi K, Beni-Hessane A, Khaloufi H. Big healthcare data: Preserving security and privacy. *Journal of Big Data.* 2018; 5(1):1–18. https://doi.org/10.1186/s40537-017-0110-7
18. Mahapatra B, Krishnamurthi R, Nayyar A. Healthcare models and algorithms for privacy and security in healthcare records, in Tanwar, S., Tyagi, S. and Kumar, N. (eds.) *Security and Privacy of Electronic Healthcare Records: Concepts, Paradigms and Solutions.* IET 2019:183. https://doi.org/10.1007/978-3-030-05487-5_11
19. Raval D, Jangale S. Cloud-based information security and privacy in healthcare. *International Journal of Computer Applications.* 2016; 150(4):11–15. https://doi.org/10.5120/ijca2016911607
20. Thapa C, Camtepe S. Precision health data: Requirements, challenges and existing techniques for data security and privacy. *Computers in Biology and Medicine* 2021;129: 104130. https://doi.org/10.1016/j.compbio med.2020.104130
21. Yigzaw KY, et al. Health data security and privacy: Challenges and solutions for the future.In Evelyn Hovenga and Heather Grain (Eds), *Roadmap to Successful Digital Health Ecosystems.* Academic Press, 2022:335–362. https://doi.org/10.1007/978-3-030-66978-9_20
22. Clayton EW, Embí PJ, Malin BA. Dobbs and the future of health data privacy for patients and healthcare organizations. *Journal of the American Medical Informatics Association.* 2023; 30(1):155–160. https://doi.org/10.1093/jamia/ocac149
23. Rai BK. PcBEHR: Patient-controlled blockchain enabled electronic health records for healthcare 4.0. *Health Services and Outcomes Research Methodology.* 2023; 23(1):80–102. https://doi.org/10.1007/s10 742-022-00279-6
24. Kumar R, Jain V, Tan GWH, Touzene A (Eds.). *Immersive Virtual and Augmented Reality in Healthcare – An IoT and Blockchain Perspective.* CRC, Taylor and Francis Group, 2023. https://doi.org/10.1201/978100 3340133
25. Gopinathan S, Suganthi M. A study of image compression and SHA 256 encryption algorithms for secure transmission. *International Journal of Computer Applications.* 2019; 181(37):9–12. https://doi.org/10.5120/ijc a2019919199
26. Velmurugadass P, et al. Enhancing blockchain security in cloud computing with IoT environment using ECIES and cryptography hash algorithm. *Materials Today: Proceedings.* 2021; 37:2653–2659. https://doi.org/10.1016/j.matpr.2020.09.637

27. Kumar R, Singh PK. A modern parallel register sharing architecture for code compilation. *International Journal of Computer Applications*. 2021; 1(16):108–113. https://doi.org/10.5120/ijca2021911540
28. Setiadi DR. PSNR vs SSIM: imperceptibility quality assessment for image steganography. *Multimedia Tools and Applications*. 2021 March; 80(6):8423–8444. https://doi.org/10.1007/s11042-020-09786-1
29. Lin W, Kuo CC. Perceptual visual quality metrics: A survey. *Journal of Visual Communication and Image Representation*. 2011 May 1; 22(4):297–312. https://doi.org/10.1016/j.jvcir.2011.02.002
30. Kumar R, Bharti V. A critical review of finger vein recognition techniques for human identification. *IEEE Conference ICIRCA*, 2–4 September, 2021. https://doi.org/10.1109/ICIRCA51532.2021.9544771

Chapter 10

# Unveiling the future

## Exploring 6G network security and data collection

*Muhammad Muzamil Aslam, Ali Tufail, Liping Du, and Sajid Ali*

### 10.1 INTRODUCTION

With the speedy advancement of communication technologies and networks, the security of network systems has received more attention. Network security is generally the data generated, obtained, or originated from the system [1]. One can compute and determine network security by analyzing data associated with network security events. Data about network security is information that is vulnerable to security risks, such as privacy, security, and confidentiality. Furthermore, gathering information on security is the initial step toward identifying hacks and security lapses. It is challenging to gather security-related data due to the quick development of 5G and next-generation 6G network systems and big data, as a lot of data is exchanged, created, and generated. The 5V features of security-related data, such as veracity, value, velocity, volume, and variety, make gathering data extremely challenging [2, 3]. Furthermore, 5G is heterogeneous, supporting D2D, M2M, H2M, M2H, and new communication technologies. 5G is comprised of various networks, including mobile cellular networks, Internet, and wireless.

Sensor networks (WSN) and mobile ad hoc networks make data collecting for security purposes challenging. To assess the security of 6G networks [4, 5] in a single network, data collection techniques for security reasons need to be improved and redesigned for large-scale networks where the network is heterogeneous. Many researchers are working on efficient and generic mechanisms related to data collection and security for heterogeneous networks.

In general, network data plays a critical role in IT, where data can be gathered from both the input and output of the system. The importance of data is due to its contribution to managing and troubleshooting network systems, detecting network interruptions, and building network traffic. The main focus of this paper is the collection of data to detect network attacks, network abnormalities, and network interferences. We focus on data packets and data flows since network attacks and intrusions are more common in

DOI: 10.1201/9781003583127-10

networks. Though other types of data, for instance, memory and CPU usage time, also help in the detection of network attacks [5–7], they were declared ineffective at examining network data packets and drifts to detect attacks on network and interruptions. A network system security level may be measured and quantified by wisdom and analyzing data related to network security measures, which explains why collecting network security-related data is critical.

Still, there is a need for academics to focus more on quantifying and describing cybersecurity-related data. In the network environment today, numerous data collection methods exist, including traffic prediction, similarity detection, sampling, and more. Prior research has collected and processed network-related data using deep learning and machine learning to benefit from the advancements in artificial intelligence [8]. There is not a thorough summary of this study. However, some are not appropriate for large-scale heterogeneous network systems because they were made for security-related data in the significant data era. This chapter presents a comprehensive evaluation of network data collection methods. We compare existing studies per functional and security objectives to collect good-quality security-related data. Afterward, we discuss issues and challenges in 6G technologies and potential future direction in this discipline. Our main contributions are as follows:

- We offer a set of prerequisites and objectives for gathering security-related data in 6G heterogeneous large-scale networks, which may be used to assess current work in this area.
- The detailed overview of the mechanisms, methods, and technologies for network data gathering by utilizing the presented prerequisites and objectives to assess their performance in gathering extremely good-quality network-related security data.
- Finally, based on the proposed literature, we present challenges and issues in future research regarding gathering data related to network security.

## 10.2 THE JOURNEY TOWARD 6G

An overview of the changes in wireless security and privacy from cellular network technology standards of the first generation (1G) to the fifth generation (5G) is given in this section.

### 10.2.1 1G–5G

The first generation (1G) network was introduced in the 1980s to offer phone services. It does not follow an authorized wireless standard and transmits data via analog transmission. This has several detrimental effects, such as ineffective transportation, delivery challenges, and a lack of security and privacy assurance. It is impossible to have private or secure phone calls or data transmission since telephone services are unsafe. A few security and privacy risks that users and entire networks must deal with are cloning, eavesdropping, and unauthorized access [9, 10]. Time division multiple access (TDMA) and code division multiple access (CDMA), which provide voice and SMS services, are the digital modulation technologies at the core of 2G networks. GSM (Global System for Mobile Communications) is the most significant and extensively used standard for 2G mobile communications [9–12].

The goal of security is the same for GSM and PSTN systems. User data protection, signal protection, authentication, anonymity, and security are the first four characteristics of the various security and privacy services it offers [5]. Temporary identifiers are employed to preserve anonymity, making ascertaining the user's genuine identity challenging. When you initially switch it on, though, you should use your actual ID; if not, please supply a temporary one. Network operators frequently employ authentication to identify a user. A "challenge-response" encryption technique serves as the authentication mechanism. Encryption is used to represent and safeguard user data, and a significant component of the encryption key is the subscriber identity module (SIM). Mobile temporary subscriber identity (TMSI) and radio path encryption are crucial technologies to safeguard user privacy [11].

Even though 1G privacy and security have significantly improved, 2G still has a lot of issues. Because users cannot authenticate to the network, there are major security difficulties and vulnerabilities in the network during user authentication. Illegal devices can impersonate base stations and other authorized network members, deceive users, and steal their personal information [8].

Moreover, it is not end-to-end encrypted. Enemies can conduct an assault since just a piece of the wireless channel is encrypted, and the fixed network is not [10]. In addition to having some privacy-related limitations, TMSI and radio path encryption are susceptible to eavesdropping and other types of attacks.

First introduced in 2000, the 3G network provided "high-speed" (at least 2 Mbps) data transmission and internet access. On the other hand, increased speed might make using more sophisticated services like video, TV streaming, and web surfing feasible, which is impossible with 1G and 2G networks. The foundation of 3G system security is 2G technology. As stated differently, it is imperative to put in place more robust security measures, such as the use of GSM and other 2G components. Authentication and Key Agreement (AKA) and two-way authentication were added to the 3G network to address further some of the 2G network's security flaws. The two-component Third Generation Partnership Project (3GPP) also offers a complete security system for access control [13]. The primary purpose of air interface security is to safeguard users and signal data transferred via wireless links. In contrast, user network authentication guarantees the physical dependability of users and both ends of the network. Regarding confidentiality, 3GPP imposes some subscriber privacy obligations on 3G users, including traceability, location, and identity confidentiality. However, 3G networks are still vulnerable to attacks via IP traffic and encryption keys. Additionally, the wireless interface that links the end device to the service network can be used to launch a variety of assaults. Threats associated with wireless interface assaults fall into the following categories: (1) unauthorized access to data, denial of service (DoS), (1) unauthorized access to data, and (3) data integrity threats [12]. Privacy concerns are frequently linked to specific sorts of attacks that aim to erase user identities and private or secret information, also known as error messages.

With uplink data speeds of up to 500 Mbit/s and downlink data rates of up to 1 Gbit/s, 4G long-range (LTE) networks were introduced in 2009. These networks decrease latency and boost spectral efficiency, which makes video chat, digital video broadcasting (DVB), and high-definition television (HD TV) possible. Coordinated multipoint transmission and receiver (CoMR), multiple input multiple output (MIMO), and orthogonal frequency division multiplexing (OFDM) are only a few of the many novel and old technologies incorporated with LTE [14]. Intelligent mobile terminals, wireless access networks, IP backbone networks, and wireless backbone networks are a few examples of 4G systems. The primary causes of security threats are thus data alteration, insertion, deletion, eavesdropping, wireless security problems, and network entity authentication problems (wired and wireless) [13]. Because consumers use mobile terminals more frequently than in the past, 4G networks are more susceptible to security and privacy problems. Dangers extend more widely, and 4G interactions get more intricate when users engage with all wireless protocols and use a range of wireless apps.

Furthermore, improvements in computational and storage power enable an increasing range of dangerous apps to be run on mobile terminals, potentially causing more harm. Examples of common occurrences are malware, viruses, and hardware platform manipulation. Furthermore, replay attacks, denial of service attacks, and eavesdropping are examples of MAC layer vulnerabilities found in the Worldwide Interoperability for Microwave Access (WiMAX) standard stemming from key management protocols' flaws. Furthermore, through the media access control (MAC) layer, LTE networks may be subject to location tracking, data integrity violations, unlawful access to user and mobile device data, and DoS assaults [15]. Faster speeds, more comprehensive systems, and more secure structures are anticipated as 5G networks get closer [6].

5G networks are primarily intended to facilitate the simultaneous supply of high-quality services for every device and link an increasing number of devices. Not only can smartphones qualify as approved devices, but other devices—like Internet of Things equipment—can also establish a connection with the network [9, 10]. 5G network security and privacy issues may be best categorized using the network architecture, particularly the design's three layers: access networks, backhaul networks, and core networks.

The 5G network's fundamental network functionalities [15]. NFV, SDN, and cloud technologies can improve network dynamism and increase risks. For instance, excessive devices and services may overload the signal load, increasing the risk of resource or denial-of-service assaults [16]. Thus far, two methods have been devised to address the issue of signal overload [17]. Using lightweight authentication and essential agreement methods is one way to enable communication between various devices. The second solution uses the group-based AKA protocol to group devices together.

The new techniques to improve 5G network performance also come with security risks [18]. For example, large MIMOs help hide both active and passive listening. When OpenFlow is used to develop SDN, the threat of malicious apps or activities increases. When a feature or service is transferred from one resource to another, NFV also poses security risks [11].

More privacy concerns must be addressed as 5G networks offer a greater variety of application scenarios and business models. Because the platform is open, users can effortlessly and frequently move private information from a closed to an open state. Doing this alters the contact's status from offline to online, substantially raising the possibility of leakage. Thus, in the coming years, we will unavoidably need to address and resolve the privacy concerns raised by 5G technology. The good news is that privacy policies have been strengthened and are improving thanks to data mining and machine-learning developments.

Numerous elements of the 5G network have been researched. They are based on artificial intelligence, such as multiple access at the MAC layer, channel coding and estimate at the physical layer, and different applications at the network layer. However, there is little support and a lack of AI applications for AI-based technologies because of the limits of traditional architecture in the early phases of the construction of 5G networks.

Smart radio and distributed AI are not supported because they are entirely dependent on AI. Furthermore, latency prevents 5G networks from responding to emergencies in "real-time," even when real-time intelligent edges are implemented (like automobile networks). However, 6G can. One-tenth the speed of 5G networks is achieved by 6G networks, which have a radio latency of 0.1 ms [19, 20]. Furthermore, only land is now covered by 5G, as certain 3D intercom tiers obstruct connections in space and underwater. Consequently, we shall outline prospective advancements in these four 6G domains in the upcoming chapters. Naturally, we also talk about the hazards to privacy and security in these domains.

## 10.3 LITERATURE REVIEW

Typically, data anomalies refer to network attacks and intrusions or interruptions. Network attacks can be detected by penetrating for irregular network data. Network security-related data is defined as information that indicates the security condition of the network system. It could be a characteristic, a fingerprint, a signature, or an attack behavior in specific [8, 12]. Time to live (TTL) is an example of data linked to network security; it specifies the maximum number of pieces that can pass before the router discards an IP address. Let us think of TTL as security-related information about network packets. It is also useful in identifying DoS attacks. The network system is being hacked if there is an unusually high TTL signal. For example, when multiple packets arrive at the same time on a network, it could be a DoS assault. To detect DoS attacks, a variety of security tools is available, including interruption detection systems (IDS) for TTL monitoring [12].

With the development of the Internet of Things big data, collecting network data is becoming critical. Malware detection [14], network management, interruption detection, network forensics, and traffic accounting are the most common uses of network data collection. Below, we go over the goals and circumstances of network data collecting in further depth.

Monitoring and identifying practice harmful actions and strategy breaches in a network is called intrusion detection. Other network security systems, such as IDS and security devices, help in network attacks and interruptions, the most common situations in network security data. Monitoring of the host state, data of the network and user behavior, and the data collection module of IDS is used. The characteristics of network activities here comprise network connection quantity, package contents, packet quantity, etc. The data-gathering unit is an essential part of the IDS paradigm. Due to the data collecting efficiency unit in IDS directly influencing the interruption detection performance, data gathering can quickly become a performance bottleneck. As a result, data collecting is critical for IDS performance [13, 15, 16]. Previous works utilized various methods to gather security-related data for interruption detection [6]. An irregularity detection method was presented based on the principle component examination (PCE). Authors further studied in [17] architectural distributed IDS to monitor and accumulate network data collectors dispersed in a network. To collect security data to record interruptions and attacks [5], a honeypot was used. Some efficient sampling methods were also proposed [20, 21] to enhance the data-gathering performance in IDS. A network managing structure (NMS) troubleshoots network problems, configures networks, and analyzes service quality. Design, network configuration, fault detection, and diagnosis are all part of the network management system. The collection of real-time network data is emphasized by a successful NMS in order to monitor relevant network resources and performance, guaranteeing the dependability and effectiveness of the network system. The main component of NMS has always been network data collecting (NDC). Debugging, optimizing, measuring, profiling, and identifying abnormalities in network systems can be better achieved with the use of enhanced technologies in network data collecting. According to data correlation, response [2, 13] proposed a network data collection model for efficient network management. A network traffic monitoring system must be installed to manage a PC room, including fault detection and network configuration. To enhance network system efficiency, collect and analyze network traffic in a smartphone [16, 1].

Traffic accounting is a new system through which internet service providers demand charges from their users before subscription internet services. In routine, it is common practice that internet access service is charged. Thus, the intranet of enterprises' Internet service providers (ISPs), and corporations have established a well-organized traffic secretarial system to charge their users. Billing information statistics, traffic collecting, user administration, and billing rule configuration are typically found in a sound traffic custodian system. However, widely used commercial transportation systems have drawbacks, including their comparatively high cost and incapacity to accommodate varying consumer demands for system functionality. Hence, nowadays, it is highly recommended that each customer gadget their traffic accounting system to meet the requirements of their functional system. He proposed a new idea that employed several network traffic collection techniques, such as Ethernet-LAN-oriented, router-oriented, firewall-oriented, and proxy server-based, which can be easily used in accounting systems [17, 19, 20, 22]. Figure 10.1 represents the 6G system foundations.

Network forensics is used to discover network irregularities, sense attacks and interruptions, and locate cyber-crimes [15]. It also involves traffic data collection because it can deliver rare data for analyzing and revealing treasured information [12]. To understand an abnormality, a forensic detective should be able to gather essential data and find out which piece of information is critical for investigation. Consequently, for network forensics investigators, efficient data collection is the basis.

Malware is a general term that refers to software that usually displays nasty running behaviors in a network system [9]. It has been demonstrated that the malicious behaviors of such software can damage a system, impact the collection of personal information, and attack a network, ultimately leading to the system's collapse. Many types of malware control malicious behaviors and operations through the network have been proposed and utilized on a wide range of systems based on abnormal data packets generated through malware software. Thus, it is highly recommended that examining network circulation can help to detect malware [20]. It is important to note that as mobile Internet usage has grown, so has the quantity of portable malware

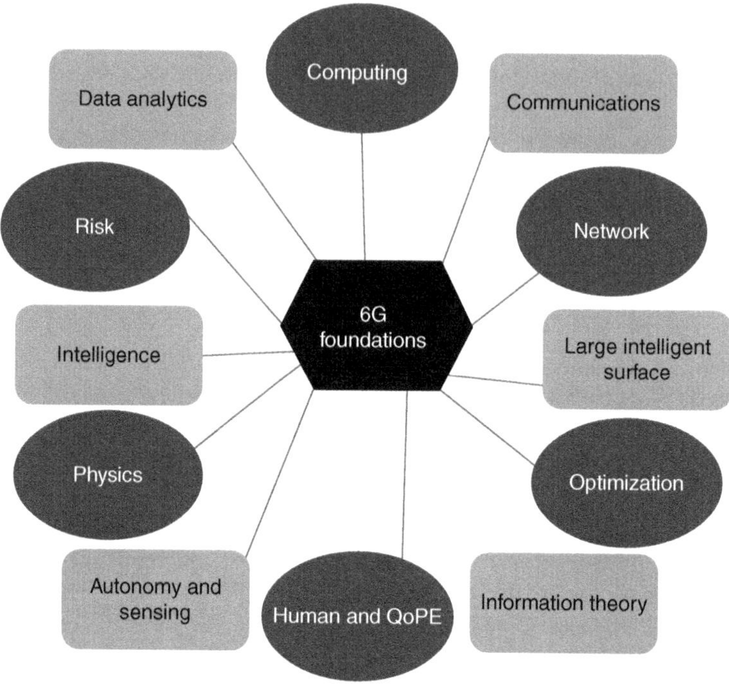


*Figure 10.1* 6G foundations.

in the system. By examining mobile communications packets, malware on mobile devices is usually found via domain name system (DNS) queries and address resolution protocol (ARP) requests. For instance, [5] described a method dubbed AS that, in tests, produced high detection rates and scalability for identifying malicious mobile applications by looking for malware network loops in the mobile Internet.

## 10.4 RESEARCH MOTIVATION

This section covers the current research on edge computing, edge intelligence, and 6G security and privacy. Next, to clarify the reasons for this research, we provide an overview of the subjects covered in this chapter and emphasize the differences.

Numerous academics have examined 6G's security and privacy while considering its revolutionary architecture and technologies. While [18, 23] concentrated on particular 6G scenarios or significant technologies, the authors carried out an extensive examination of the security and privacy of 6G networks' physical layer, connection layer, and service layer. In particular, it concentrated on security defensive solutions based on cutting-edge technologies and novel attack vectors for 6G radio technology. We investigate the cyber risk of an ultra-scale multiple-input multiple-output (MIMO) system operating at 6G THz at the physical layer and ubiquitous intelligence throughout the network. This work also addresses novel data protection methods, artificial intelligence security, quantum secure communication, and real-time adaptive security. While [24, 25] concentrated on particular 6G scenarios or significant technologies, the authors of [26] carried out an extensive examination of the security and privacy of 6G networks' physical layer, connection layer, and service layer [27]. In particular, it concentrated on security protection solutions based on cutting-edge technologies and novel attack vectors for 6G radio technology. We investigate the cyber risk of an ultra-scale MIMO system operating at 6G THz at the physical layer and ubiquitous intelligence throughout the network. This work also addresses novel data protection methods, artificial intelligence security, quantum secure communication, and real-time adaptive security. Additionally, this article offers two examples demonstrating how to adjust the security/privacy protection level to strike a balance between privacy and service quality. Furthermore, it is anticipated that in the 6G era, edge computing and FL would achieve an innovative equilibrium.

Research [7] examines security vectors and solutions for MEC security and privacy using the recently published MEC standard ETSI architecture. The integration and use of the suggested solution are the main topics of this essay. Additionally, the primary environment for utilizing edge computing is now Internet of Things (IoT) networks; relevant security and privacy-related works have been evaluated in [21], [15]. The authors [21] focused on DoS, side channel, malware injection, and authorization assaults as the four common threats against edge computing infrastructure. It also describes the workings, solutions, and sources of today's problems. Research on signal processing, resource allocation, eavesdropping coding, and other physical layer security techniques is done [15, 27]. The most frequent security and privacy risks to 6G networks are discussed in the review article currently available, along with countermeasures that leverage edge computing and edge intelligence. Increased security and privacy risks may jeopardize the benefits of 6G networks due to the application of MEC and artificial intelligence.

A review of relevant literature and current challenges in FL-based network optimization approaches is available in [22]. Another decentralized method is blockchain, which tracks and verifies any changes to data via a follower system and ledger. The topics of recent review studies [23], and [18] are generic 6G techniques as well as network security and privacy. The security and confidentiality of the edge network have become a focus for researchers since meeting the 6G KPI requirements for processing, caching, and intelligence is crucial [21]. While edge computing and intelligence have drawbacks, most recent research has been on how these strategies enhance security and privacy protection. Studies that discuss edge computing services' security concerns are few [7, 21].

Regarding edge service security and privacy concerns as well as potential solutions, there are no survey studies available. To address security and privacy from a systems viewpoint, edge computing, edge caching, and edge intelligence should also be examined, considering the close relationship and coexistence of satisfying user needs. Moreover, in addition to well-known security and privacy attacks such as denial-of-service attacks, malware, and eavesdropping, network edge applications such as collaborative computing, caching, and intelligence have triggered many new security and privacy attacks such as data leakage—toxic and unreliable players. Traditional core network threats like data misuse might worsen for edge networks as 6G access networks are increasingly susceptible to attacks from ISPs and outside vendors. As a result, 6G advantage networks have new requirements, methods, and security and privacy concerns.

The study aims to assess the advancements in security and privacy attack defense technologies and 6G network edge computing, caching, and intelligent service technologies. This work focuses on 6G access networks, where the design of security countermeasures must consider emerging technologies like O-RAN, FL, and blockchain. Furthermore, this study aims to clarify how FL and blockchain may enhance edge network security and address current issues. An explanation of how edge computing, edge caching, and edge intelligence work together to preserve security and privacy will also be provided.

## 10.5 REQUIREMENTS

We studied network security-related data in this section. Given the present research work, there are security and functional requirements. As per requirements, present research on security-related data concerning security and function can be discussed. The following requirements are for security-related data.

### 10.5.1 Security requirements

This chapter discusses 6G network security concerns. The remaining security flaws from 5G to 6G are explained by pre-6G security. The security threats brought on by cutting-edge 6G technology and architectural changes are also covered in this section. The security challenges associated with 5G network software technologies, including network slicing, multi-access edge computing (MEC), network function virtualization (NFV), and software-defined networking (SDN), will persist in 6G systems. Assaults against SDN controllers, assaults on northbound and southbound interfaces, and

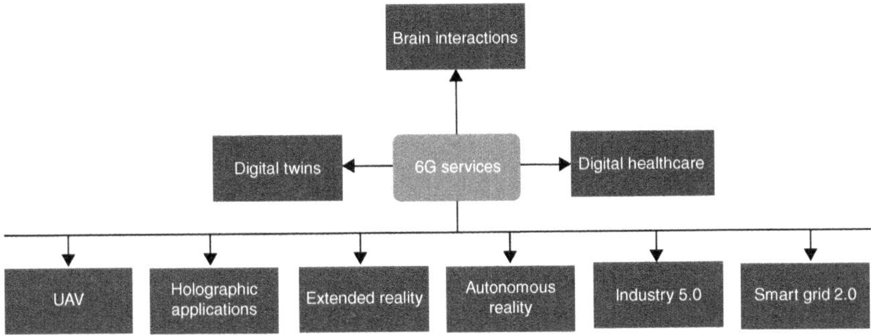


*Figure 10.2* 6G applications and services.

vulnerabilities present in the platforms used to deploy SDN controllers and applications are SDN's most frequent security risks [7]. NFV-related security vulnerabilities include attacks against hypervisors, virtual machines (VMs), virtual network functions (VNFs), VNF managers, and NFV coordinators [5]. Due to the extreme fragmentation of 6G networks, MEC is susceptible to physical security risks, decentralized denial of service (DDoS), and man-in-the-middle assaults [8]. Potential network slicing concerns include denial of service assaults and information theft by hacking slicing [9]. The promised dynamics and complete automation of 6G networks are not achievable due to attacks on network software technologies. 6G aims to achieve the Internet of Everything (IoE), a network of billions of devices. Figure 10.2. is a visual representation of 6G applications.

Basic device security designs based on SIM cards are not feasible for IoE deployments in 6G, especially for small devices like in-body sensors. More extensive networks have extremely inefficient central management and distribution [10]. Attackers mainly target IoT devices with little resources since they cannot afford sophisticated encryption methods that provide robust security [11]. This tool can be used as a springboard for attacks and is readily hacked. Privacy concerns are raised by the hyper-connected Internet of Everything 6G applications that capture data. Data theft will impact identity, location, and privacy via resource-constrained IoT devices. In healthcare, education, and industry, on-premise 5G network deployments are frequently used to provide services. Enabling smaller networks like in-body networks, drone swarms, and environmental sensor networks, as well as longer battery life, 6G further broadens this idea [4]. These tiny networks can be integrated with a wide area network or function alone when needed. Unlike native 5G networks, numerous stakeholders in a native 6G network implement varying degrees of embedded security. 6G networks with inadequate local security present a possibility for attackers to start assaults.

Next, get on the network where the compromised network has acquired credibility. These specifications concern network system security and quality during cyber security data collection. It should prevent data loss and ensure data truth during data gathering as follows.

- Should warranty the safety of gathered data.
- Must be able to prevent data leakage.
- Client privacy should be taken care of during the data collection process.
- It should have access control capability, which can ensure that users are willing to access data and support access for capable users.
- Should have the capability of verifying the authenticity and integrity of collected data.

### 10.5.2 Function requirements

- Gather necessary security-related data in various contexts.
- Capable ofstoring gathered data on a storage medium.
- Ability to control and manage data.
- Capable of loading data collected information.
- Scalable and flexible when gathering data.
- Capability to know when and where to gather security-related data.
- Capability of transferring data to another system.
- Cheap in terms of storage resources and computation resources for data gathering.
- Unable to create new data, which may affect already-present collected data.
- Keep the original network system.
- Should not disturb the original network system.
- Should be generic and universal.
- Must be able to support different applications.
- Should be adaptive and automatic.

## 10.6 OBJECTIVES

We highlighted many objectives based on functional and security requirements. There are the following objectives.

- Integrity: sSurety of loss of information and data
- Authentication: Ensure the authenticity of data collection node and reliability of gathered data
- Confidentiality: Information and data are unable to transfer to illegal networks and are ensured by a cryptographic system

- Self-protection: Providing safety in data collection and preventing system disturbances
- Non-repudiation: The location and time of these behaviors and one's collection non-repudiation.
- Privacy protection: Other sources of collected information must be secure for the prevention of privacy leakage
- Stability: In this data-gathering system relevant to network security, the mechanism should be normal
- Cost: Cost should be appropriate for applications, and data collection cost should be in consideration
- Scalability: It describes the scalability quality for security-related network data-gathering technology
- Applicability: In this network, data collection mechanisms or techniques are to be used in the actual network for gathering the network security-related data
- Flexibility: It shows that data related to network security is switchable between various network contexts and reserves suitable functional delay boundaries
- Adaptability: It studies how gathering procedures may fix to various network situations and contexts
- Generality: It explains that security-related data-gathering techniques apply to several scenarios and can gather various data.
- Efficiency: It shows that the data collection technique is capable of collecting necessary data efficiently without interference in the performance of regular network

## 10.7 KEY CHALLENGES AND FUTURE DISCUSSION

This section presents several critical challenges related to data collection in security. Some of such challenges are given below. Every generation of cellular networks aims to meet issues posed by new applications and business models by defining or upgrading at least one component of the security architecture, such as new key management and authentication. Based on previous 6G white papers and ongoing studies, this section provides an overview of the 6G roadmap and new developments in the three levels (physical layer, connection/network layer, and service/application layer) of 6G enabling technologies [13, 14].

We describe security needs and possible remedies—especially in security architecture—through modifications to those elements. The wheel of 6G's usual 10-year development cycle is set in motion.

Drawing on existing research (e.g., [1, 2, 10]), Figure 10.3 outlines the expected evolution of 6G in a roadmap that includes comparisons of spectrum consumption with earlier generations and key performance indicators (KPIs). Significantly faster network speeds (>1 Tbps), lower latency (<1

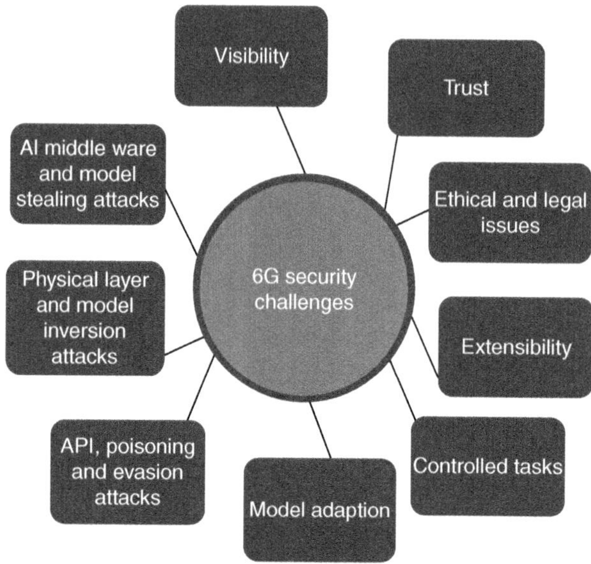


*Figure 10.3* 6G security challenges.

ms), and 10–100 times more energy efficiency are what 6G will have to offer over 5G. Figure 10.3 represents the attack threats and challenges in 6G technology. The long-term objectives of the Internet of Senses and the United Nations Sustainable Development Goals (SDGs) for a sustainable and carbon-neutral future are also considered when designing 6G [15] [16]. Three key air interface modifications are targeted by 6G to meet KPI requirements: (1) moving communications to higher frequency bands, (2) establishing reconfigurable surfaces to create intelligent radio environments, and (3) doing away with conventional cell structures such as cell-free numerous inputs and outputs combined (MIMO) [13, 14, 16], and so on. In order to provide native 6G communications (6G THz), 6G will explore new high-frequency spectrum and photonic signals (300 GHz–3 THz) [2, 10, and [17]. Higher frequencies in earlier network generations are no longer applied or used, with the exception of using compatible frequencies (6G mmWave). To enhance 6G communications, additional key technologies include multi-user MIMO, holographic radio beamforming, VLC, and orbital angular momentum [22]. There will be several notable differences between 5G and 6G network architecture at the connection (network) level. Initially, network as a service (NaaS) and network automation may be realized with 6G. Thanks to NaaS, users or businesses can customize networks according to their needs. A new network architecture is necessary for this per-user basis paradigm. Key technologies to accomplish the goal include cloudization, deep slicing/function virtualization, end-to-end softwarization, and intent-based networking.

Second, the quick transition of 5G infrastructure deployments to open source and cloud-based networks, especially for core/RAN network components, heralds the arrival of the "full openness" age of 6G. Because features and improvements may be added at scale, the development model can help realize the potential of 6G [19]. The main driving force behind this trend is the desire of leading vendors (like AT&T) not to rely just on one equipment manufacturer. The first wireless cellular network to be fully AI-powered may be 6G. The idea of "connected things" in 5G will be transferred to "connected intelligence" in 6G when AI is expected to manage the majority of network functions and nodes [20]. [21] states that time-sensitive networking (TSN) or deterministic networking (DetNet), which promises zero data loss and guaranteed latency, can be a significant 6G upgrade to meet the criterion of 6G ultra-reliable and low-latency communications (uRLLC). Long-distance, high-mobility communications (LDHMC) and highly low-power communications (ELPC) are two new features made possible by 6G's enhanced network capabilities. It also expands the ability of the three primary 5G services (uRRLC, eMBB, and mMTC). Notably, novel applications such as holographic telepresence, tactile/haptic Internet, smart medical nano-robot, fully autonomous driving, and extended reality/digital twin are expected to debut with 6G [2, 3]. Extended reality (XR) is the term for the integration of virtual reality (VR), augmented reality (AR), and mixed reality (MR) to provide real-time human–machine interactions in real and virtual integrated environments. For many upcoming businesses, including manufacturing, education, and training—mainly training for hazardous situations too costly or harmful to do on the field—XR will be a crucial technology in 6G. Similarly, the digital twin expands the use of virtual reality technologies (VR, AR, and MR) to replicate real-world counterparts (copies of actual items and their actions in real-time), for example, by simulating metropolitan landscapes for use in planning. An Internet network with extremely low latency, high availability, dependability, and security that can provide real-time human–machine interaction is known as the tactile/haptic Internet. Lastly, future 6G technologies that need high reliability and ultra-low latency include holographic telepresence and autonomous driving. To accommodate these time-sensitive applications, additional processing power will be added to 6G communications at each hop in addition to the edge. It is anticipated that AI technologies will be crucial in increasing the nodes' level of intelligence and autonomy [6, 7, 18, 23].

- Intelligence: It represents that intelligence cannot fulfill users' intelligence needs. Developing automatic and intelligent data collection systems is still a key challenge. There is no sufficient work on machine learning to create such an intelligence system.

- Heterogeneity: As 6G is coming shortly, it is not easy to deploy suitable data collection in a huge-scale data heterogeneous network system. However, several network systems form this network, such as wireless sensor networks, satellite communication, and the Internet. Still, there are some challenges in the collection of network security data. First, there is a lack of comprehensive, complete, and extendable security relevant data report techniques that may be useable in various network contexts. Secondly, there is an insufficient data collection method that may transfer adaptively between various network contexts. A severe challenge for heterogeneous network systems is in designing an impressive security-related data collection mechanism.
- Adaptability: For various application requirements, collected data needs to be varied. We require a specific type of data recognition in mass data to solve the data collection validity issue, specifically in 6G. Still, there is insufficient work on adaptability objectives. Sufficient data mining, pattern-learning techniques, or machine-learning methods can solve such a challenge.
- Security and privacy: With 6G technology, much more attention will be paid to security and information privacy. There are still several security issues in the upcoming 6G technology for data collection mechanism techniques. However, security surety, privacy, and data leakage protection during network security-related data collection are critical challenges. Such issues can be resolved by more work on security techniques and cryptography.

In future discussions, we can conclude that, for network security data collection, intelligent data collection will get much more attention from researchers. Because of enhancements in machine learning, artificial intelligence, and deep learning may play an essential role in different fields to gain intelligent data and processing. In the future, data collection must be joined with artificial intelligence to collect intelligence data. Network systems with the capability of machine learning or deep learning may affect the enhancement of security, efficiency of data collection, and effectiveness, which is suitable. Because of hardware development, it is currently not well-known. With the enhancement of technologies such as big data, smart cities, intelligent hospitals, IoT [18, 21, 23, 24], 6G research work, and terahertz communication, there is improved technology that may reduce collected data and keep data accuracy results. Data fusion and intelligent sampling will gain more attention from researchers in data collection.

Reliance management of security-related data collection is a hot research field that ensures the variety of data collected and completes the data collection process, including trustworthiness, storage, processing, transmission, analysis, etc. Reliance management is helpful in automatic decision-making and plays a crucial role in challenging network contexts. It helps overcome the perception of uncertainty to take strict action.

The protection and security of data collection are also hot research topics for researchers in 6G technology. Such a challenge may suppose that different kinds of data exist for various meanings, not only for security. In this regard, to see challenges, active data collection may be helpful for network security measurement. Network system applications may be regenerated by such a type of technology. They may get network attack and interruption information, but such a data collection mechanism is unsuitable for effective network performance.

In previous research, researchers studied centralized reliance management, which depends on a trusted party. Such a solution is not sufficient in a 6G heterogeneous network system. There is a requirement for a new trust management solution in future 6G technology of data collection techniques.

## 10.8 Conclusion

Information technology plays an essential role in our daily lives, and we connect ourselves with several types of applications in the physical layers. Data usage is present in this regard, and several security threats are present. This chapter studied security-related data collection ideas, objectives, issues, and requirements. As 5G technology is in progress and the world is looking toward 6G technology, putting 6G technology in mind we discussed security issues and critical data collection challenges. In our chapter, we also presented future discussions about data collection technologies. Finally, we concluded that researchers need to pay more attention to security issues in such technologies.

## 10.9 AUTHOR'S CONTRIBUTION

All authors equally contribute to this chapter.

## 10.10 FUNDING

There is no external funding for this chapter.

## 10.11 CONFLICTS OF INTEREST

There is no conflict of interest between all authors.

## REFERENCES

[1] N. A. Azeez, T. M. Bada, S. Misra, A. Adewumi, C. Van der Vyver, and R. Ahuja, "Intrusion Detection and Prevention Systems: An Updated Review,"

In N. Sharma, A. Chakrabarti, and V. Balas (eds.), *Data Management, Analytics and Innovation. Advances in Intelligent Systems and Computing*, vol. 1042, Springer, pp. 685–696, 2020. doi: 10.1007/978-981-32-9949-8_48.

[2] M. M. Aslam, J. Zhang, B. Qureshi, and Z. Ahmed, "Beyond 6G-Consensus Traffic Management in CRN, Applications, Architecture and Key Challenges," *2021 IEEE 11th International Conference on Electronics Information and Emergency Communication*, pp. 182–185, 2021 [Online]. doi: 10.1109/ICEIEC51955.2021.9463832.

[3] M. M. Aslam, L. Du, Z. Ahmed, and M. N. Irshad, "A Deep Learning Based Power Control and Consensus Performance of Spectrum Sharing in CRN-Network," *Wireless Communication and Mobile Computing*, no. submitted, vol. 2021, pp. 1–17, 2019.

[4] M. M. Aslam, M. N. Irshad, H. Azeem, and M. Ikram, "Cost Effective & Energy Efficient Intelligent Smart Home System Based on IoT," *Afyon Kocatepe University International Journal of Engineering Technology and Applied Sciences*, vol. 3, pp. 10–20, 2020.

[5] M. M. Aslam, L. Du, X. Zhang, Y. Chen, and B. Qureshi, "Sixth Generation (6G) Cognitive Radio Network (CRN) Application, Requirements, Security Issues, and Key Challenges," *Wireless Communication and Mobile Computing*, vol. 2021, pp. 1–29, 2021.

[6] M. N. Irshad, L. Du, I. A. Khoso, T. Bin Javed, and M. M. Aslam, "A Hybrid Solution of SDN Architecture for 5G Mobile Communication to Improve Data Rate Transmission," *2019 28th Wireless and Optical Communications Conference*, WOCC 2019 Proceedings, pp. 1–5, 2019. doi: 10.1109/WOCC.2019.8770664.

[7] M. M. Aslam, A. Tufail, K. H. Kim, R. A. A. H. M. Apong, and M. T. Raza, "A Comprehensive Study on Cyber Attacks in Communication Networks in Water Purification and Distribution Plants: Challenges, Vulnerabilities, and Future Prospects," *Sensors 2023*, vol. 23, no. 18, p. 7999, September 2023. doi: 10.3390/S23187999.

[8] H. Rodrigues Dias Filgueiras et al., "Wireless and Optical Convergent Access Technologies Toward 6G," *IEEE Access*, vol. 11, pp. 9232–9259, 2023. doi: 10.1109/ACCESS.2023.3239807.

[9] C. X. Wang et al., "On the Road to 6G: Visions, Requirements, Key Technologies, and Testbeds," *IEEE Communications Surveys and Tutorials*, vol. 25, no. 2, pp. 905–974, 2023. doi: 10.1109/COMST.2023.3249835.

[10] X. Mu, J. Xu, Y. Liu, and L. Hanzo, "Reconfigurable Intelligent Surface-Aided Near-Field Communications for 6G: Opportunities and Challenges," *IEEE Vehicular Technology Magazine*, vol. 19, pp. 65–74, 2024. doi: 10.1109/MVT.2023.3345608.

[11] Z. Zhang et al., "Quality-of-Experience Evaluation for Digital Twins in 6G Network Environments," *IEEE Transactions on Broadcasting*, vol. 70, no. 3, pp. 995–1007, 2024. doi: 10.1109/TBC.2023.3345656.

[12] O. Abbasi, A. Yadav, H. Yanikomeroglu, N. D. Dao, G. Senarath, and P. Zhu, "HAPS for 6G Networks: Potential Use Cases, Open Challenges, and

Possible Solutions," *IEEE Wireless Communications*, vol. 31, no. 3, pp. 324–331, January 2023. doi: 10.1109/MWC.012.2200365.

[13] N. Chukhno et al., "Models, Methods, and Solutions for Multicasting in 5G/6G mmWave and Sub-THz Systems," *IEEE Communications Surveys & Tutorials*, vol. 26, no. 1, pp. 119–159, 2023. doi: 10.1109/COMST.2023.3319354.

[14] H. Poddar, S. Ju, D. Shakya, and T. S. Rappaport, "A Tutorial on NYUSIM: Sub-Terahertz and Millimeter-Wave Channel Simulator for 5G, 6G and Beyond," *IEEE Communications Surveys & Tutorials*, vol. 26, no. 2, pp. 824–857, December 2023. doi: 10.1109/COMST.2023.3344671.

[15] R. Kumar, R. C. Singh, and V. Jain, *Modeling for Sustainable Development: A Multidisciplinary Approach*, Nova Science Publishers, 2023. doi: https://doi.org/10.52305/HAXA0362.

[16] X. Lin, L. Kundu, C. Dick, E. Obiodu, T. Mostak, and M. Flaxman, "6G Digital Twin Networks: From Theory to Practice," *IEEE Communications Magazine*, vol. 61, no. 11, pp. 72–78, November 2023. doi: 10.1109/MCOM.001.2200830.

[17] N. N. Dao, "Internet of Wearable Things: Advancements and Benefits from 6G Technologies," *Future Generation Computer Systems*, vol. 138, pp. 172–184, January 2023. doi: 10.1016/J.FUTURE.2022.07.006.

[18] M. M. Aslam, Z. Ahmed, L. Du, M. Z. Hassan, S. Ali, and M. Nasir, "An Overview of Recent Advances of Resilient Consensus for Multiagent Systems Under Attacks," *Computational Intelligence and Neuroscience*, vol. 2022, 2022. doi: 10.1155/2022/6732343.

[19] V. Dala Pegorara Souto et al., "Emerging MIMO Technologies for 6G Networks," *Sensors 2023*, vol. 23, no. 4, p. 1921, February 2023. doi: 10.3390/S23041921.

[20] Z. Chen, Z. Zhang, and Z. Yang, "Big AI Models for 6G Wireless Networks: Opportunities, Challenges, and Research Directions," in *IEEE Wireless Communications*, vol. 31, no. 5, pp. 164–172, October 2024, doi: 10.1109/MWC.015.2300404

[21] A. Alkhateeb, S. Jiang, and G. Charan, "Real-Time Digital Twins: Vision and Research Directions for 6G and Beyond," *IEEE Communications Magazine*, vol. 61, no. 11, pp. 128–134, November 2023. doi: 10.1109/MCOM.001.2200866.

[22] A. T. Jawad, R. Maaloul, and L. Chaari, "A Comprehensive Survey on 6G and Beyond: Enabling Technologies, Opportunities of Machine Learning and Challenges," *Computer Networks*, vol. 237, p. 110085, December 2023. doi: 10.1016/J.COMNET.2023.110085.

[23] R. Kumar, V. Jain, W. Y. Leong, and S. Teyarachakul. *Convergence of IoT, Blockchain and Computational Intelligence in Smart Cities*. Taylor & Francis, CRC Press, 2023, ISBN: 9781032404240.

[24] W. Y. Leong, Y. Z. Leong and W. S. Leong, "Smart Manufacturing Technology for Environmental, Social, and Governance (ESG) Sustainability," *2023 IEEE 5th Eurasia Conference on IOT, Communication and Engineering (ECICE)*, Yunlin, Taiwan, pp. 1–6, 2023. doi: 10.1109/ECICE59523.2023.10383150.

[25] R. Jain, S. Chaudhary, and R. Kumar, "Green Approach for Next Generation Computing: A Survey", *International Journal of Advanced Engineering Research and Science (IJAERS)*, vol. 4, no. 1, pp. 78–82, 2017.
[26] R. Rana and R. Kumar, "Performance Analysis of AODV in Presence of Malicious Node", *Acta Electronica Malaysia (AEM)*, vol. 3, no. 1, pp. 1–5, 2019.
[27] R. Kumar, *Information and Communication Technologies*, Laxmi Publications, 2009.

Chapter 11

# Network security and data privacy in the 6G environment

## Impacts and challenges

*Navjot Singh Talwandi, Shanu Khare, Payal Thakur, and Rajendra Kumar*

## 11.1 INTRODUCTION TO 6G NETWORKS

The evolution from fifth generation (5G) to sixth generation (6G) is an exciting development in the field of wireless communication technology. While 5G has already brought about significant improvements in terms of speed, latency, and capacity compared to its predecessors, 6G promises to take these advancements even further [1]. 6G is expected to offer significantly higher data rates than 5G, with some estimates suggesting that it could provide peak data speeds of up to 1 terabit per second, which is around 100 times faster than what 5G is currently achieving. This will enable new applications such as real-time holographical communications, ultra-high definition video streaming, and prompt downloads of massive files. Additionally, with increased speed, 6G is also expected to have much lower latency than 5G, potentially reducing end-to-end delay to just a few microseconds. This will be particularly beneficial for time-critical applications such as autonomous vehicles, remote surgery, and industrial automation [2]. Another key feature of 6G will be its ability to support vast numbers of connected devices simultaneously, making it ideal for use cases involving the Internet of Things (IoT) and blockchain in peer-to-peer communication. It is anticipated that 6G networks will be able to handle trillions of connections at a time, enabling seamless connectivity between humans, machines, and objects. Also, 6G networks will likely be more energy efficient than 5G. The technologies, such as artificial intelligence (AI) and machine learning along with augmented and virtual reality will ensure it. These technologies will allow network resources to be allocated more efficiently, resulting in reduced power consumption and longer battery life for devices.

The 6G networks may also incorporate advanced security features, including quantum encryption, ensuring the confidentiality, integrity, and availability of sensitive information transmitted over the network [3]. All in all, the transition from 5G to 6G represents a major step forward in the

DOI: 10.1201/9781003583127-11

evolution of wireless communication technology by offering unprecedented levels of speed, latency, capacity, energy efficiency, and security. 6G technology is set to transform the way we live, work, and communicate for years to come.

## 11.2 FOUNDATIONS OF NETWORK SECURITY AND DATA PRIVACY

Here are some of the key features and promises of 6G technology:

**Ultra-high data rates:** As mentioned earlier, 6G is expected to deliver significantly higher data rates than 5G, with peak data speeds of up to 1 TBPS or more. This would enable new applications, such as real-time holographic communications, immersive virtual reality experiences, and high-fidelity telepresence.

**Extremely low latency:** 6G networks promise to reduce end-to-end latency to just a few microseconds, making them ideal for time-critical applications, such as autonomous driving, remote surgery, and industrial automation.

**Massive connectivity:** 6G networks will be capable of supporting vast numbers of connected devices simultaneously, enabling seamless connectivity between humans, machines, and objects. This makes it well suited for use cases involving the Internet of Things (IoT) and machine-type communications.

**Energy efficiency:** Advances in AI and machine learning will make 6G networks more energy efficient than their 5G counterparts, resulting in reduced power consumption and longer battery life for devices.

**Security:** 6G networks may incorporate advanced security features, including quantum encryption, to ensure the confidentiality, integrity, and availability of sensitive information transmitted over the airwaves.

**Global coverage:** 6G networks aim to provide global coverage, ensuring consistent connectivity regardless of location. This is especially important for rural areas, where access to high-speed internet is often limited.

**Intelligent networking:** With the help of AI and machine-learning algorithms, 6G networks will be able to adapt dynamically to changing traffic patterns and user demands, providing optimal performance at all times.

**Integrated satellite communications:** 6G networks may integrate satellite communications into the core network infrastructure, providing ubiquitous connectivity across land, sea, and air.

## 11.3 SECURITY ARCHITECTURE IN 6G NETWORKS

### 11.3.1 Design principles for secure 6G systems

Security is of paramount importance in next-generation networks due to the increasing reliance on digital services and the growing volume of sensitive data being transmitted over wireless networks. With the advent of 6G networks, there will be an exponential increase in the number of connected devices, leading to an expanded attack surface for cybercriminals. Therefore, securing these networks against unauthorized access, eavesdropping, and other malicious activities is crucial [4]. Next-generation networks must incorporate robust security measures to protect users' privacy and prevent data breaches. One approach to achieving this goal is through the deployment of advanced cryptographic techniques such as post-quantum cryptography, which provides protection against attacks by quantum computers. Additionally, secure authentication mechanisms should be implemented to verify the identity of devices connecting to the network, preventing impersonation, and man-in-the-middle attacks.

Network slicing, a key feature of 6G networks, allows multiple logical networks to operate within a single physical infrastructure. Each slice can be customized to meet specific requirements, including security policies. Implementing stringent security protocols for each slice ensures that sensitive data remains isolated and protected from unauthorized access. Moreover, AI and machine-learning (ML) techniques can be employed to detect and respond to threats in real-time. By analyzing network behavior and identifying anomalies, AI and ML systems can quickly identify and mitigate potential security risks before they become critical.

In summary, security is a vital component of next-generation networks. As these networks continue to evolve and expand, so too do the associated security challenges [5]. By implementing robust security measures, such as advanced cryptographic techniques, secure authentication, network slicing, and AI/ML- powered threat detection, network operators can safeguard user privacy and ensure the confidentiality, integrity, and availability of sensitive data transmitted over the network.

### 11.3.2 Integration of security in network architecture

Designing secure 6G systems requires adherence to several principles that prioritizes security while delivering high-performance networking capabilities. Some of these design principles include the following:

**End-to-end security:** Security should be integrated throughout the entire network architecture, from endpoints to the core. Encryption, authentication, and authorization mechanisms should be applied consistently across all layers of the network stack.

**Zero-trust architecture:** A zero-trust model assumes that any device or entity attempting to access the network is potentially hostile until proven otherwise. Access control policies based on least privilege and multi-factor authentication should be enforced to limit exposure to security threats.

**Privacy preservation:** User privacy should be preserved by minimizing the collection and storage of personal data. Differential privacy techniques and homomorphic encryption can be used to analyze aggregate data without revealing individual user identities.

**Quantum resistance:** Given the emergence of quantum computing, 6G systems should employ quantum-resistant cryptographic techniques that resist attacks by quantum computers. Post-quantum cryptography standards should be adopted wherever possible.

**Real-time threat detection:** Advanced analytics using AI and ML should be deployed to monitor network behavior continuously, detecting potential security threats in real-time. Anomaly detection algorithms should be trained to recognize normal network activity and flag abnormal behaviors as potential threats.

**Software defined networking (SDN):** SDN enables centralized management and programmability of network functions, simplifying security policy implementation and maintenance. SDN can also facilitate rapid response to emerging security threats through automated patching and software updates.

## 11.4 ENCRYPTION AND AUTHENTICATION IN 6G

### 11.4.1 Advancements in cryptographic techniques

Advancements in cryptographic techniques play a pivotal role in enhancing the security of modern communication networks, including 6G systems. Recent developments in cryptography focus on improving resilience against both classical and quantum attacks, addressing issues related to scalability, and maintaining compatibility with legacy systems.

Traditional public-key cryptography methods rely on mathematical problems that are difficult for classical computers to solve but can be solved relatively easily by quantum computers. Post-quantum cryptography employs alternative hardness assumptions, such as lattice problems or coding theory, to construct cryptosystems that remain secure in the presence of large-scale quantum computers. Homomorphic encryption enables computation on encrypted data without requiring decryption first. This technique preserves privacy and confidentiality when processing sensitive information, enabling scenarios like outsourcing computations to cloud servers without exposing raw data [6]. Multi-party computation (MPC) allows multiple parties to jointly perform computations on private inputs without sharing them directly. MPC maintains privacy and secrecy during collaborative tasks, such as data mining or machine learning. Functional encryption extends traditional encryption schemes by granting selective functionalities to authorized entities without disclosing full plain text contents. For instance, attribute-based encryption permits fine-grained access control based on predefined attributes associated with users or devices. Transport Layer Security (TLS) 1.3 is the latest version of the widely adopted transport layer security protocol responsible for establishing secure communication channels between client and server applications. TLS 1.3 introduces enhancements such as improved security, faster connection setup times, and better resistance to downgrade attacks [7].

### 11.4.2 Multi-layered authentication protocols

Multi-layered authentication protocols represent a powerful approach to bolstering security in contemporary communication networks. By deploying multiple sequential stages of authentication checks, these protocols effectively thwart various types of intrusion attempts and reinforce user identity verification. Layered authentication typically involves combining different factors of authentication, such as something you know (passwords), something you possess (smartcards or tokens), something you are (biometrics), or somewhere you are (geolocation).

The primary objective of multi-layered authentication protocols is to establish strong confidence in user identities while impeding unauthorized access attempts. Such protocols generally comprise three main layers:

**Initial layer – preliminary authentication:** At the initial stage, standard authentication approaches, such as password-based login credentials, are checked. If successful, the subsequent layers are initiated; otherwise, the attempted access is denied.

**Secondary layer – stronger factor check:** Upon passing the preliminary check, stronger forms of authentication come into play, usually entailing one or more non-reusable factors. Examples include smartcard swipes, OTP generation via mobile apps, or fingerprint scans. Only after satisfying these secondary checks does access get granted temporarily.

**Final layer – contextual analysis:** The final layer incorporates contextual analysis to assess whether the authenticated session aligns with typical usage patterns and expectations. Behavioral biometric profiling, geographical location tracking, and historical transaction records serve as valuable sources of insight during this phase. Suspicious deviations trigger additional scrutiny, possibly prompting reauthentication requests or blocking questionable transactions altogether.

Multi-layered authentication protocols find extensive applicability across numerous domains, ranging from financial institutions and healthcare providers to government agencies and enterprise networks [8]. They cater to heightened security needs by integrating disparate authentication techniques, fortifying user validation, and deterring sophisticated cyberattacks. Furthermore, many modern implementations leverage machine-learning algorithms to optimize decision-making and risk assessment processes, fostering intelligent adaptability amid dynamic threat landscapes.

## 11.5 EDGE COMPUTING SECURITY IN 6G

### 11.5.1 Security challenges in edge computing environments

Edge computing is a distributed computing paradigm that brings computation and data storage closer to the location where it is needed, thus reducing latency, bandwidth usage, and cost. While edge computing offers numerous benefits, such as improved performance, reduced network congestion, and enhanced privacy, it also introduces new security challenges that must be addressed to ensure the confidentiality, integrity, and availability of sensitive data and applications [9].

One major challenge in edge computing environments is securing the communication between edge devices and the cloud or other centralized systems. Since these devices are often connected over untrusted networks, they are vulnerable to eavesdropping, man-in-the-middle attacks, and other forms of tampering. To mitigate these risks, organizations can use encryption, authentication, and access control mechanisms to protect their communications.

Another challenge is ensuring the security and integrity of the software running on edge devices. These devices may have limited resources, making them susceptible to malware, viruses, and other types of cyber threats. Additionally, since edge devices are typically deployed in remote locations, patching and updating them can be challenging, increasing the risk of vulnerabilities remaining unpatched for extended periods. Therefore, implementing robust endpoint protection measures, including antivirus software, intrusion detection/prevention systems (IDS/IPS), and regular updates, is essential [10].

Physical security is another concern in edge computing environments, especially given the decentralized nature of these systems. Physical access to an edge device can compromise its security, allowing attackers to steal sensitive data, install malicious firmware, or disrupt operations. Organizations should implement appropriate physical security controls, such as locks, alarms, surveillance cameras, and environmental sensors, to prevent unauthorized access to their edge devices.

Finally, managing identities and access privileges across multiple edge devices can be complex and error-prone, leading to potential security breaches. Implementing identity and access management (IAM) solutions that provide role-based access control (RBAC) and multi-factor authentication (MFA) can help address this challenge by limiting user access based on job responsibilities and requiring additional verification factors before granting access [11].

In summary, while edge computing offers significant advantages in terms of performance, scalability, and cost savings, it also presents unique security challenges. Addressing these challenges requires a comprehensive approach that includes secure communication protocols, robust endpoint protection, strong physical security controls, and effective IAM practices. By taking a proactive stance towards security, organizations can fully realize the benefits of edge computing while minimizing the associated risks.

### 11.5.2 AI and machine learning for edge security

AI and ML technologies offer promising opportunities for enhancing security in edge computing environments. With the ability to analyze vast amounts of data quickly and accurately, AI and ML algorithms can detect anomalies, identify patterns, and make real-time decisions to improve threat detection, response times, and overall system resilience.

For instance, ML models trained on historical data can predict normal behavior patterns of edge devices and flag any deviations from those patterns as suspicious activities. This allows for early identification of potential security threats, enabling organizations to take prompt action to contain them before they escalate into more serious incidents. Similarly, AI algorithms can automate the process of analyzing log files, identifying indicators of compromise (IOCs), and generating alerts for further investigation [12].

Moreover, AI and ML can enhance the effectiveness of existing security measures, such as firewalls, IDS/IPS, and anti-malware solutions, by providing adaptive and dynamic responses tailored to specific threats. For example, ML-powered intrusion prevention systems can automatically adjust rules and policies based on real-time analysis of traffic patterns, improving their accuracy and reducing false positives.

However, deploying AI and ML in edge computing environments also poses some challenges. One critical issue is the need to balance the tradeoff between model accuracy and resource consumption, particularly regarding processing power, memory, and energy requirements. Running sophisticated ML models on resource-constrained edge devices could lead to slower response times, higher battery drain, and increased costs. As such, optimizing ML models for edge computing is crucial to ensure efficient and effective deployment.

Additionally, protecting the training data used to develop AI and ML models is vital, as adversaries could exploit biases, weaknesses, or errors in the training data to evade detection or launch targeted attacks. Ensuring data privacy and security throughout the entire lifecycle of AI and ML development, deployment, and maintenance is paramount [12].

To summarize, leveraging AI and ML for edge security provides several benefits, including improved threat detection, automated decision-making, and enhanced efficiency. However, addressing the challenges related to resource constraints, model optimization, and data privacy is necessary to ensure successful implementation. By combining human expertise with advanced AI and ML capabilities, organizations can build more intelligent, adaptable, and resilient edge computing environments capable of defending against increasingly sophisticated cyber threats.

## 11.6 BLOCKCHAIN APPLICATIONS FOR NETWORK SECURITY

### 11.6.1 Decentralized security solutions

In recent years, there has been growing interest in decentralized security solutions to address the unique challenges of edge computing environments. Unlike traditional centralized approaches, which rely on a single point of control and trust, decentralized security architectures distribute responsibility and authority among multiple nodes or entities within a network. Decentralization enables greater flexibility, fault tolerance, and resistance to single points of failure, thereby strengthening overall system resilience. Blockchain technology is one notable example of a decentralized security solution that has gained traction in various industries. By using consensus algorithms, cryptographic techniques, and immutable ledgers, blockchain ensures transparency, accountability, and security without relying on intermediaries or trusted third parties [13].

Blockchain-enabled security solutions can benefit edge computing environments by facilitating peer-to-peer communication, enforcing smart contracts, and verifying digital identities. For instance, decentralized identity platforms built on blockchain enable users to maintain ownership and control of their personal information, eliminating the need for central authorities or service providers to store sensitive data. This reduces the surface area for attacks and lowers the risk of large-scale data breaches. Similarly, decentralized autonomous organizations (DAOs) powered by smart contracts can facilitate collaborative defense strategies, shared threat intelligence, and collective bargaining power among participants in an edge computing ecosystem. DAOs allow self-governing communities to establish rules, enforce agreements, and manage resources transparently, promoting collaboration and mutual aid in responding to emerging security threats [14].

### 11.6.2 Smart contracts and immutable ledgers

Smart contracts and immutable ledgers are two core components of decentralized security solutions, offering novel ways to create trust, enforce rules, and verify transactions without relying on intermediaries. A smart contract is a self-executing program stored on a blockchain that automatically executes predefined actions when certain conditions are met. It acts as a digital agreement between parties, eliminating the need for manual intervention or validation by trusted third parties. By encoding business logic directly into code, smart contracts increase transparency, reduce ambiguity, and minimize disputes, thereby fostering trust and confidence among stakeholders. They can streamline processes, lower transaction costs, and accelerate time-to-market, ultimately benefiting both consumers and producers.

Immutable ledgers, meanwhile, serve as a tamper-proof record of all transactions executed within a decentralized system. Once written onto the ledger, data cannot be altered or deleted, creating a permanent and auditable history of events. This feature enhances data integrity, non-repudiation, and traceability, deterring fraudulent activity and bolstering accountability [15]. Together, smart contracts and immutable ledgers form the foundation of innovative security solutions designed to safeguard edge computing environments. Use cases include decentralized identity management, permissionless access control, and distributed file sharing. For instance, decentralized identity platforms leverage smart contracts to manage issuance, revocation, and verification of digital credentials, giving users full control over their personal information. Permissionless access control systems employ smart contracts to dynamically evaluate access requests and determine whether to grant or deny permissions based on predefined criteria, ensuring fine-grained, context-aware authorization policies. Distributed file sharing networks harness immutable ledgers to track changes, monitor access rights, and resolve conflicts, promoting transparency and fairness among peers. While smart contracts and immutable ledgers hold great promise for enhancing edge computing security, they also introduce new challenges. Programming errors, coding bugs, and logical flaws in smart contracts can lead to unintended consequences, security vulnerabilities, or financial losses. Furthermore, reconciling competing interests, establishing clear jurisdiction boundaries, and complying with regulations become more complicated in cross-border scenarios involving diverse actors and varying legal regimes [16]. Addressing these issues demands ongoing research, development, and refinement of best practices, tools, and methodologies for building secure, reliable, and compliant smart contracts and immutable ledgers. Nevertheless, embracing these technologies represents a strategic step forward toward realizing the vision of a safer, more open, and highly interconnected world enabled by edge computing.

## 11.7 PRIVACY-PRESERVING TECHNOLOGIES IN 6G

The advent of sixth-generation (6G) wireless networks promises unprecedented connectivity, speed, and capacity, opening up new avenues for economic growth, social progress, and technological innovation. At the same time, 6G raises concerns about privacy and data protection due to its pervasive sensing, ubiquitous monitoring, and massive data aggregation capabilities. To address these concerns, researchers are exploring privacy-preserving technologies specifically tailored for 6G networks [17].

Privacy-enhancing techniques under investigation include homomorphic encryption, differential privacy, secure multiparty computation, zero-knowledge proofs, and federated learning. Homomorphic encryption enables computations on encrypted data without decryption, preserving confidentiality during processing. Differential privacy adds carefully calibrated noise to statistical queries to obfuscate individual contributions while maintaining aggregate insights, preventing unwanted inferences about sensitive attributes. Secure multiparty computation permits joint calculations on private inputs without exposing underlying data, supporting cooperative analytics tasks without compromising confidentiality. Zero-knowledge proofs enable users to demonstrate possession of secret knowledge without revealing the actual content, enabling secure authentication and authorization mechanisms. Lastly, federated learning distributes model training across multiple nodes, retaining raw data locally, and transferring only learned parameters, thereby avoiding direct exposure of sensitive information [18].

Integrating these privacy-preserving technologies into 6G networks holds immense potential for balancing the tension between data utility and privacy, unlocking value creation while respecting individual rights. Applications span vertical sectors ranging from healthcare, finance, manufacturing, transportation, agriculture, entertainment, education, and beyond.

Healthcare providers can utilize privacy-preserving techniques to exchange medical records, coordinate care plans, and conduct clinical trials without violating patient confidentiality. Financial institutions can perform credit scoring, fraud detection, and investment analyses on protected customer data, enhancing services without infringing upon financial privacy. Manufacturers can share production secrets, quality metrics, and supply chain information confidentially, driving operational improvements and product innovations. Transport operators can collect granular mobility patterns, travel preferences, and location histories privately, informing urban planning initiatives and public policy decisions. Agricultural firms can extract valuable crop yield forecasts, soil condition assessments, and weather predictions from farmers' proprietary data, boosting productivity sustainably. Entertainment companies can generate custom playlists, recommend movies, and curate online experiences based on user tastes, behaviors, and preferences without disclosing intimate details. Educational institutes can personalize learning pathways, gauge student engagement levels, and measure teaching efficacies using anonymous feedback, encouraging continuous improvement and excellence [19].

In sum, privacy-preserving technologies represent a critical component of future 6G networks, striking a delicate equilibrium between data utilization and privacy protection. Embracing these advances paves the way for sustainable socioeconomic advancement grounded in responsible data management principles, fostering trust, inclusiveness, and equitable outcomes for all stakeholders.

## 11.8 THREAT INTELLIGENCE AND DETECTION IN 6G NETWORKS

Sixth-generation (6G) networks, characterized by ultrafast speeds, near-zero latencies, and hyperconnectivity, call for advanced threat intelligence and detection mechanisms to counteract ever-evolving cybersecurity threats. Harnessing AI and ML-driven analytics, 6G networks aim to anticipate, identify, and neutralize nefarious activities before causing harm, thereby safeguarding users' privacy, data integrity, and system reliability.

Threat intelligence involves gathering, synthesizing, and disseminating timely, accurate, and relevant information concerning potential or current cyber threats targeting 6G networks. Sources range from internal logs, external feeds, industry reports, dark web scanning, honey pots, and crowdsourced submissions. Integrating disparate sources fuels holistic situational awareness, guiding informed decision-making and prioritizing defensive investments.

Detection entails continuously scrutinizing 6G network traffic, configurations, and endpoints to discern abnormalities, anomalous trends, and malicious intent. Advanced AI/ML algorithms excel at pattern recognition, correlation mining, and outlier segmentation, unearthing subtle cues suggestive of lurking dangers. Anomaly-based methods compare observed behaviors against expected baselines derived from historical norms, spotlighting irregularities warranting deeper inspection. Misuse-based approaches match observed patterns with known signatures of past assaults, triggering alerts when matches exceed preset thresholds. Hybrid schemes blend elements from both categories, capitalizing on complementary strengths and synergies [20].

Upon detection, incident responders classify severity, scope, and impact, determining optimal remedial courses of action. Mitigation steps depend on scenario particulars but generally involve isolating affected segments, eradicating root causes, restoring functionality, fortifying defenses, and conducting post-incident reviews. Rapid containment limits collateral damage, expedites recovery, and reinforces consumer trust in 6G networks. In addition to conventional protective measures, 6G networks incorporate quantum-resistant cryptography, post-quantum encryption standards, and advanced key management protocols to harden infrastructure against futuristic quantum computers wielding devastating cracking abilities. Robust security orchestration, automation, and response (SOAR) platforms seamlessly integrate heterogeneous tools, technologies, and workflows, enhancing operational visibility, streamlining workflow execution, and slashing response times.

Overall, integrating cutting-edge threat intelligence and detection mechanisms constitutes a cornerstone of 6G networks' design philosophy, assuring safe, secure, and stable telecommunications environs amid burgeoning complexity and sophistication. Effectively tackling cyber threats necessitates vigilant preparedness, swift reaction, and iterative enhancement, cultivating a vibrant culture rooted in continual learning, active collaboration, and collective responsibility.

## 11.9 SECURING INTERNET OF THINGS (IOT) IN 6G

The advent of 6G technology is expected to bring about a new era of connectivity, with an unprecedented number of devices becoming connected to the internet. This proliferation of IoT devices will create numerous opportunities for businesses and individuals alike, but it also poses significant security challenges. To ensure the secure deployment of IoT devices in 6G networks, several measures must be taken. Firstly, there should be robust authentication mechanisms in place to verify the identity of devices connecting to the network. This can include multi-factor authentication methods, such as biometric or token-based systems [21].

Secondly, encryption techniques should be employed to protect data transmitted between devices and the network. End-to-end encryption can help prevent eavesdropping and tampering, ensuring that sensitive information remains confidential. Thirdly, regular software updates and patches are crucial to address vulnerabilities and maintain the security of IoT devices. Manufacturers should provide timely updates and encourage users to install them promptly. Fourthly, access control policies need to be implemented to limit the exposure of devices to potential threats. This includes segmenting networks and restricting device communication based on their roles and functions.

Lastly, monitoring and anomaly detection tools should be deployed to detect any suspicious activity or behavior patterns. Artificial intelligence and machine-learning algorithms can analyze vast amounts of data and identify unusual patterns, enabling quick response times and minimizing damage caused by cyber-attacks.

## 11.10 REGULATORY COMPLIANCE AND 6G SECURITY

### 11.10.1 Legal and regulatory frameworks for 6G security

As 6G networks become increasingly complex, legal and regulatory frameworks play a critical role in ensuring their security. Governments and international organizations must establish clear guidelines and regulations that define acceptable behaviors, practices, and responsibilities related to 6G security. These frameworks should cover various aspects, including privacy protection, data sovereignty, liability allocation, incident reporting, and interoperability standards [22]. Regulations should require manufacturers to implement specific security features and protocols to minimize risks associated with IoT devices. Additionally, they should mandate regular audits and assessments of 6G networks' security posture [23]. Penalties for non-compliance should be severe enough to discourage malicious actors from exploiting vulnerabilities.

International cooperation is essential for establishing consistent legal and regulatory frameworks across different regions. Collaborative efforts among governments, industry stakeholders, and standardization bodies can promote harmonized approaches to 6G security while avoiding conflicting requirements and fragmented markets. Moreover, public awareness campaigns and education initiatives should inform consumers about the importance of 6G security and empower them to make informed decisions when using connected devices. By fostering a culture of security awareness and responsibility, we can build a safer and more resilient 6G ecosystem. In summary, effective legal and regulatory frameworks for 6G security require comprehensive guidelines, strict enforcement mechanisms, international collaboration, and public engagement. Through coordinated efforts, we can create a secure and reliable 6G landscape that benefits society at large.

### 11.10.2 Compliance challenges and solutions

Achieving compliance with emerging 6G security regulations presents unique challenges for both private and public sector entities. Navigating complex regulatory landscapes, understanding evolving best practices, and allocating resources effectively can be difficult tasks. However, adopting a strategic approach to compliance can help mitigate these challenges.

Firstly, organizations should establish dedicated teams responsible for managing 6G security compliance. These teams should have expertise in relevant areas, such as cybersecurity, risk management, and legal matters. They should work closely with other departments to develop holistic strategies that align with business objectives and regulatory requirements. Secondly, continuous training and upskilling programs can help employees stay abreast of changing regulations and threat environments. Providing staff with adequate knowledge and skills enables them to contribute effectively towards maintaining compliance and enhancing overall security posture [24]. Thirdly, leveraging advanced technologies like artificial intelligence, machine learning, and automation can streamline compliance processes and reduce manual effort. For instance, AI-powered analytics platforms can monitor networks in real time, detect anomalies, and generate reports automatically, simplifying audit trails and reducing human error. Finally, engaging third-party experts and consultants can offer valuable insights into best practices and emerging trends in 6G security compliance. External advisors can help evaluate existing controls, recommend improvements, and support implementation efforts, thereby accelerating progress towards full compliance.

In conclusion, addressing compliance challenges requires a proactive and integrated approach involving specialized personnel, ongoing training, technological innovation, and external expertise. Embracing this strategy empowers organizations to navigate complex regulatory landscapes confidently and strengthen their overall security posture in preparation for the 6G era.

## 11.11 DATA GOVERNANCE AND COMPLIANCE IN 6G NETWORKS

### 11.11.1 Managing and protecting sensitive data

In the context of 6G networks, managing and protecting sensitive data becomes paramount due to increased connectivity and data volumes. To ensure data integrity, confidentiality, and availability, organizations must adopt robust data management and protection strategies. Implementing strong access controls is vital for limiting unauthorized access to sensitive data. Role-based access control (RBAC), attribute-based access control (ABAC), and mandatory access control (MAC) are some examples of effective access control models that can help enforce granular permissions based on user attributes, roles, or security levels.

Encryption plays a crucial role in safeguarding data during transmission and storage. Advanced cryptographic techniques, such as homomorphic encryption and zero-knowledge proofs, enable computations on encrypted data without decrypting it, thus preserving privacy and security. Data masking, pseudonymization, and anonymization techniques can further enhance data protection by obscuring personally identifiable information (PII). These methods allow legitimate use cases while preventing unauthorized disclosure or misuse of sensitive data. Additionally, deploying data loss prevention (DLP) solutions can help detect and prevent accidental or intentional data leaks. DLP tools monitor network traffic, inspect content, and apply policies consistently to prevent exfiltration attempts [25].

Organizations should regularly review and update their data retention policies to strike a balance between operational needs and privacy concerns. Implementing automated deletion schedules ensures that outdated or unnecessary data does not accumulate over time, reducing the attack surface area.

Lastly, educating employees about proper data handling procedures and raising awareness around social engineering tactics helps foster a culture of data stewardship within the organization. Regular phishing simulations, interactive workshops, and targeted trainings can equip staff members with the necessary knowledge and skills to manage and protect sensitive data effectively. In summary, successful data management and protection in 6G networks necessitate a combination of technical safeguards, organizational policies, employee education, and periodic reviews to maintain an optimal level of security and privacy. Adhering to these principles contributes to building trust and confidence among customers, partners, and regulators, ultimately driving long-term success in the digital age.

### 11.11.2 Compliance with global data protection regulations

With increasing global data protection regulations, such as GDPR, CCPA, LGPD, and others, achieving compliance has become a top priority for organizations operating in multiple jurisdictions. Meeting stringent data protection requirements involves understanding regional differences, adapting internal policies, and implementing appropriate technical measures. A key aspect of complying with global data protection regulations is conducting thorough gap analyses to identify discrepancies between current practices and desired outcomes. Based on these findings, tailored action plans can be developed to close gaps and achieve compliance incrementally.

Standardization and centralization of data protection policies can facilitate efficient management of diverse regulatory demands. Developing common frameworks and templates allows for easier adaptation to local requirements while maintaining consistency throughout the enterprise. Collaborating with trusted partners and service providers who share similar values and commitment to data protection can alleviate the burden of navigating complex regulatory landscapes. Joint ventures, partnership agreements, and vendor evaluations should incorporate rigorous assessments of each party's ability to meet compliance obligations. Leveraging cloud services can offer advantages in terms of scalability, flexibility, and built-in compliance capabilities. Many cloud providers invest heavily in meeting various data protection requirements, which may benefit smaller enterprises seeking cost-effective ways to comply with multiple regulations simultaneously.

Monitoring and reporting mechanisms should be established to demonstrate accountability and continuously improve data protection practices.

Periodic audits, self-assessments, and benchmarking activities can reveal areas requiring improvement and help prioritize investments accordingly. Ultimately, staying updated on emerging trends and developments in global data protection regulations requires ongoing vigilance and adaptability. Organizations that embrace this mindset and actively seek innovative solutions will position themselves favorably in today's rapidly evolving digital environment.

## 11.12 HUMAN-CENTRIC SECURITY IN 6G

As 6G networks emerge, human-centric security gains prominence as a core principle guiding design, development, and operation. Human-centric security aims to prioritize individual privacy, autonomy, and wellbeing while providing seamless and intuitive experiences for end-users. One important facet of human-centric security is incorporating user-friendly consent mechanisms that allow people to easily understand and control how their personal data is used. This might involve designing graphical interfaces that convey complex concepts in simple language, employing gamification elements to incentivize participation, or utilizing voice recognition technologies to create natural interactions.

Another component of human-centric security is recognizing and accommodating diverse user groups and scenarios. Customizable security settings, inclusive design, and accessible features cater to varying skill levels, cultural backgrounds, and physical abilities, promoting equitable access and inclusiveness. Privacy-preserving technologies, such as differential privacy, federated learning, and homomorphic encryption, represent additional pillars of human-centric security in 6G. These innovations enable computation on encrypted data without exposing underlying sensitive information, facilitating collaborative analysis and decision-making while respecting individual privacy rights.

Security incidents affecting mental health and emotional wellbeing should also be addressed through human-centric approaches. Proactively identifying stressors, offering psychological support, and developing empathetic responses to adversities can cultivate positive relationships between users and technology, leading to improved overall satisfaction and loyalty. Education and awareness remain crucial components of human-centric security. Engaging users in continuous learning journeys, presenting relatable case studies, and integrating experiential learning methods empower individuals to make informed choices regarding their online safety and privacy [26].

In essence, human-centric security represents a paradigm shift towards placing people at the heart of 6G networks' design, construction, and maintenance. Combining technical ingenuity, ethical considerations, and socioemotional factors, human-centric security promises to engender greater trust, agency, and enjoyment in our ever-evolving digital world.

## 11.13 CHALLENGES AND LIMITATIONS OF 6G SECURITY

The development of 6G technology is expected to bring about significant advancements in terms of speed, connectivity, and overall network performance. However, with these improvements come new security challenges and limitations that must be addressed. One major challenge is the increased attack surface due to the vast number of connected devices and networks. This makes it easier for malicious actors to exploit vulnerabilities and launch attacks on the system. Additionally, as data rates increase, there is a greater risk of eavesdropping, interception, and manipulation of sensitive information transmitted over the air-waves. Another limitation of 6G security is the need for advanced cryptographic techniques that can keep up with the high-speed transmission rates. Traditional encryption methods may not be able to handle the sheer volume of data being transferred, leading to potential security breaches. Moreover, the use of AI and machine-learning algorithms in 6G systems introduces new threats, such as adversarial machine-learning attacks where an attacker intentionally manipulates input data to mislead AI models.

Furthermore, ensuring user privacy in 6G will also pose unique challenges. With more sophisticated tracking technologies like location-based services and facial recognition becoming prevalent, users' personal information could easily fall into the wrong hands. Therefore, robust privacy protection mechanisms should be implemented to prevent unauthorized access or disclosure of private data. Lastly, regulatory compliance and international cooperation are essential components of 6G security. As 6G networks transcend national borders, global collaboration between governments, industry players, and standardization bodies becomes crucial in establishing consistent security policies and regulations. Without proper coordination, disparities in legal frameworks across countries could hinder progress towards achieving secure 6G infrastructure.

## 11.14 COLLABORATION AND INFORMATION SHARING IN 6G SECURITY

### 11.14.1 Industry collaboration for security standards

To address the complex security challenges posed by 6G technology, strong industry collaboration is necessary to establish effective security standards. By working together, various stakeholders from academia, government agencies, telecom operators, device manufacturers, and software developers can pool their resources and expertise to develop comprehensive security measures that protect both the infrastructure and end-users. Collaborative efforts among these entities enable the sharing of best practices, identification of common threats, and creation of uniform security protocols. For instance, through joint research initiatives, organizations can collaborate on developing innovative cryptographic techniques capable of securing high-speed data transmissions while maintaining low latency requirements. Furthermore, they can work together to define security specifications for emerging technologies such as edge computing, IoT devices, and AI-powered applications, thereby ensuring seamless integration within the broader 6G ecosystem.

Industry alliances and consortiums play a critical role in driving this collaboration. These groups provide platforms for members to engage in discussions around key issues, exchange knowledge, and align their strategies towards building secure 6G networks. Examples include the 5G Public Private Partnership (5G PPP), which has been instrumental in shaping European Union (EU) policy regarding 5G security, and the Next G Alliance, established by the USA to foster innovation and leadership in 6G. Moreover, cross-border partnerships are vital in creating globally accepted security standards that facilitate interoperability and consistency across different regions. International organizations like the International Telecommunication Union (ITU) and the Third Generation Partnership Project (3GPP) serve as forums for nations to discuss and agree upon common guidelines and specifications related to 6G security.

In conclusion, fostering industry collaboration for security standards in 6G is crucial in addressing the intricate challenges associated with this next-generation technology. Through collective efforts, stakeholders can create a secure and trustworthy environment that benefits businesses and consumers alike, ultimately paving the way for widespread adoption and growth of 6G worldwide.

### 11.14.2 Information sharing for collective security

An essential aspect of ensuring robust 6G security lies in promoting information sharing for collective defense. By enabling open communication channels among stakeholders, including telecom operators, vendors, regulators, and researchers, valuable insights can be exchanged regarding threat intelligence, security incidents, and best practices. Such collaboration allows the entire ecosystem to benefit from shared experiences and learnings, thus strengthening overall resilience against cyberattacks. One approach to facilitating information sharing is through establishment of trusted platforms or communities dedicated to disseminating real-time updates on emerging threats and vulnerabilities. These platforms can help streamline the process of reporting and responding to security incidents, allowing quicker mitigation actions and reducing potential damage. They can also encourage collaboration on developing countermeasures, such as patches or configuration changes, that can be widely adopted across the industry.

Additionally, implementing automated information exchange mechanisms can further enhance the efficiency and effectiveness of collective security efforts. AI and ML tools can analyze large volumes of data generated by diverse sources, identify patterns indicative of suspicious activities, and alert relevant parties accordingly. This proactive approach enables faster response times and minimizes the impact of targeted attacks. International cooperation plays a critical role in fostering a culture of information sharing for 6G security. Cross-border collaboration agreements between governments, regulatory authorities, and industry associations can facilitate the exchange of sensitive data without compromising national security interests. Standardized data formats and APIs can ensure compatibility and ease of integration between different systems, making it simpler for participants to contribute and consume shared information. However, it is important to balance the benefits of information sharing with appropriate safeguards to protect confidentiality, integrity, and availability of the data exchanged. Implementing stringent access controls, encryption, and anonymization techniques can minimize risks associated with data breaches or unauthorized disclosures. Regular audits and monitoring of information-sharing platforms can help maintain security hygiene and detect any anomalies that might indicate potential insider threats or external attacks. In summary, information sharing for collective security is a powerful strategy to bolster 6G network defenses. By encouraging transparent communication, leveraging advanced analytics, and fostering international collaboration, the industry can build a stronger and more resilient security posture capable of addressing evolving threats in the rapidly changing landscape of 6G technology.

## 11.15 FUTURE TRENDS IN NETWORK SECURITY AND DATA PRIVACY IN 6G

In the era of 6G, we can expect several future trends in network security and data privacy aimed at addressing the increasing complexity and scale of cybersecurity threats. Some notable developments include the following:

**Quantum-resistant cryptography:** As quantum computers become more powerful, traditional encryption algorithms may become vulnerable to attacks.

Thus, transitioning to quantum-resistant cryptographic techniques will be crucial in protecting sensitive information transmitted over 6G networks. Researchers are already exploring novel approaches based on lattice theory, multivariate polynomials, and code-based cryptography to ensure long-term security.

**Zero-trust architecture:** Adopting zero-trust principles will gain traction in 6G environments, focusing on continuous verification and authentication of every user, device, and application requesting access to resources. This model reduces reliance on perimeter defenses and focuses instead on verifying identities, assessing risks, and enforcing granular access control policies throughout the network.

**Sensor-driven threat detection:** Leveraging advances in sensor technologies, 6G networks can incorporate distributed sensing capabilities to monitor network behavior and detect unusual patterns that might indicate security breaches or malicious activities. Real-time analysis of contextual data from multiple sensors can improve accuracy and speed in identifying threats, enabling prompt responses before substantial damage occurs.

**Privacy-preserving techniques:** Enhanced data privacy will be achieved through innovations such as homomorphic encryption, differential privacy, and secure multiparty computation. These techniques allow processing encrypted data directly, preventing exposure of raw information during computations while still delivering meaningful results.

**AI-assisted security management:** Integrating AI and ML algorithms in network management systems can significantly improve security operations by automating tasks like threat detection, incident response, and remediation. Intelligent security bots powered by natural language processing and computer vision can interact with human operators, providing actionable recommendations and simplified workflows for managing complex security scenarios.

**Blockchain-enabled decentralized identity:** Utilizing blockchain technology can offer secure, tamper-proof digital identity solutions for individuals, devices, and applications in 6G networks. Decentralized identity systems empower users to manage their own credentials, eliminating single points of failure and enhancing privacy protections.

**Legal and ethical frameworks:** Establishing clear legal and ethical guidelines for handling data and preserving user privacy rights will be paramount in the age of 6G. Governments, industry leaders, and civil society organizations must work together to craft balanced policies that promote innovation while safeguarding individual liberties and societal values. By embracing these future trends, the 6G community can effectively tackle emerging security challenges and ensure a safe, reliable, and trustworthy networking experience for everyone involved.

## 11.16 CONCLUSION: NAVIGATING THE IMPACTS AND CHALLENGES OF 6G SECURITY

In conclusion, navigating the impacts and challenges of 6G security requires concerted effort from all stakeholders – governments, industries, academic institutions, and consumers. Embracing cutting-edge technological advancements brings immense opportunities for enhanced connectivity, productivity, and socioeconomic prosperity; however, it also presents unprecedented security concerns that demand timely attention and resolution.

Encourage close collaboration amongst various industry players, including telecom operators, equipment vendors, software providers, and research institutes, to develop cohesive security standards and share best practices. Facilitate open communication channels for sharing threat intelligence, security incidents, and lessons learned across the ecosystem to enhance overall resilience against cyberattacks. Drive investments in research and development of novel security technologies, such as quantum-resistant cryptography, AI-assisted security management, and decentralized identity systems, to stay ahead of evolving threats. Develop robust yet adaptable regulatory frameworks that strike an optimal balance between fostering innovation and ensuring adequate security measures are in place. Educate and equip consumers with the knowledge and tools required to make informed decisions about their digital footprint and online safety. By adhering to these guiding principles, we can collectively overcome the challenges posed by 6G security and unlock its transformative potential for businesses, societies, and individuals alike. Together, let us shape a secure, inclusive, and thriving digital future built upon the foundation of 6G technology.

## REFERENCES


1. Hakeem, S., Hussein, H., Kim, H., 2022. Security Requirements and Challenges of 6G Technologies and Applications. *Sensors (Basel, Switzerland)*, 22. https://doi.org/10.3390/s22051969.
2. Je, D., Jung, J., Choi, S., 2021. Toward 6G Security: Technology Trends, Threats, and Solutions. *IEEE Communications Standards Magazine*, 5, pp. 64–71. https://doi.org/10.1109/MCOMSTD.011.2000065.
3. Liu, J., Yu, Y., Li, K., Gao, L., 2021. Post-Quantum Secure Ring Signatures for Security and Privacy in the Cybertwin-Driven 6G. *IEEE Internet of Things Journal*, 8, pp. 16290–16300. https://doi.org/10.1109/jiot.2021.3102385.
4. Xu, Q., Su, Z., Li, R., 2022. Security and Privacy in Artificial Intelligence-Enabled 6G. *IEEE Network*, 36, pp. 188–196. https://doi.org/10.1109/MNET.117.2100730.
5. Soleymani, S., Goudarzi, S., Anisi, M., Movahedi, Z., Jindal, A., Kama, N., 2022. PACMAN: Privacy-Preserving Authentication Scheme for Managing Cybertwin-Based 6G Networking. *IEEE Transactions on Industrial Informatics*, 18, pp. 4902–4911. https://doi.org/10.1109/TII.2021.3121505.
6. Porambage, P., Gür, G., Osorio, D., Liyanage, M., Gurtov, A., Ylianttila, M., 2021. The Roadmap to 6G Security and Privacy. *IEEE Open Journal of*

*the Communications Society*, 2, pp. 1094–1122. https://doi.org/10.1109/OJCOMS.2021.3078081.

7. Porambage, P., Gür, G., Osorio, D., Livanage, M., Ylianttila, M., 2021. 6G Security Challenges and Potential Solutions. *2021 Joint European Conference on Networks and Communications 6G Summit (EuCNC/6G Summit)*, pp. 622–627. https://doi.org/10.1109/EuCNC/6GSummit51104.2021.9482609.
8. Ylianttila, M., Kantola, R., Gurtov, A., Mucchi, L., Oppermann, I., Yan, Z., Nguyen, T., Liu, F., Hewa, T., Liyanage, M., Ijaz, A., Partala, J., Abbas, R., Hecker, A., Jayousi, S., Martinelli, A., Caputo, S., Bechtold, J., Morales, I., Stoica, A., Abreu, G., Shahabuddin, S., Panayirci, E., Haas, H., Kumar, T., Ozparlak, B., Roning, J., 2020. 6G White Paper: Research Challenges for Trust, Security and Privacy. *arXiv*, abs/2004.11665.
9. Wang, M., Zhu, T., Zhang, T., Zhang, J., Yu, S., Zhou, W., 2020. Security and Privacy in 6G Networks: New Areas and New Challenges. *Digital Communications and Networks*, 6, pp. 281–291. https://doi.org/10.1016/j.dcan.2020.07.003.
10. Shirwaikar, R., Faisal, A., Singh, A., Shanbhag, D., 2021. A Review on Privacy and Security in 6G Networks. *2021 International Conference on Forensics, Analytics, Big Data, Security (FABS)*, 1, pp. 1–6. https://doi.org/10.1109/FABS52071.2021.9702654.
11. Osorio, D., Ahmad, I., Sánchez, J., Gurtov, A., Scholliers, J., Kutila, M., Porambage, P., 2022. Towards 6G-Enabled Internet of Vehicles: Security and Privacy. *IEEE Open Journal of the Communications Society*, 3, pp. 82–105. https://doi.org/10.1109/ojcoms.2022.3143098.
12. Chaudhry, S., Irshad, A., Khan, M., Khan, S., Nosheen, S., Alzubi, A., Zikria, Y., 2023. A Lightweight Authentication Scheme for 6G-IoT Enabled Maritime Transport System. *IEEE Transactions on Intelligent Transportation Systems*, 24, pp. 2401–2410. https://doi.org/10.1109/TITS.2021.3134643.
13. Rajendra Kumar, R. C. Singh, Rohit Khokher, Vishal Jain, 2023. The Catalyst for Clean and Green Energy Using Blockchain Technology. In *Modeling for Sustainable Development: A Multidisciplinary Approach*, Nova Science Publishers Inc., pp. 23–39.
14. Stergiou, C., Psannis, K., Gupta, B., 2021. IoT-Based Big Data Secure Management in the Fog Over a 6G Wireless Network. *IEEE Internet of Things Journal*, 8, pp. 5164–5171. https://doi.org/10.1109/JIOT.2020.3033131.
15. Nguyen, V., Lin, P., Cheng, B., Hwang, R., Lin, Y., 2021. Security and Privacy for 6G: A Survey on Prospective Technologies and Challenges. *IEEE Communications Surveys & Tutorials*, 23, pp. 2384–2428. https://doi.org/10.1109/COMST.2021.3108618.
16. Garzon, S., Yildiz, H., Küpper, A., 2021. Decentralized Identifiers and Self-Sovereign Identity in 6G. *IEEE Network*, 36, pp. 142–148. https://doi.org/10.1109/MNET.009.2100736.
17. Mao, B., Kawamoto, Y., Kato, N., 2020. AI-Based Joint Optimization of QoS and Security for 6G Energy Harvesting Internet of Things. *IEEE Internet of Things Journal*, 7, pp. 7032–7042. https://doi.org/10.1109/jiot.2020.2982417.

18. Yang, P., Xiao, Y., Xiao, M., Li, S., 2019. 6G Wireless Communications: Vision and Potential Techniques. *IEEE Network*, 33, pp. 70–75. https://doi.org/10.1109/MNET.2019.1800418.
19. Mathew, A., 2021. Edge Computing and Its Convergence with Blockchain in 6G: Security Challenges. *International Journal of Computer Science and Mobile Computing*, 10, pp. 8–14. https://doi.org/10.47760/ijcsmc.2021.v10i08.002.
20. Gorrepati, U., Zavarsky, P., Ruhl, R., 2021. Privacy Protection in LTE and 5G Networks. *2021 2nd International Conference on Secure Cyber Computing and Communications (ICSCCC)*, pp. 382–387. https://doi.org/10.1109/ICSCCC51823.2021.9478109.
21. Ayaz, F., Sheng, Z., Tian, D., Nekovee, M., Saeed, N., 2022. Blockchain-Empowered AI for 6G-Enabled Internet of Vehicles. *Electronics*, 11. https://doi.org/10.3390/electronics11203339.
22. Ahmad, I., Kumar, T., Liyanage, M., Okwuibe, J., Ylianttila, M., Gurtov, A., 2018. Overview of 5G Security Challenges and Solutions. *IEEE Communications Standards Magazine*, 2, pp. 36–43. https://doi.org/10.1109/MCOMSTD.2018.1700063.
23. Lin, J., Srivastava, G., Zhang, Y., Djenouri, Y., Aloqaily, M., 2021. Privacy-Preserving Multiobjective Sanitization Model in 6G IoT Environments. *IEEE Internet of Things Journal*, 8, pp. 5340–5349. https://doi.org/10.1109/JIOT.2020.3032896.
24. Dang, S., Amin, O., Shihada, B., Alouini, M., 2019. What Should 6G Be? *Nature Electronics*, 3, pp. 20–29. https://doi.org/10.1038/s41928-019-0355-6.
25. Nguyen, T., Tran, N., Lovén, L., Partala, J., Kechadi, M., Pirttikangas, S., 2020. Privacy-Aware Blockchain Innovation for 6G: Challenges and Opportunities. *2020 2nd 6G Wireless Summit (6G SUMMIT)*, pp. 1–5. https://doi.org/10.1109/6GSUMMIT49458.2020.9083832.
26. Kumar, R., 2005. *Human Computer Interaction*, Firewall Media.

Chapter 12

# 6G networks

## Pioneering advanced communication techniques for call centers and beyond

*Danish Ather, Rajneesh Kler, Tanveer Baig Z, G. Prakash Babu, Ajay Rastogi, and Nazia Ahmed*

### 12.1 INTRODUCTION

An attempt to untangle what makes 6G so unique, and to have a closer look into the technology's potential implementation areas is the core aim of the current work. It is done so by pursuing a broad overview of technical advancement 6G and its advantages and disadvantages, which are emergent after the advancement. This leads to the fact that when the world is undergoing the telecommunications revolution, the next one – 6G – is properly designed to be a successor of the networks starting to appear worldwide. This imagining of substandard of connectivity is anticipated to give least possible peak data rates that are several orders greater than 5G, together with the ability to support the novel services and deployment areas and provide unprecedented levels of reliability and latency as low as the sub-millisecond range. The 6G capability is not just performing a miracle in terms of speed but it also enables the public to have a fully interconnected digital world in which the barrier between digital and physical reality become so narrow that people feel them being interlinked. Apart from elevating the performance of mobile internet, which is the first service anticipated with the 6G technology, it is also expected to be the key to the symbiotic integration of emerging technologies like the Internet of Things (IoT), holophonic telepresence, autonomous systems, artificial intelligence, and so on. Guiding us to the age of near-absolute connectivity which means enabling an intelligence environment all around us and making our network services as if they do not exist in everyday life. The 6G connection is supposed to be integrated at the societal level and that is why it is expected to witness more innovations in the digital world to strengthen sectors like smart cities, eHealth, and Industry 5.0 [1–4].

Evidently, the 6G networks are the most significant in the history of mobile networks' technology to extend the edge of the communication platforms. By using the exceptional high-frequency ranges and through unique architectures, for example, beamforming, and massively increasing the number of MIMO antenna elements, the 6G networks are expected to


DOI: 10.1201/9781003583127-12

give not only fast data rates but also effective and secure methods of interaction. The 6G with power to cover a large amount of data will enable the transmission of data in real times, which will support cloud-based AI services and smart gadgets, therefore driving how the data is consumed and processed. Among industrial and business community, the 6G will raise a jumping board for novel business models and changing service ecosystems. Because of the speed of communications, information exchange will be fast, the remote operation and management of machines will be evident with near-zero delays, and advanced call center operations will lead to better customer service.

Besides, online customers can apply for all kinds of services by means of 6G technology to give them personalized and interactive video support, which will increase the effectiveness of customer service and customer satisfaction. What can be mostly emphasized while implementing 6G networks in society is that the innovation will lead to outstanding gains in various aspects of our economic life. It is worth noting that in terms of education, for example, 6G could help in the creation of an immersive learning environment adding virtual and augmented reality in the process. On another instance, environmental monitoring would be the prime benefactor of 6G through its fibrous network of sensors generating the most informative evidence of resource management and climate change issues [5, 6].

Thus, the chapter targets institutions, industries, academia, and policymaking sectors as stakeholders to provide a comprehensive understanding and exploitation potential of 6G for societal development purposes. Among the objectives is also a thorough dive into 6G technical advancements. A detailed examination of proposed architectural modification, frequency bands and technological enablers that characterize 6G is carried out. A scrutiny of the AI merging and network management as well as the service delivery, which is among the most defining features within the 6G network is also discussed. The chapter also proposes training and working with various stakeholders for developing a 6G future. The research categorizes potential applications that will be the focus of the discussion on the need for an integrated approach in response to the future opportunities and challenges. It also hopes to promote a debate on the normative and regulatory issues associated with implementing the 6G networks. Research is presented in a detailed manner, covering both qualitative and quantitative techniques. At last, the chapter predominately concentrates on technical aspects and use cases, but it also tackles the socio-economic impacts of 6G [7, 8].

## 12.2 EVOLUTION OF COMMUNICATION NETWORKS: FROM 1G TO 6G

The evolution is summarized in Figure 12.1.

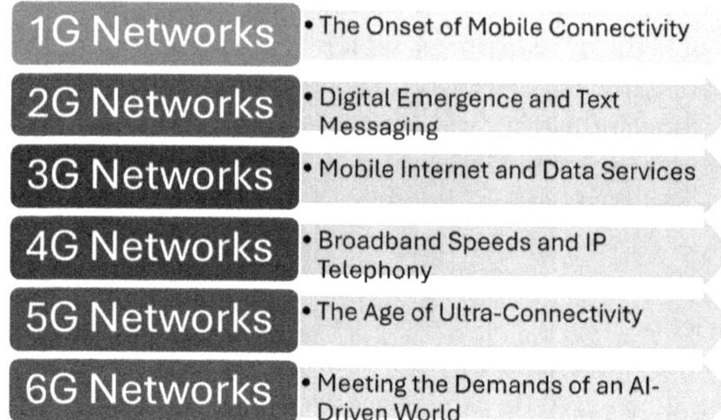


*Figure 12.1* Evolution of communication networks: from 1G to 6G.

### 12.2.1 The onset of mobile connectivity

First-generation (1G) period began in the 1980s with mono frequency generation, comprised of analogue systems such as the advanced mobile phone system (AMPS). During this time, there were car phones and giant, clumsy handsets. They were mostly used by a certain social category of people who were high-cost payers, and the network coverage was limited. Since 1G was analog, it was more likely that calls could be wiretapped and often the conversations were not so clear and broke off many times because there were few channels, and no connection ran between the cell towers while the calls were in progress.

### 12.2.2 Digital emergence and text messaging

The second-generation network, the GSM was a key system, which was launched in the early 1990s. It was the one that pioneered the migration from analog modulation, which was the initial signal generation technology of mobile phones, to digital modulation. This time, it had change encryptions, SMS text messages, and voice mail that had a great effect on privacy as well as the addition of more functions. In fact, although there was narrowband data, it could still send texts but not multimedia content because users' growing appetite was not met as well [9].

### 12.2.3 Mobile internet and data services

3G networks was the time when International Telecommunication Union (ITU) was putting effort into the standards for IMT-2000, which played an important role in unifying the 2G systems, depleting the fragmentation of the 2G networks and providing a basis for the delivery of broadband content. It increased internet speeds by several factors and made live video calling and data-intensive mobile services available. This was the beginning for Internet-connected smartphones. Certainly, gapless go-ahead of the theory was put into practice by intermittent coverage and due to different reasons the earlier 3G network was low quality, so it did not meet the large-scale high data speed requirement [10].

### 12.2.4 Broadband speeds and IP telephony

The 4G era was launched in the late 2000s, 4G (IP telephony) standards like LTE were designed for data and not for voice, with IP telephony (VoLTE) introduced later. This era of technology, thus, was all about high-speed mobile broadband networks, which via huge bit rate leap enabled services like HD video streaming, high-quality video calls, and online gaming. Nonetheless, the data consumption increase has led to a sharp increase in the demand for much higher speeds and lower latencies, things that 4G cannot handle as far as the network congestion and latency are concerned.

### 12.2.5 The age of ultra-connectivity

Subsequently the roll-out of 5G was entrusted to address the emerging needs of a society that is becoming deeply connected with a raised number of devices having increased throughput and URLLC. The platform formation was intended to be multipurpose to be applied to a range of applications, from enhanced mobile broadband (eMBB) up to mission critical communication. True as it seems, the 5G broadband deployment is slow, depending on infrastructural difficulties and economic issues. These obstacles have impeded the full achievement of the technology's capabilities [11–13].

### 12.2.6 The promise of pervasive computing

The 6G network designing is envisaged as a forward-thinking adaptive response to currently unimagined needs of the changes in society and technology in the future. Extending over and beyond connectivity, 6G means to offer pervasive computing and communication with almost no lags, thus making way for an era where applications that demand extreme reliability, like surgeries performed remotely, can be effectively carried out, and enhancing the feeling of oneness within extended reality (XR) environments [14]. The 6G network is anticipated to operate with the signal rates from sub-THz frequencies that are capable to transform the huge bandwidth along with the propagation challenges requiring both network structure modification and antenna technology development.

### 12.2.7 Driving a real-time AI world

The world is standing before a situation where not only data workloads become a first rate task for the network but also they must be supported by the systems that will allow nearly-real-time decision-making capabilities. Numerous applications of AI-driven 6G networks are visualized with not only support of these applications but also utilizing AI for optimization for network, predictive maintenance, and personalize services meanwhile letting the network environment be parallel to a self-learning and self-adapting network, which could greatly improve the efficiency and users' experience.

### 12.2.8 Motivation

The aim of providing 6G networks indicates that a stage of moving up in the arena of communication technologies is being realized, which is going to be the basis of rethinking the landscape of communication for call centers among other things. Yet, this evolution is driven by the rising needs of citizens and businesses for predominantly faster, more reliable, and sacred communication channels capable of carrying loads of data traffic. This arises from the constantly changing and without boundaries requirements of digital societies. The implication of the digital era for industries as well as everyday activities has led to the outdatedness of the current networks becoming clearer and, therefore, a new advanced network is much needed. The 6G is supposed to be much more than an improvement over 5G and provides some of the fundamental enhances data speed, latency, and brings new possibilities with it. However, the future of 6G lies in the hidden urge of escaping the limitations of existing technologies towards the uncharted territories of technological creativity while fostering the utilization of emerging technologies like artificial intelligence, Internet of Things (IoT), and augmented reality. Call centers now have access to various advanced features such as real-time data analytics, VR-based training, and AI-powered platforms for customer interactions, which serve as the key to service delivery. The call centers are not the only domain area that 6G is intending to cover. It also aims to automate smart cities, autopilot vehicles, telemedicine, and so forth. This would lead us to think about a future where digital interactions are more engaging, simple, and sure.

This determination is enhanced with the aim to bridge the digital divide by creating a fair and just access to broadband connectivity for different regions and socioeconomic groups across the nation. 6G's ability to attain wide coverage, including in far-flung areas and not only the ones that are served, but it is also an indication of how 6G will be able to usher in equal opportunities for all and steer away the digital divide which prevails in the current era. The goal of 6G network creation is no longer just the technological development but also social capacitation, whose essence is to construct the connected society where everyone has equal opportunity. While there is ample potential for 6G to shape our future, the way to a perfect implementation difficult since major investment in infrastructure is required to make 6G a reality. Even though these obstacles do exist, the mutual desire of all the stakeholders from the business side, the academia, and the government, together, reinforces the collective vision of utilizing 6G as a basis for ramping up a new age of communication, eradicating the bounds of the past and doing unimaginably great things with the ones in the time to come.

## 12.3 THEORETICAL FOUNDATIONS OF 6G NETWORKS

### 12.3.1 Fundamental principles of 6G technologies

The telecommunication network of the future, 6G, is conceived, not only to leverage and expand on what was achieved by its previous generations, but also to respond to the ever-growing need for connectivity in a highly interconnected world. Ignoring the presentation of theoretical basis of 6G, this means creating an everywhere, no-misconception, and exceptionally large networking infrastructure. This entails the applications of networks not only throughout the media, business, and environment but also, integrating networks into almost every part of daily lives through the overall communication of everyone on Earth. This can be visualised in Figure 6.2.

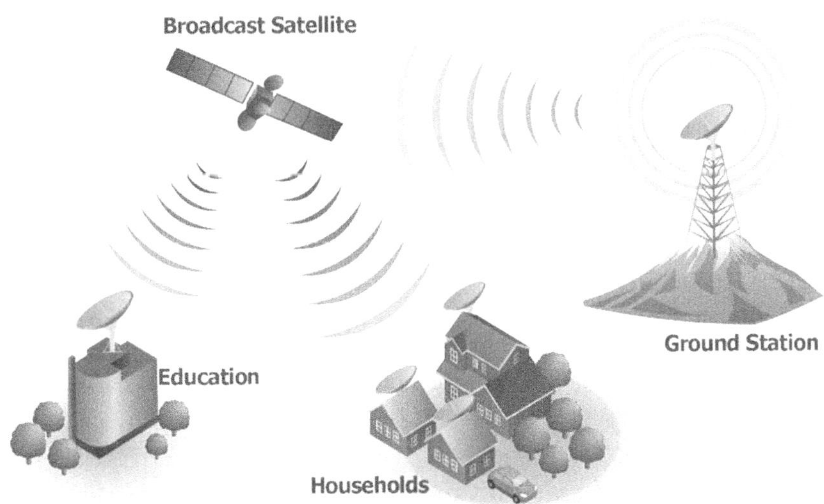


*Figure 12.2* A schematic representation of a satellite broadcasting system.

The figure illustrates a signal transmission from an orbiting satellite to the ground station. The signalling frames are communicated from the ground station towers and simultaneously accepted by the dishes linked to schools and residential quarters, which are employed for multiple functions [15, 16].

### 12.3.2 THz-band frequencies' exploitation

One of the most important improvements which supports 6G is investigating the terahertz (THz) range, home to the frequency of the range from 0.1 to 10 THz. This spectrum is like a wide territory ranging from microwave to the infrared regions, possessing many untapped portions that can be used for transmitting terabits-per-second data rates expected for 6G technology. Although the THz frequency might bring some problems, for example, it faces very high propagation loss and is very sensitive to the atmospheric conditions. Mastering these requires creativity and new materials like plasmonic for antennas and new modulation schemes, which will make terahertz communications technologies.

### 12.3.3 AI integration and machine learning

At the level of the networks, AI and machine learning will be needed at a great scale than before due to the complexity and dynamics of 6G networks. The AI algorithms will be the main driver in key areas like network operations, resource allocation, and security. They are expected to employ the network on patterns of usage and adapt real time for usage changes and mobility of the devices. Likewise, AI will help in predictive maintenance through data analysis and will forecast and prevent network issues that keep the service down.

### 12.3.4 Utilizing the powerful abilities of big data analytics

Big data analytics, which aim to integrate data that can be analysed and from the massive volumes of data, meaningful conclusions are produced for network users and IoT devices. It will also provide an avenue for the establishment of more complex technologies like smart cities' real-time environmental monitoring, traffic management, and provision of timely and targeted healthcare services. Big data analytics similarly will be instrumental in focusing on the protection of the network discovering anomalies, which might be indications of cyber threats.

### 12.3.5 Advanced solution of slicing and virtualization

Network slicing will go through a refinement as 6G comes into existence and it will provide more user-defined virtual networks that are dynamic and optimized in accordance with specific criteria such as latency, bandwidth, security, and reliability. Having that will guarantee a variety of different types of service, from the vehicle networks that require the low-latency, ultra-reliability, to the massive IoT deployments, that can operate at the same infrastructure on the same physical network layer. Techniques of virtualization are improved to make these slices to be distinct, flexible, and adjustable through real-time usage.

### 12.3.6 Embracing edge computing

Edge computing is an essential element in 6G networks as it is moving the data processing closer to where it is produced and used. This distribution provides quick response, network traffic resilience, and may be a constituent of AR and local AI done on the premises. The computation resources that are deployed by 6G networks, enables a move to the intelligent and reactive devices that may be able to do the sophisticated tasks without constant communication remote servers.

### 12.3.7 Quantum communications in place for the security maximization

Conceptual inventions of quantum communication for 6G have brought a revolutionary way of securing the network communication at the very foundation, taking advantage of QKD that offers a means of arranging secure communication, which logically is insusceptible to any computational and technological developments such as quantum computing. Even though quantum networking technology for the 6G implementation has several practical drawbacks, the quantum research is striving to set up a secure channel for the communication, which will protect against the cyber threats.

### 12.3.8 6G is what sets it apart from its pioneer

While 6G is not only about the leap in the same generations, since the speed and latency enhance, it also involves qualitative breakthroughs. This triggers a shift in the nature of networks and shows how they are becoming an integral part of life. While most of 5G capabilities have not been used in everyday life yet, 6G is the technology that will enable new communication techniques such as holographic presence and high-precision positioning systems. In fact, the strong complementarity of AI, quantum technologies, and advanced materials science make the 6G generation much more intelligent, complex, but at the same time, more versatile network.

Through a theoretical inspection of 6G, one can imagine a future where wireless networks will not be more speedy or efficient but will work together with our societies, so the life of a person may become easier, and new services not even thought of would become reality. It is the implementation of these technologies coupled with a fresh vision of the whole architecture and operations, underpinned by the stated principles, that is a prerequisite for this vision to be realized.

## 12.4 6G AND CALL CENTER OPERATIONS: A DETAILED PERSPECTIVE

### 12.4.1 Transforming call center infrastructure with 6G

The 6G deployment into call center operations is predicted to remodel and usher into service delivery sector a completely new era that is unfathomably clear through robust network properties. 6G is expected to have even less latency (5×) than current 5G and transmission speeds of up to 1 terabit per second (Tbps). This gigantic transition will bring 20 Gbps speed and 10 ms latency from 5G, which were not possible before because of bandwidth limitations so now call centres can implement advanced technologies which were not possible before.

### 12.4.2 Real-time language translation

A call center which deals with international calls that requires immediate translation must have agents who are fluent to avoid any miscommunication. The system can quickly interpret and transmit languages end to end with a latency of 0.1 ms, which means that you do not even notice the delay during the conversation. This feature, joined with 6G's ability to perform intense high-speed data processing, leaves room for the use of complicated neural network algorithms for translation, which necessitates enormous computational power. This leads to the development of a live and continuous dialogue that is not affected by the diversity of languages being used, and thus improves the quality of services. This can be seen is Figure 12.3.

### 12.4.3 AI-powered customer service bots

In a 6G-driven environment, AI robots would take a blink of the eye to respond to customer questions and give them answers in milliseconds. Take, for example, a call from the customer who has an issue. The artificial intelligence system would quickly go through terabytes of data, such as the customer's history, product information, and possible solutions before making any recommendations, which happen in a second. The speed in which this processing happens makes the bot able to provide accurate and automated answers, leading to significant improvement in the call resolution time from before which it would take up to minutes or even hours.

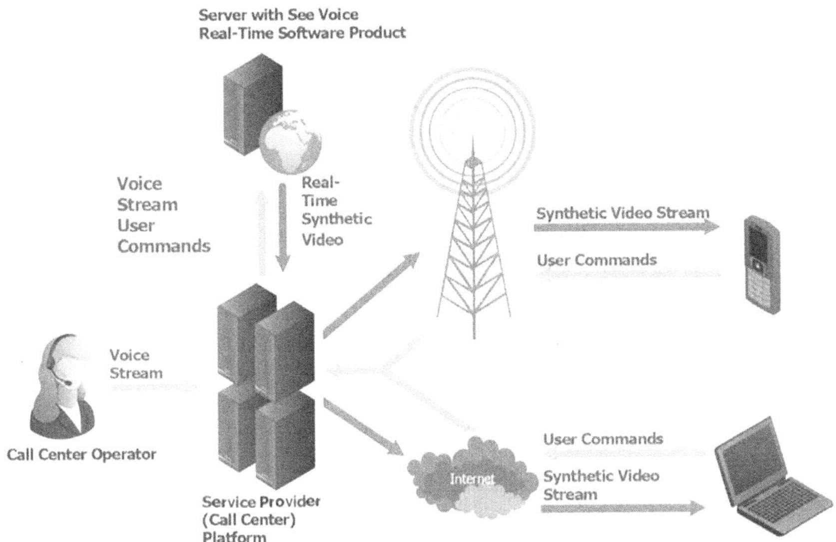


*Figure 12.3* 6G and call center operations: a detailed perspective.

### 12.4.4 Immersive VR customer support

Applying 6G to achieve low latency and high bandwidth VR support sessions for residents through a call center would also be possible. Another instance is where virtual reality session that requires a video resolution of 4 K taping at 60 frames per second (fps) needs a bandwidth ending up at 50 Mbps. And by 6G, speeds of 1 Tbps and higher were achieved effectively, allowing multiple users to attend each session without having to worry about the quality of the interaction, thereby, customers will have more dedicated time to engage fully in the platform. Such a solution is supposed to lead time reduction for response time from the current 50% to the guided instructions [14, 17].

### 12.4.5 Impact on customer experience and operator efficiency

- **Enhanced customer experience:** The introduction of 6G will not only be manifested by increasing customer satisfaction rates but also by ensuring that each customer will be satisfied. To illustrate, the skills to trim down minutes to seconds help the customer satisfaction score as well as lower the rates of abandonment which are currently at an astounding rate of 5 to 10% for the longer wait time.

- **Improved operator efficiency:** As with 6G, the reduction in operators is anticipated to be powerful. AI bots that deal with repetitive queries can cut down operator task volume by 30% or so which means they are able to narrow their attention to the more time consuming and complex problems. Concurrently, the AHT for calls would decrease by up to 50% due to the expedient access of data and proper integration of support systems.
- **Predictive and proactive service:** Through analysis of customer data in the form of real-time information, the call centers which leverage the 6G network can indicate possible issues before they occur, and, in this way, the number of requests can be decreased by 20% or even more. Then again, proactive customer service can increase customer retention up to 5–10% because when customers' satisfaction with a product or service that meets their needs is considered, they become more willing to remain faithful to a particular brand.

The incorporation of 6G into call center operations is no longer only for the purpose of enhanced services but rather a symbol of the development of the service industry that will become more interactive, personalized, and modernized. Efficient data-to-data communication brought by speed and latency of 6G to call centers can help both in improving customer service quality and rethinking the digital customer interaction standards. Various benefits as discussed are listed in Table 12.1

It can be observed that 6G is demonstrated to be highly beneficial in contribution to improve call center operations due to the significant reduction in speed, latency, and efficiency, thereby resulting in the best customer experience and the operational excellence. Visually Figure 12.4 represents the benefits at a glance.

The figure shows how 5G and 6G technologies differ as far as the data transmission speed (Gbps) represented by bars and the latency (ms) which is depicted by line plot. The significant increase in data traffic speed by almost 500%, compared to 3G, and approximately 100%, in comparison to 4G, is not an understatement, particularly with 6G technology's dramatic reduction in latency. Further, the other parameters are shown in Figure 12.5.

A clear enrichment of call center operations as the 6G is introduced to the world of network technology can be observed. This shows a reduction of 50% of times to resolve calls, so it means that the usual time to solve a customer problem may now be achieved in half of the time. Besides that, the presence of the AI-powered chatbot is forecasted to reduce the maintenance operator workload by 30% signifying that the call center agents may perform their tasks throughout a shorter period, thus boosting productivity. Besides this, the figure points to a 20% decrease in inbound calls, could be a resultant effect of a powerful problem resolution and automated service, which could eliminate direct customer–agent conversation. Lastly, it

*Table 12.1* Benefits of 6G as compared to 5G

| *Feature* | *5G* | *6G* | *Improvement/Impact* |
|---|---|---|---|
| Data transmission speed | Up to 20 Gbps | Up to 1 Tbps | 50× increase, enabling faster data processing |
| Latency | ~10 ms | ~0.1 ms | 100× decrease, allowing for real-time interactions |
| Real-time language translation | Limited by latency | Instantaneous | Eliminates communication barriers |
| AI-powered customer service bots | Constrained by speed | Terabytes of data processed in milliseconds | Reduces resolution times from minutes to seconds |
| Immersive VR customer support | Bandwidth limitations | Supports multiple 4 K sessions simultaneously | Reduces problem resolution times by over 50% |
| Call resolution time | Several minutes | Seconds | Dramatic decrease, enhancing customer satisfaction |
| Operator efficiency improvement | – | Up to 30% reduction in routine inquiries workload | Operators can focus on complex issues |
| Average handling time (AHT) | – | Up to 50% decrease | Increases call center throughput and service quality |
| Inbound call reduction | – | Predictive service can reduce calls by up to 20% | Lower operational cost, improved customer satisfaction |
| Customer retention rate improvement | – | 5–10% improvement due to proactive service | Higher customer loyalty and brand satisfaction |

projects a 7.5% betterment in client retention rate triggered by the superior grade service delivered in the inevitable 6G network. These set of indicators, put together, send out a message of 6G's potential in revolutionizing call center operations, upon which the customers and service providers would emerge better off.

## 12.5 6G IN ACTION: SOME CASE STUDIES

### 12.5.1 Deployment in call centers

In Japan, one of the leading multinational telecommunications firms has performed a trial manufacture of 6G technologies' deployment within its

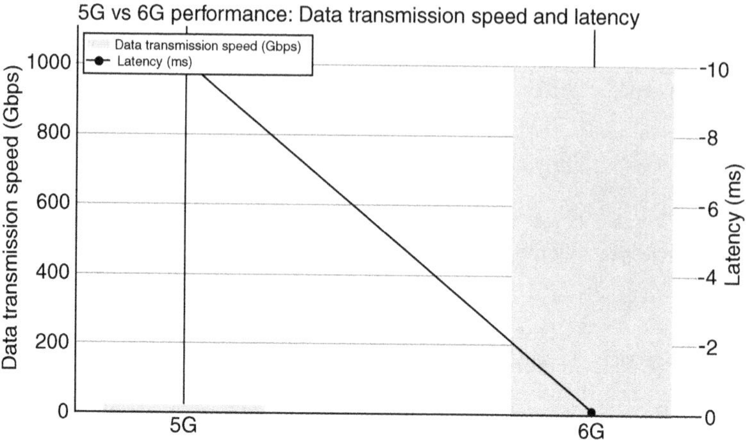


*Figure 12.4* A comparative graph showcasing the performance metrics for 5G and 6G wireless technologies.

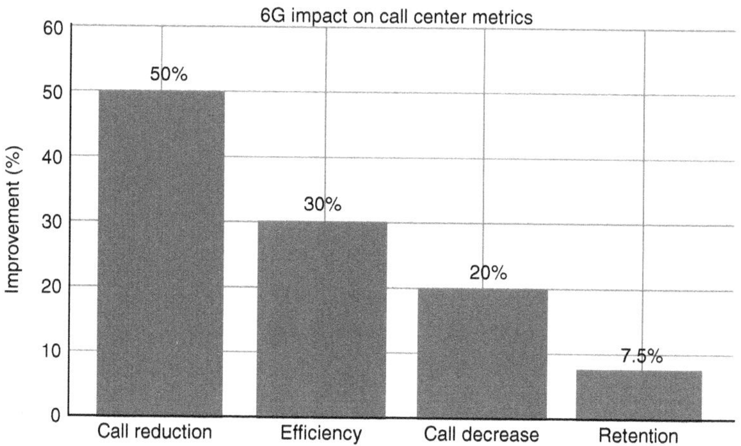


*Figure 12.5* 6G impact on call centers.

customer service call centers at the first go. The project revolved around 6G's potential to integrate real-time multilingual translating services and, to help with problem-solving, utilize VR through the capability of low latency and high bandwidth. It showed indeed fantastic improvement – 60% increase in call resolution efficiency and 40% rise in positive customers' feedback. Principal insights being the need to develop a coherent approach to 6G technology implementation together with existing business relationship systems and citing the growth of staff education and enhancement as a proof of concept. The lessons learned from this case came through the establishment of rigorous pre-deployment testing phases and the implementation of gradual integration method by means of which there was an opportunity for iterative feedback and the refinement of the system processes.

### 12.5.2 Healthcare sector innovation

The second example is where a group of Japanese hospital entities that used 6G connectivity to perform remote diagnostics and patient tracking. This 6G deployment was based on its ability to stream complex data with relationships between different medical devices in real-time and evaluate the situation from a distance. The improvements that were done includes patients' wating time reduction, the quicker diagnosis allowing a better and faster treatment and preventing contagious diseases' exposure risk are visible. The lessons learned shed light on the necessity of having strong cybersecurity measures that also reaffirmed the importance of partnership as a tool to bridge the gap between the tech companies and health service providers.

### 12.5.3 Smart city development

In the setting of smart city domestic, several organizations in Japan merged with technical firms to develop 6G incorporated transaction management systems. These systems have maximized the advantages of 6G's high-speed data transfer to real time monitoring and analysis of traffic and therefore improving the management of traffic congestion through consistently reducing average commute times. These complements put forward the fact that there is a huge need to increase the infrastructure and the windows people open for the community during the design phases let us obtain even better outcomes. Top practices advised prioritizing scalability in the initial design that allows the system to be compatible with different designs in the future. Moreover, it was essential that the technology was vertically integrated to be adaptable to other smart city applications [18].

### 12.5.4 Manufacturing sector transformation

An automotive production company, which used 6G technology, aiding both the manufacturing robots and assembly line equipment to connect and work together is another example to refer to. The higher bandwidth and low latency permitted the performance of immediate analysis and necessary adjustments, which optimized the production process in terms of efficiency and eliminating useless output. The Modbus protocol was successfully implemented, which resulted in the following lessons learned: reliable network communication is indeed a critical aspect and cooperating closely between the network engineers and process experts is a very important feature. A best practice referred to as the implementation of dual systems in maintenance of the network operation also appeared, which calls for the network to be up and running even in the case of disruptions.

### 12.3.5 Challenges and considerations in the deployment of 6G networks

The pathway to 6G networks' realization remains bleak and complex as it covers techno-legal and ethical aspects. Indeed, there are several technical difficulties in making the 6G happen. The foreseeable use of THz frequency range inevitably leads to a complete revolution of the current networking infrastructure due to the intricate propagation obstacles it poses. Small cells are the only option to overcome THz frequency limitations but these need to be placed in urban environments, where there are huge costs of installation and potentially adverse effects on the environment. Besides this, having an additional spacecraft network for obtaining global coverage necessitates attention to be paid to limiting debris in space and assigning slots in orbit. Also, technical obstacles are present with the development of the devices which should be rendered energy-efficient without causing extra energy costs or environmental pollution and thus becoming the extremal point for current semiconductor and materials science.

In 6G networks, the security has become a huge subject since artificial intelligence (AI) and the Internet of Things (IoT) are expected to be integrated deeper into life. These devices are never offline and are interconnected, creating a network through which attackers can organize a more threatening cyber-assault. Privacy dangers are not lower; the capacity of networks to have access to the massive amount of data gives more chance for the data to be collected, processed, and then be used in an improper way. Ensuring data encryption and secure communication protocols that ordinary quantum attacks can break interpose another complication.

From the ethical point of view, 6G is going to seep into our everyday life; however, it raises the issues of surveillance and data ownership. The possibility of monitoring 24/7 at the hands of diversified devices can lead to invasive surveillance if there are not a set of strict moral codes to abide by. Network management and data processing through AI systems present also ethical concerns. Concerns arise when AI, like the decisions affecting priority network operations or data management, are not transparent and accountable, with humans in charge having clear mechanisms for oversight to avoid discrimination or unfair action. Regulatory aspect is the biggest problem for international community in unifying the spectrum allocations for the 6G, which may require coordination between the member countries to eliminate the interference. Here, countries and regions must find the frequencies at the negotiating table, but not colliding with the existing allocations used for services, for example, weather forecasting and satellite communications, which will be disrupted by new 6G networks.

The process of designing and imposing legal frameworks that can evolve to the fast-paced revolution of technologies implied in the 6G network is sure to become an arduous and permanent responsibility, which requires global cooperation and promptness. And then we must take social economy factor into account. The urban–rural digital divide could get worse if the planning and roll-out of 6G networks gives preference to urban or economically well-off areas at the expense of rural or low-income regions. Ensuring equal availability of 6G benefits for everyone will yet be a difficult task for which governmental entities and entrepreneurs will have to strive. As parting words, the stages of 6G deployment entails thoughtful procedures to prevent numerous technical and non-technical issues associated with it. Multidisciplinary collaborations between technologists, policymakers, ethicists, and industry players are indispensable for a purposeful generation of the 6G transition that is technically robust and benefits society as well. The degree of success of 6G will not be decided on the grounds of the technical marvels alone but also on how the adherence to principles of security, privacy, and ethics is done and how it contributes to closing the digital divide.

## 12.6 CONCLUSION

The chapter has had so many highlights and raised the most diverse issues, which are critical to the upcoming creation of 6G networks. The leap from the first generation (1G) to the sixth (6G) generation of mobile communications is a remarkable progress of the field, seen as an immense combat that was won. To every generation, technological progress has always been accompanied by a considerable increase in speed, capacity, and functionality. Yet, 5G and its successors only anticipate a large step forward in the system design. It predicts a period characterized by ubiquitous computing, where computing will be made of high bandwidth, reliable, and moment-to-moment connections in support of innovative applications in diverse fields such as healthcare, education, industry, and consumer technology.

6G will be heavily influenced by theoretical foundations of the technology, especially its usage of terahertz frequencies, AI networks, and advanced network slicing, which are the key elements of the paradigm shift. The future marketplace will be heavily reliant on technologies that are designed to greatly enhance operational efficiency and will specifically emphasize the need for better call resolution time, increased operator efficiency, reduced inbound call traffic, and higher customer retention rates. A conceptual framework that articulates these case studies, for example, the developments of Japanese call centers, paves the way for practical and tangible implementations of 6G technological innovations.

Nevertheless, the omnipresent hope about 6G's possibilities is damped by the awareness of the tremendous barriers that lie ahead. Challenges in a network rollout and regulation like robust technical architecture, privacy and ethics concerning the data use, and regulation and spectrum allocation are all issues to be dealt with. The dawn of an era where 6G technology is imminent, presents the realization that the full potential is limitless but attaining this comes with a deep complex process that will require collaborative efforts from several fields of expertise. For 6G technology to become a reality, we must ensure that it proceeds while benefiting the entire world equitably, ethically, and sustainably. It should hence be an imperative that the development of 6G is done so that it reflects these principles as this will safeguard the potential benefits of 6G globally.

In general, 6G technology seems to be the most exciting technological road to an improved communication connecting people worldwide at an unbelievable level of efficiency and service. The dividing line at the same moment allows for the integration of parallel processes of technological as well as socio-economic solutions. Tomorrow of 6G will come in because of our collective inspiring actions today, which should definitely be approached globally with great care, comprehensive planning and cooperation in order to put this concept into practice.

## 12.7 FUTURE SCOPE

With the arrival of 6G networks to the communications world, a paradigm shift has emerged not only in traditional call centers but in our ability to mend all the details of a digital life. This next stage wireless technology is on the point of delivering unprecedented speed, reliability, and capacity features, making it possible for users to exchange data and multimedia in real-time without any lag, accelerating and even disrupting the ways people work and have fun. The high-tech implications of 6G for call centers are not less than these: zero latency, interactive audio and video of the highest quality, and services provided by the most advanced AI obtainable so customers' convenience and satisfaction can substantially be improved. Also, state-of-the-art technologies like holographic calls and augmented reality (AR) client assistance can interact with existing clients no matter where they are located and will provide a more personalized and live interaction. Despite the fact that the 6G networks are a great tool that humankind can use to further its development, the road towards 6G is full of problems. The design setup and launching of such an intricate technology worldwide carries with it the risk of the mission becoming expensive. Besides, there is high consumption of energy that needs to be released in the workings of these networks, this has made many consider sustainability and environmental conservation. But the issue of digital equity needs to be considered as well, as the rollout of 6G could make the digital divide even greater for citizens living in any emerging markets or economically poor nations.

The grasp of 6G networks from the perspective of law inflicts a pile of legal and ethical issues as well. Data protection and security are preliminarily being challenged by increasingly connected devices and services, which leads to an advanced state of complexity. It is the case that governments and the regulatory agencies will have to grapple with such issues with the aim to ensure protection of rights of individuals and at the same time, create an environment of innovation. Undoubtedly, 6G would create opportunities to disrupt not only the telecommunication sector but also various other industries. In healthcare, e.g., telemedicine may be applied in remote surgeries and data monitoring of patients' health from anywhere on Earth now. In the transportation side, autonomous vehicles can share information with each other and manage the roads through infrastructure whereas helping with safety as well as efficiency. And now applications in the city, agriculture, education, and entertainment, are also equally hopeful, those are the solutions which are smart, more efficient and responsive to such needs of day-to-day problems.

Nevertheless, 6G like everything has two sides to it, and thus the question of surveillance and invasion of privacy arises given the increase of embedded devices and services. Data collection and processing will be on the remarkable growth and so stringent policies should be put in place to ensure privacy is not compromised. The fact that individuals might gain from unauthorized usage of such a powerful tech cannot be ignored, thus emphasizing the importance of appropriate moral standards and oversight. Inter-working and standards are another major obstacle. Being designed for the 6G network, devices and networks must be accessible all around the world and interact with each other. International standards possess paramount importance across all aspects. Nonetheless, it is very noteworthy that arriving at an understanding and harmony among countries in a world that is so divergent in many respects needs lots of time. The starting costs of investments for implementation and deployment of 6G networks will be very high, but the potential economic rewards could be great according to the economic forecast. The emergence of new sectors and employment is the most probable consequence, while others will undergo a process of transformation. The opportunity for economic development and innovation is wide open and interesting, however, the advantages need to be in tandem with the costs and disruptions to laid off workers.

Briefly, 6G networks are envisaged to dispel all barriers standing in the way of further evolution of telecommunications and thus to ramp-up development of a myriad of industries. Nevertheless, they will also have to encounter a bunch of technical, ethical, and regulatory issues. Combating these problems must involve cooperation from the government, industry, and civil society. The possible capabilities of 6G to radically upgrade global transmission systems, foster economic development, and positively influence living conditions are immense; however, achieving these goals will require some difficulties being surmounted. On the dawn of a new technological era, we cannot be certain of the consequences of 6G technological achievements that will be felt in many lives for years to come.

## REFERENCES

[1] T. B. Ahammed, R. Patgiri, and S. Nayak, "A Vision on the Artificial Intelligence for 6G Communication," *ICT Express*, vol. 9, no. 2, pp. 197–210. 2023. doi: 10.1016/j.icte.2022.05.005.

[2] C. De Alwis, A. Kalla, Q. V. Pham, P. Kumar, K. Dev, W. J. Hwang, and M. Liyanage, "Survey on 6G Frontiers: Trends, Applications, Requirements, Technologies and Future Research," *IEEE Open Journal of the Communications Society*, vol. 2, pp. 836–886, 2021. doi: 10.1109/OJCOMS.2021.3071496.

[3] P. Porambage, G. Gur, D. P. M. Osorio, M. Liyanage, A. Gurtov, and M. Ylianttila, "The Roadmap to 6G Security and Privacy," *IEEE Open Journal of the Communications Society*, vol. 2, pp. 1094–1122, 2021. doi: 10.1109/OJCOMS.2021.3078081.

[4] M. Banafaa, I. Shayea, J. Din, M. H. Azmi, A. Alashbi, Y. I. Daradkeh, and A. Alhammadi, "6G Mobile Communication Technology: Requirements, Targets, Applications, Challenges, Advantages, and Opportunities," *Alexandria Engineering Journal*, vol. 64, pp. 245–274, 2023. doi: 10.1016/j.aej.2022.08.017.

[5] D. Ather, R. Singh, and R. S. Shukla, "Routing Protocol for Heterogeneous Networks in Vehicular Ad-Hoc Network for Larger Coverage Area," *Engineered Science*, vol. 17, pp. 266–273, 2022.

[6] M. Tahira, D. Ather, and A. K. Saxena, "Modeling and Evaluation of Heterogeneous Networks for VANETs," *2018 International Conference on System Modeling & Advancement in Research Trends (SMART)*, Moradabad, India, 2018, pp. 150–153. doi: 10.1109/SYSMART.2018.8746981.

[7] C. X. Wang et al., "On the Road to 6G: Visions, Requirements, Key Technologies, and Testbeds," *IEEE Communications Surveys and Tutorials*, vol. 25, no. 2, pp. 905–974, 2023. doi: 10.1109/COMST.2023.3249835.

[8] H. Viswanathan and P. E. Mogensen, "Communications in the 6G Era," *IEEE Access*, vol. 8, pp. 57063–57074, 2020. doi: 10.1109/ACCESS.2020.2981745.

[9] S. A. Abdel Hakeem, H. H. Hussein, and H. W. Kim, "Vision and Research Directions of 6G Technologies and Applications," *Journal of King Saud University – Computer and Information Sciences*, vol. 34, no. 6, pp. 2419–2442, 2022. doi: 10.1016/j.jksuci.2022.03.019.

[10] Z. Qadir, K. N. Le, N. Saeed, and H. S. Munawar, "Towards 6G Internet of Things: Recent Advances, Use Cases, and Open Challenges," *ICT Express*, vol. 9, no. 3, pp. 296–312, 2023. doi: 10.1016/j.icte.2022.06.006.

[11] A. Talpur and M. Gurusamy, "On Attack-Resilient Service Placement and Availability in Edge-Enabled IoV Networks," *IEEE Transactions on Intelligent Transportation Systems*, vol. 24, no. 6, pp. 6244–6256, 2023. doi: 10.1109/TITS.2023.3249830.
[12] D. Ather, R. Singh, and R. S. Shukla. "Routing Protocol for Heterogeneous Networks in Vehicular Ad-Hoc Network for Larger Coverage Area," *Engineered Science*, vol. 17, pp. 266–273, 2022.
[13] A. Gupta, and D. Ather. "A Comparative Study of Various Routing Protocol for VANETs," *MATRIX Academic International Online Journal of Engineering and Technology*, vol. 4, no. 1, pp. 12–18, 2016.
[14] R. Kumar, V. Jain, G. W. H. Tan, and A. Touzene (Eds.). *Immersive Virtual and Augmented Reality in Healthcare – An IoT and Blockchain Perspective*. CRC, Taylor and Francis Group, 2023. https://doi.org/10.1201/9781003340133.
[15] G. K. Surbhi Gupta and D. Ather, "BEG: Bin Oriented Energy Efficient Gathering of Data in WSN," *International Journal of Engineering, Applied and Management Sciences Paradigms (IJEAM)*, vol. 54, no. 3, p. 418, 2019.
[16] R. S. Shukla and Danish Ather. "Simulation Based Protocols Comparison for Vehicular Ad-Hoc Network Routing," In *2021 10th International Conference on System Modeling & Advancement in Research Trends (SMART)*, pp. 198–203, IEEE, 2021.
[17] R. S. Shukla, D. Ather, and R. Singh. "An Efficient Route Maintenance Routing Algorithm for VANETS," *International Journal of Recent Technology and Engineering (IJRTE)*, vol. 8, no. 4, p. 4921, 2019.
[18] R. Kumar, V. Jain, W. Y. Leong, and S. Teyarachakul. *Convergence of IoT, Blockchain and Computational Intelligence in Smart Cities*. Taylor & Francis, CRC Press, 2023, ISBN: 9781032404240.

Chapter 13

# Enhancing distributed denial of service cloud security threats using artificial neural networks in 6G networks

*Renu Devi, Naresh Kumari, and Sumeet Gill*

## 13.1 INTRODUCTION

Over the Internet, cloud computing provides computing resources, like storage, processing power, and software. These resources are available on-demand, without the need for local hardware or infrastructure. Cloud computing plays an important role in the implementation of 6G architecture [1]. A fundamental driver behind 6G, is its ability to deliver an ultra-fast and reliable connectivity, specifically suited to support a diverse range of applications and services with stringent requirements [2]. In the architecture of 6G, cloud computing entails combining cloud-based services and infrastructure with the wireless network infrastructure. This integration facilitates the transfer of resource-intensive tasks and data storage to remote cloud servers, alleviating the load on local devices and enhancing overall system performance.

### 13.1.1 Advantages of cloud computing in 6G networks

Cloud computing facilitates smooth scalability, enabling the network to allocate resources dynamically in response to demand fluctuations. Through this approach, resources are effectively utilized and applications are provided with optimal performance. The cloud computing platform enables 6G networks to adapt to the changing requirements and demands of its users. An advantage of a cloud infrastructure is its ability to be quickly deployed and innovated as well as easily reconfigured and customized to support different types of applications. Cloud-based services ensure continuity of service and high availability. Cloud infrastructure provides redundancy and backup mechanisms to improve the availability and reliability of 6G networks while minimizing the possibility of service interruptions [3].

### 13.1.2 Cloud service providers

Cloud computing refers to the use of a shared and mostly virtual environment. To ensure the security of their data, cloud service providers rely


DOI: 10.1201/9781003583127-13

on consumers comprehending the consequences of their data being stored within the provider's data center. A user must understand how its cloud provider controls the user's access to its cloud services. Although federal managers retain responsibility for their data in the cloud, their responsibility for its protection has been transferred to the vendor, and therefore the responsibility for its protection has been left to the cloud vendor [4]. To decide which vendor is most appropriate when the agency is prepared to make a choice, understanding how the service provider has operated historically in the federal environment may be helpful. As compared to smaller or less established companies that provide cloud services, it is expected that large-scale cloud providers will offer greater security.

### 13.1.3 Cloud service models in 6G networks

**Software as a service (SaaS):** The consumer is provided with the ability to use the application modules of the provider, which run on a cloud infrastructure hosted by the provider. A thin client interface like a browser makes the apps accessible from different devices. Users have access to limited program configuration settings for the applications they are using, the cloud infrastructure underlying their applications is not managed or controlled by the consumer, including the networks, operating systems, storage, or specific application capabilities.

**Platform as a service (PaaS):** In this instance, the cloud provider gives users the option to use programming languages and tools that the provider supports to install apps that they have developed or purchased onto the cloud infrastructure. Consumers are not responsible for managing or controlling the cloud infrastructure supporting the application hosting environment. This encompasses networks, servers, operating systems, and storage. In spite of this, they may be able to control some aspects of the deployed applications or even the configuration of the application hosting environment.

**Infrastructure as a service (IaaS):** As a user, one has the ability to provision processing resources, storage capabilities, networks, and other computing resources. This allows one to run arbitrary software, such as operating systems and applications, and deploy and manage these resources as you see fit. Typically, cloud infrastructure is not managed or controlled by consumers. In the cloud architecture, they do, however, have control over storage, operating systems, and installed apps. They may even have restricted control over specific networking elements, including host firewalls.

### 13.1.4 Cloud deployment models

**Private cloud:** The cloud infrastructure is solely operated by the company; it can be located on-site or off-site, and it can be managed internally or by a third party.

**Community cloud:** In addition to supporting the mission, security requirements, policy, and compliance requirements of a specific community, the cloud infrastructure is utilized by multiple organizations. There are several ways in which an organization can manage its infrastructure, including either using it on-premises or off-premises, and either by themselves or by a third party.

**Public cloud:** An enterprise that offers cloud-based services owns the cloud infrastructure, which is shared by a sizable industry association or the general public.

**Hybrid cloud:** The cloud architecture is made up of two or more clouds, which can be either private or public. Each cloud maintains its unique identity and is connected via proprietary or standardized technologies, which makes data and applications more portable.

### 13.1.5 An overview of sixth-generation technology

The major disadvantages of the previous mobile networks are as follows:

1G: This service has a very low coverage area; the number of users is limited and the voice quality is poor.
2G: Mobile applications are limited in terms of functionality, data rates are low, and the number of users is limited.
3G: The cost of the spectrum license and the installation of infrastructure are high.
4G: Smartphone prices have increased, and spectrum licenses have become more expensive.
5G: It is not easy to deploy, there are security concerns, and the gadgets are very expensive.

The research has already started on sixth-generation technology in 2017. The research on 6G has begun in numerous countries, including Finland, Japan, the United States, South Korea, and China. The term 6G refers to a communications network of the sixth generation. In 6G, the speed is measured in terabits per second [5].

Cloudification, in terms of cloud services, is the primary function that is involved in transferring or migrating the services to the cloud. The availability of on-demand services has increased in recent years, as many cloud service providers, such as Google and Microsoft Azure, are offering their services via the Internet [6]. The services provided by these companies are unique. In today's world, it would be extremely difficult for an organization to function without cloud computing services [7]. According to the predictions, most organizations will migrate to the cloud by 2025. As a result, there is no need to invest and maintain hardware infrastructure, and the setup is quick and simple. Cloud services handle extensive volumes of data and require large-scale storage capabilities. Therefore, cloud services must have high-speed connectivity to efficiently manage this vast amount of data. To handle the volume of traffic resulting from 5G and 6G networks, cloudification is a requirement. By cloudifying the 5G/6G networks, the speed and capacity of the 5G/6G networks would be 1000 times greater than those of previous technology. The advantages of cloudification extend primarily to operators, enterprises, and consumers.

**Cloudification in 5G/6G involves the following key technologies:**

**Software-defined networking (SDN):** It refers to the utilization of software-based techniques to manage and control network infrastructure within a cloud environment. In cloudification, SDN enables dynamic and flexible management of network resources, allowing for programmable control over network behavior. Essentially, SDN in cloudification facilitates the automation and optimization of network operations to meet the evolving demands of cloud-based services and applications.

**Virtualization:** A virtual software-defined network is formed by combining several physical networks or dividing a physical network into multiple independent networks.

**Network slicing:** To maximize revenue and minimize operational expenses, multiple independent networks can be operated within the same infrastructure environment.

**Self-managed network (SMN):** This network is primarily focused on identifying and repairing network problems.

### 13.1.6 Cloud service problems in 6G networks

According to predictions, forthcoming 6G networks are expected to elevate voice quality, extend coverage to larger areas, and operate with minimal power consumption. Additionally, advancements may enable non-technical individuals to program robots for personal applications easily. In today's world, automation has become increasingly popular. There is a need for automation in every sector, including space, industries, health sectors, and transportation. In the next generation, everything will be connected to the cloud through sensors in cities, homes, factories, vehicles, and industries. For these applications to be supported, high data transmission rates are required. Apart from these advancements, cloud services are encountering several challenges, as depicted in Figure 13.1.

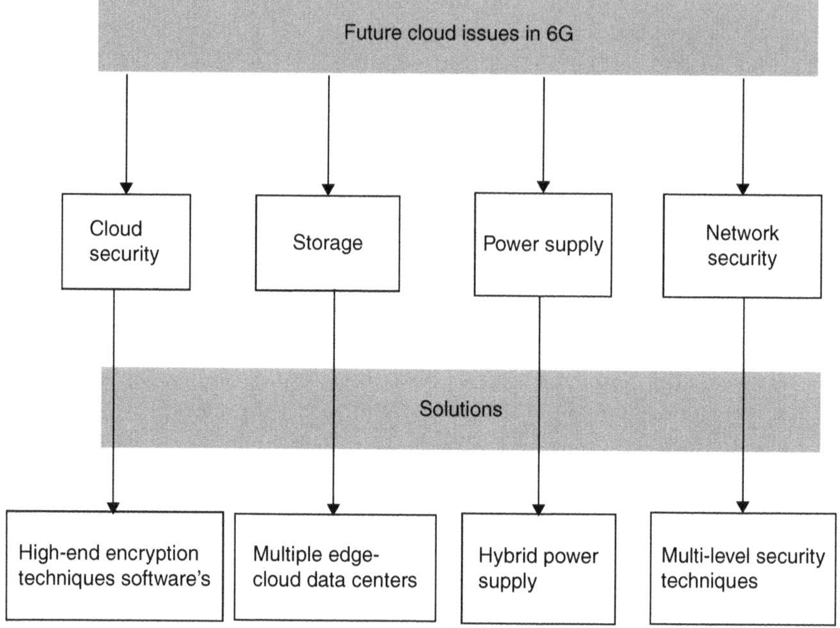


*Figure 13.1* Cloud issues in 6G networks.

#### *13.1.6.1 Storage issues*

By 2030, in the realm of 6G cloud computing, all services are expected to be interconnected, with the majority of organizations having migrated their data centers to the cloud. There is a requirement for massive storage capacities in this case. To store IoT networked device data in bulk, many edge cloud centers need to be installed. Consequently, the number of cloud data centers will increase significantly in the future.

#### *13.1.6.2 A study of the energy consumption of cloud services*

Second, there is the issue of managing the myriad of cloud edge centers, which require significant processing power. As such, to optimize this, there is a need to install more wind farms and solar farms (renewable sources of energy) to achieve the goal. By transmitting data to the cloud, all network devices will consume more energy. In the current year, the cloud services consume 2000 terawatt-hours of electricity. By 2030, this will reach 8000 terawatt-hours.

### 13.1.7 Cloud security issues in 6G networks based on unique identification number

This section presents a few of the potential threats that could occur in a 6G network. Over the past few years, the number of attacks on the cloud has increased significantly. Attackers steal bank data using malware virus. DDOS attacks are also among the most annoying. An attack occurs in which the attacker utilizes their knowledge and prevents users from gaining access to cloud services. This indicates a lack of trust and privacy. Phishing occurs when an attacker uses a phone, e-mail, or SMS in an attempt to steal sensitive information. SaaS (software as a service) cloud models consist of third-party providers delivering software application products over the Internet and managing them remotely. There are many examples of SaaS cloud computing, such as Dropbox, Spotify, Google Apps, and others. Presently, there is a widespread migration of applications to the SaaS cloud, a development that is captivating global markets. Since SaaS applications have been targeted by attackers. A large number of individuals worldwide have previously been affected by the WannaCry malware. As a result of this attack, various sectors, including the police department, health department, and IT software, are affected, indicating a lack of cyber security [8]. It is the cloud provider's responsibility to secure their services. Then the cloud providers will be able to gain the trust of their customers if this is the case. The following are some of the main concerns associated with cloud security relating to IP addresses of devices:

- Unauthorized access
- IP spoofing
- Denial of service attacks (DoS attacks)
- Data leakage
- Man-in-the-middle (MITM) attacks
- Identity and access management (IAM) risks

6G ecosystem partners, such as network operators and device manufacturers, need to work closely together to research and develop new security technologies. DDoS represents a specific type of DoS security threat. This paper focuses particularly on addressing DDoS security threats within the cloud environment.

A DDoS attack happens when there is a breach in security. In this threat, users who have been authorized for access to the cloud are prevented from using its devices, systems, or other resources [9]. There is a network of zombies that are operated remotely, are well arranged and widely distributed and conduct DDoS attacks [10]. The attacker initiates the attack with the assistance of zombies, also known as secondary victims.

##### *13.1.7.1 DDoS attacks are generally categorized into three categories*

Volume-based/bandwidth-based attacks: Through this technique, the attacker attempts to overwhelm the user with a large volume of garbage data, which consumes the network's bandwidth and resources [11].

Protocol attacks: A flaw in multiple network protocols is exploited in this attack in an attempt to overload the target's resources [12].

Application layer attacks: In this attack, certain online applications are targeted, and HTTP requests are sent that exceed the capacity of the applications [13].

### 13.1.8 Application of machine learning in 6G network

A crucial role is expected to be played by this technology in 6G networks [14, 15]. Here are some potential applications of ML technology:

**The optimization of networks**
**Identification of anomalies**
**Predictive maintenance**
**Intelligent network management**

**Enhanced security:** By detecting security issues and addressing them quickly, ML algorithms can enhance the safety of network. Using machine learning, network security measures can be automatically activated in response to unusual traffic patterns and devices.

For machine-learning algorithms to be implemented in 6G networks, significant investments in artificial intelligence and machine-learning infrastructure will be required, as well as consideration of concerns regarding data privacy and security [16, 17].

### 13.1.9 Integration of ML in the security of cloud data

The use of ML techniques enhances security in different domains. These methods identify and learn from vast amounts of data using artificial neural networks, leading to better and more trustworthy security solutions. In terms of cloud security, automated threat detection, and enhancement of cloud security are the key benefits of ML implementation. By utilizing machine learning, we can minimize the need for manual monitoring and analysis and identify security incidents more quickly, respond to them promptly, and reduce them. Due to these facts, ML minimizes the possibility of human error by automating and automating various security tasks, which can result in security breaches and the corresponding costs that come with them. It is possible to increase the effectiveness of overall security by implementing automated systems powered by machine learning [18]. To

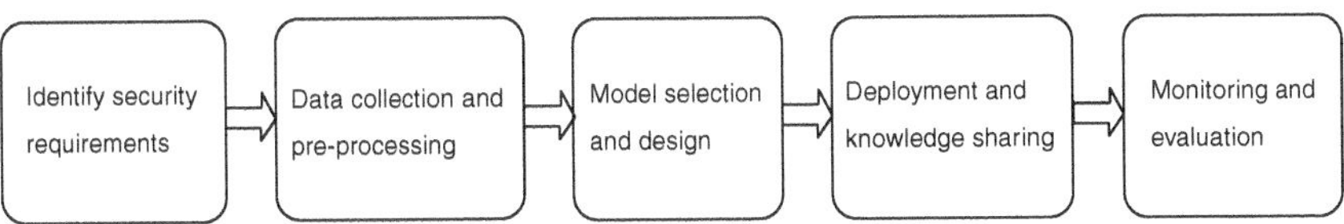


*Figure 13.2* Important steps for integrating ML in cloud data security.

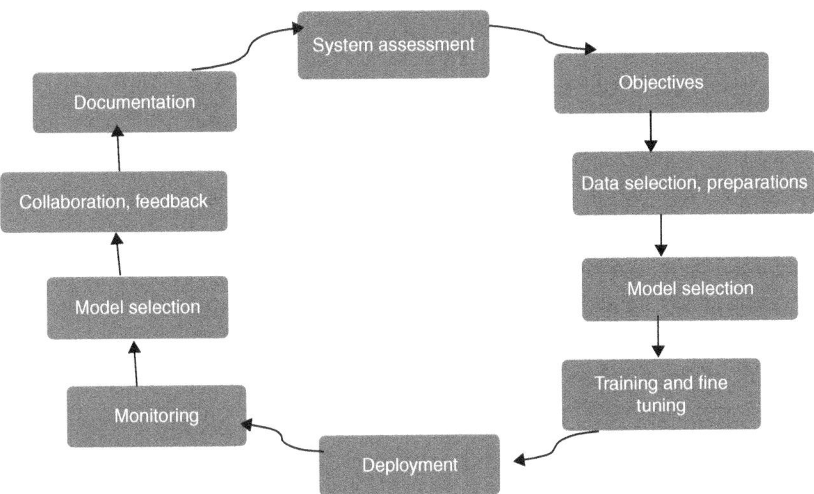


*Figure 13.3* An overview of the key phases of integrating ML into existing security systems,

detect and classify threats more accurately, machine-learning models learn patterns and relationships in data. Apart from the aforementioned benefits, there exist additional non-technical advantages that need consideration. These include qualitative aspects such as enhanced reputation, increased customer trust, and gaining a competitive edge.

Machine-learning models and algorithms are important in improving data security in the cloud [19]. Machine-learning algorithms such as "SVM", "LINEAR REGRESSION", and "RANDOM FOREST" are commonly used in cloud security.

Important steps involved in integrating machine learning within cloud security are provided in Figure 13.2. By adhering to these procedures, companies can effectively incorporate machine learning into their strategies for securing cloud data [20]. However, a more comprehensive assessment would be necessary for enterprises aiming to incorporate machine learning into their current security structures and solutions (Figure 13.3). The first step towards implementation involves evaluating the positive and negative factors of the existing system. This method would make it easier to identify the tasks and domains that benefit from the incorporation of machine learning. ML can be used to enhance areas and tasks that are very closely associated with predetermined targets that have been set by the organization. For the optimization of strategies, it is essential to have clearly defined objectives, so that machine learning can be applied to specific tasks or challenges within the cloud security system [21]. The latter directly points to selecting appropriate models for the tasks and purposes identified. A large number of pre-processed and well-chosen data are needed for the model's training. Developing a successful model depends in part on the selection of the dataset and the preparation of the data [22]. Having chosen the data, it is necessary to adjust the hyper-parameters, perform cross-validation, and optimize the model to fit the particular situation for security in the cloud. It is essential to have appropriate mechanisms planned for monitoring and providing feedback at this stage of integration.

## 13.2 RELATED WORK

The term "DDoS security threats" describes attacks aimed at overwhelming target networks with excessive traffic. Researchers have developed numerous methods for detecting DDoS attacks, with the primary focus always on accuracy. Many researchers are actively engaged in addressing this threat. In this section, we have outlined some existing research efforts concerning cloud security and denial of service security threats.

Amitha et al. [23] introduced a hybrid model combining "radial basis function (RBF) and long short-term memory (LSTM)" networks for identifying and enhancing DDoS attacks. This approach is designed to provide a high level of data security in the cloud. An overview of DDoS attacks and their impact on cloud computing is provided by Deshmukh and Devadkar [24]. They also highlighted factors to be considered when selecting DDoS defense mechanisms.

El-Sofany et al. [25] presented a comprehensive study on DoS attacks, analyzed and evaluated different techniques for detecting DDoS attacks, and presented an experimental study on cloud computing applications and their impact on them. Masdari and Jalali [14] described a detailed description of the different types of DoS attacks for the environment of cloud computing. Memon et al. [26] propose a solution that can help the Pakistan Information Technology (IT) services overcome attacks caused by distributed denial of service.

In Pandithurai et al. [27], the forecasting of DDoS attacks in a cloud setting is demonstrated through the utilization of a honey badger optimization algorithm incorporating feature selection and Bi-LSTMs. In Thirumaran et al., the deep learning (DL) approach is used by the author of this research study for developing an intrusion detection system (IDS) in cloud computing. Two IDS network datasets are tested for feature selection with the implemented hierarchical long short-term memory (HLSTM) method: Bot-IoT and NSL-KDD (network security lab knowledge discovery and data mining). According to the conclusions reached in this paper, an intrusion detection network can provide a high level of security and performance while being economical [28]

Emil Selvan et al. established a deep neuro-fuzzy network (DNFN) based on fractional anti corona virus optimization (FACVO) to identify DDoS attacks. The features are merged by the RV coefficient in conjunction with a "deep quantum neural network" (deep QNN), and then the information is augmented and the fusion is completed. An algorithm created by FACVO that can identify and target DDoS attacks is the basis for the DNFN. It can be concluded therefore that the proposed approach achieved the following testing accuracy, test performance, test result precision, test result reliability, and test result accuracy values of 0.9304, 0.9088, 0.9293, and 0.8745 concerning the NSL-KDD dataset without attack, and 0.8991, 0.9015, and 0.8648 concerning the BoT-IoT dataset without attack [29].

Thangasamy et al. (2022) proposed a method for identifying DDoS assaults based on hybrid long-short-term memory (LSTM) model and deep belief network (DBN) feature extraction. The method uses the NSL-KDD dataset. Particle swarm optimization (PSO) is combined with the hybrid LSTM method to reduce prediction error in particle swarm optimization (PSO) according to this deep belief network method, IP packets are extracted to extract the features and it uses the PSO-LSTM model to identify DDoS attacks. Further, DDoS attacks are detected and predicted accurately. Based on the comparison of accuracy, recall, f-measure, and precision of the proposed PSO-LSTM architecture with other classification techniques, the PSO-LSTM architecture outperforms standard support vector machine (SVM) techniques [30].

In Farhat et al. (2023), with the CICIDS2017 benchmark dataset and multiple machine learning (ML) algorithms, this paper proposes an attack detection system for cloud environments. XG Boost (associated with the influence gradient algorithm) has been implemented by the authors to demonstrate the effectiveness of the model, which demonstrates 99.11% detection accuracy when it comes to detecting intrusions [31].

Aashmi et al. (2022) propose an enhanced hybrid method that integrates the Spearman coefficient, a mitigation strategy, and multivariate correlation analysis. The distribution of resources only for authenticated requests greatly helps to increase the security of the network. A comparison of the results of the analysis with the normalized information has revealed that the detection accuracy of 99% for both DoS attacks and DDoS attacks is very encouraging when compared with baseline measurements [32].

In 2023, Bano et al. introduce a machine-learning framework aimed at identifying, monitoring, and offering preventive measures for DDoS attacks. Their study assesses the effectiveness of four commonly utilized algorithms: naive Bayes, decision tree, random forest, and SVM [33].

In Azizpour et al., a biological genetic algorithm combined with deep-learning methods is proposed to identify potential network traffic anomalies. Maximum vote-based attack detection validation is performed using the available samples, and the results are analyzed. Suspicious traffic is applied to CNN, DT, and KNN for input. Lastly, they calculated precision, accuracy, recall, and error as part of their evaluation method (NADA). In the simulation of NADA, they found that compared to alternative methods, the metrics mentioned above improved on average by approximately 7% compared to those examined by alternative methods [34].

Kelechi et al. proposed framework for safeguarding data in cloud environments addresses prevalent security challenges, such as shielding data from breaches and thwarting unauthorized users who pose risks to cloud security. This model not only enhances the effectiveness and benefits of security in cloud computing but also elevate the encryption of cloud data [35].

Kumar et al. categorize DDoS research papers into various domains including management of networks, methods of deep learning, machine learning, management of resource, load distribution, and fuzzy approach. "Accuracy, precision, recall, and F1 score" are compared among 41 distinct suggested approaches, with accompanying comparative graphs. SaDE-ELM emerges as the top performer across all datasets, while SVM demonstrates the poorest performance [36].

## 13.3 ARTIFICIAL NEURAL NETWORKS

At the advanced level, a neural network can be described as a computational framework that processes input information and generates output by utilizing interconnected nodes organized into layers. This model determines which input signal to send to the next layer by filtering data passing through the network. A neural network determines which neurons will be activated, and if no specific neurons are present, the network develops as a purely linear model. An ANN model based on feed-forward propagation is presented as a new approach to enhance cloud data security. Three stages of this approach training, testing, and validation are involved. To train an artificial neural network (ANN), labeled data is utilized, whose characteristics of the input data are identified, and associated results are made available for learning. Back-propagation is used to adjust the weights and biases after the input data is transmitted to the network and the results are compared to the predicted results. Several tasks can be solved using the ANN method, such as the detection of cloud data security threats. Its effectiveness is dependent on several criteria, such as the type and volume of data for training, the network architecture selected, and the optimization method applied to modify the weights.

### 13.3.1 Dataset

The dataset used for this experiment is created by us. The dataset comprises passwords of different lengths. Each password includes a combination of elements such as uppercase letters, lowercase letters, special characters, and numerical values. The "input data and the output labels" constitute the two main components of an ANN dataset. "The dataset is separated into three subsets: training, testing, and validation". The training subset of the data is used to train an ANN model. A large amount of the dataset is used in the training procedure for optimizing the network's weights and biases. MATLAB has been used to create the training graphs for this training, as seen in Figures 13.1 to 13.3. The validation subset is used to modify the hyper-parameters of the ANN model during the training phase. An ANN model's final performance is evaluated using a test subset. In addition, the model's generalization performance is evaluated unbiased on unseen data.

### 13.3.2 Application layer in ANN

ANN models contain multiple layers of neurons. Each neuron represents an output value in the application layer. The role of this layer is to convert the knowledge gained by the preceding layers into a format appropriate to solve the current issue. It gives the final outputs of this model.

### 13.3.3 Validation layer in ANN

In the training of an "artificial neural network" (ANN), a segment of the labeled dataset is set aside for validation purposes. This validation set, distinct from the training data, serves the sole purpose of assessing the model's performance during training, rather than contributing to the actual training process. Its role is to monitor the model's ability to generalize, fine-tune hyper-parameters, and optimize the model. Regression analysis is employed to evaluate the model's performance throughout training. Figures 13.4 to 13.6 present the outcomes of this neural network validation process. These figures include regression plots that depict the network's predictions versus the actual targets across both training and testing datasets.

## 13.4 PROPOSED SIMULATION

A multilayer feed-forward network is the basic architecture of the system, based on the concept of back-propagation. There are fully interconnected neurons in each layer [37]. Each unit's input patterns are calculated by adding the previous layer's output patterns and multiplying them by the weight matrix [38]. For this network to be trained, the inputs and expected outputs must be known in advance. In the following step, we obtain

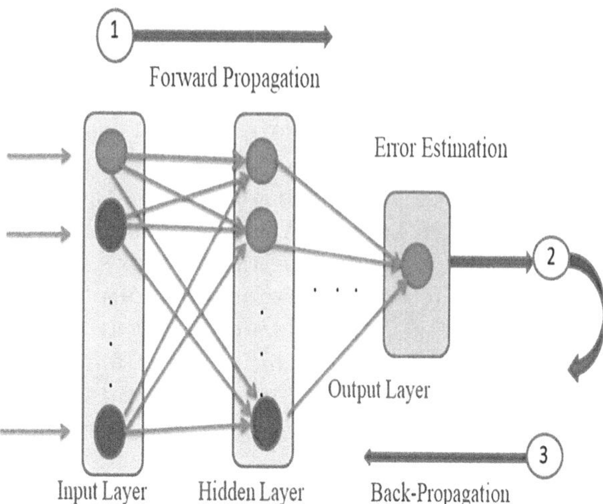


*Figure 13.4* Feed-forward back-propagation algorithm of artificial neural networks.

network parameters such as output, error values, and weight matrix after the training has been completed [39]. A back-propagation model is used to reduce the error values during the training process, and this is the fundamental idea underlying our study. Figure 13.4 is the diagram of the FFBP algorithm of ANN.

## 13.5 EXPERIMENTAL SETUP

The dataset that is used for this study is created by us. This dataset contains the unique identification numbers of various devices. In general, two primary components are input and output data. An effective training procedure typically involves partitioning the dataset. Usually, three partitions exist, one is training, the second is validation, and last is testing. One of the most significant portions of the dataset is utilized as a training subset. The MATLAB program is used to generate the graphs that illustrate the training, validation, and testing processes. Validation subsets are designed to assist with measuring the performance of ANN models on unknown data and facilitate conclusions regarding adjustments during the training phase. An ANN model's final performance is evaluated using test subsets.

In this experimental setup, the unique identification number of two smartphones has been trained using the MATLAB tool. The first step in training is to consider the two unique identification numbers of smartphones. After memorizing the pattern of unique identification numbers of two smartphones and tablets together, we have taken the next step to memorize

*Table 13.1* Binary converted input data for training of unique identification number of two smartphones

| | | | | | | | | | | |
|---|---|---|---|---|---|---|---|---|---|---|
| 0 | 1 | 0 | 1 | 0 | 0 | 1 | 0 | 0 | 1 | 0 |
| 0 | 1 | 0 | 1 | 0 | 0 | 1 | 0 | 0 | 1 | 0 |
| 1 | 1 | 0 | 1 | 0 | 0 | 0 | 1 | 1 | 1 | 0 |
| 1 | 1 | 0 | 1 | 0 | 0 | 0 | 1 | 1 | 1 | 0 |
| 0 | 0 | 0 | 1 | 0 | 0 | 1 | 1 | 0 | 1 | 0 |
| 0 | 0 | 0 | 1 | 0 | 1 | 0 | 0 | 1 | 0 | 0 |
| 1 | 0 | 0 | 0 | 0 | 0 | 1 | 0 | 0 | 1 | 1 |
| 1 | 0 | 0 | 0 | 0 | 1 | 0 | 0 | 0 | 1 | 1 |
| 0 | 0 | 0 | 0 | 0 | 0 | 1 | 1 | 0 | 1 | 0 |
| 0 | 0 | 0 | 1 | 0 | 0 | 1 | 1 | 0 | 1 | 1 |
| 1 | 0 | 1 | 0 | 1 | 1 | 0 | 1 | 0 | 0 | 0 |
| 1 | 0 | 1 | 0 | 0 | 0 | 0 | 0 | 1 | 0 | 1 |
| 1 | 1 | 0 | 0 | 1 | 0 | 0 | 1 | 0 | 1 | 0 |
| 0 | 0 | 1 | 1 | 0 | 0 | 0 | 1 | 0 | 1 | 1 |
| 0 | 0 | 0 | 0 | 1 | 0 | 0 | 0 | 1 | 0 | 0 |
| 0 | 1 | 0 | 0 | 1 | 0 | 0 | 0 | 1 | 1 | 0 |

*Table 13.2* Binary converted input data for training of unique identification numbers of four devices (two smartphones and tablets)

| | | | | | | | | | | | |
|---|---|---|---|---|---|---|---|---|---|---|---|
| 0 | 0 | 0 | 0 | 0 | 0 | 0 | 0 | 0 | 1 | 0 | 1 |
| 0 | 0 | 0 | 0 | 0 | 0 | 0 | 0 | 0 | 1 | 0 | 1 |
| 0 | 0 | 0 | 0 | 0 | 0 | 0 | 0 | 0 | 0 | 0 | 0 |
| 0 | 1 | 0 | 1 | 0 | 0 | 1 | 0 | 0 | 0 | 1 | 1 |
| ... | ... | ... | ... | ... | .... | .... | ... | ... | ... | .... | ..... |
| 0 | 1 | 0 | 0 | 0 | 0 | 1 | 1 | 0 | 1 | 1 | 1 |
| 0 | 0 | 0 | 1 | 0 | 0 | 0 | 0 | 1 | 0 | 1 | 0 |
| 0 | 0 | 1 | 0 | 0 | 0 | 0 | 0 | 1 | 0 | 1 | 0 |
| 0 | 0 | 0 | 1 | 0 | 0 | 0 | 0 | 1 | 0 | 1 | 0 |

the unique identification numbers of two smartphones, tablets, and desktop computers at the same time. For these trainings, we have used 42 neurons, one hidden layer, TRAINBR training functions; TANSIG transfers functions, PURELIN transfer functions, and some basic parameters to achieve a decent result. Tables 13.1, 13.2, and 13.3 represent the binary-converted input data for these trainings.

## 13.6 RESULTS

The simulation has been conducted using the MATLAB software. There are two sections to the dataset. Overall, 30% of the data is used for testing and validation, while 70% of the data is used for training. These simulation findings are shown in Figures 13.5 to 13.7.

*Table 13.3* Binary converted input data for training of six devices' unique identification numbers (two smartphones, tablets, and laptops)

| | | | | | | | | | | | |
|---|---|---|---|---|---|---|---|---|---|---|---|
| 0 | 0 | 0 | 0 | 0 | 0 | 0 | 0 | 0 | 1 | 0 | 1 |
| 0 | 0 | 0 | 0 | 0 | 0 | 0 | 0 | 0 | 1 | 0 | 1 |
| 0 | 0 | 0 | 0 | 0 | 0 | 0 | 0 | 0 | 0 | 0 | 0 |
| 0 | 1 | 0 | 1 | 0 | 0 | 1 | 0 | 0 | 0 | 1 | 1 |
| ... | ... | ... | ... | ... | .... | .... | ... | ... | ... | .... | ..... |
| 0 | 1 | 0 | 0 | 0 | 0 | 1 | 1 | 0 | 1 | 1 | 1 |
| 0 | 0 | 0 | 1 | 0 | 0 | 0 | 0 | 1 | 0 | 1 | 0 |
| 0 | 0 | 1 | 0 | 0 | 0 | 0 | 0 | 1 | 0 | 1 | 0 |
| 0 | 0 | 0 | 1 | 0 | 0 | 0 | 0 | 1 | 0 | 1 | 0 |

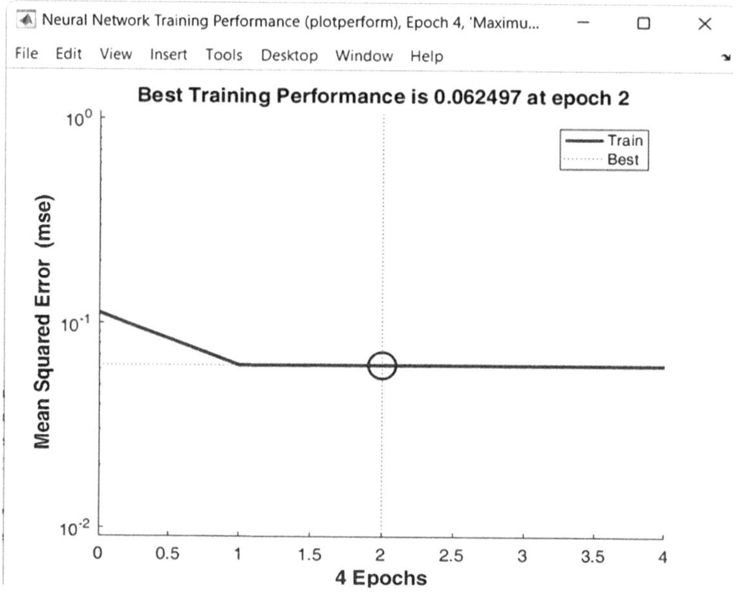

*Figure 13.5* Performance curve of training of unique identification numbers of two smartphones.

The performance curves of these trainings are displayed in Figures 13.8 to 13.10. In Figure 13.8, the training curve represents a training line that converges to approximate zero, and the training performance of this curve is 0.062497 at 2 epochs out of 4. Figure 13.9 shows the best performance of the training of four devices' unique identification numbers. The performance of this training is 0.12802 at epoch 3 out of 11 epochs. In Figure 13.10, we have shown the training and testing line, which converges

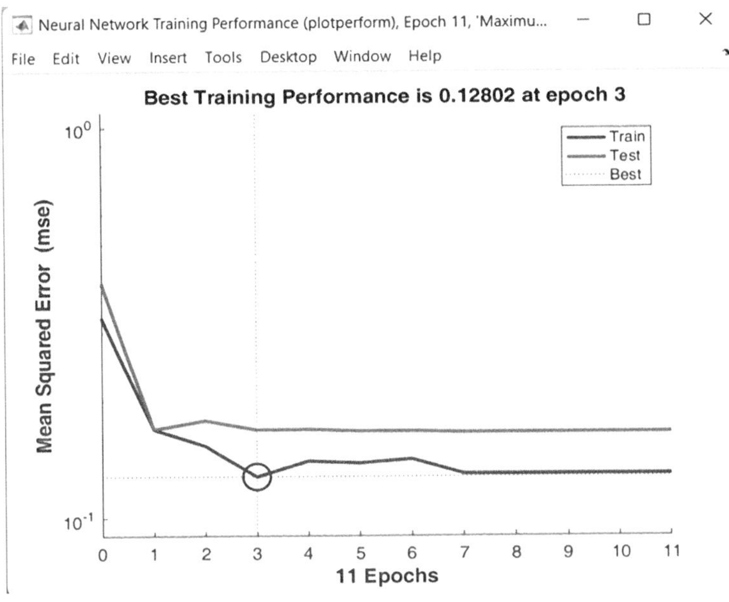


*Figure 13.6* Performance curve of training of unique identification numbers of two smartphones and tablets.

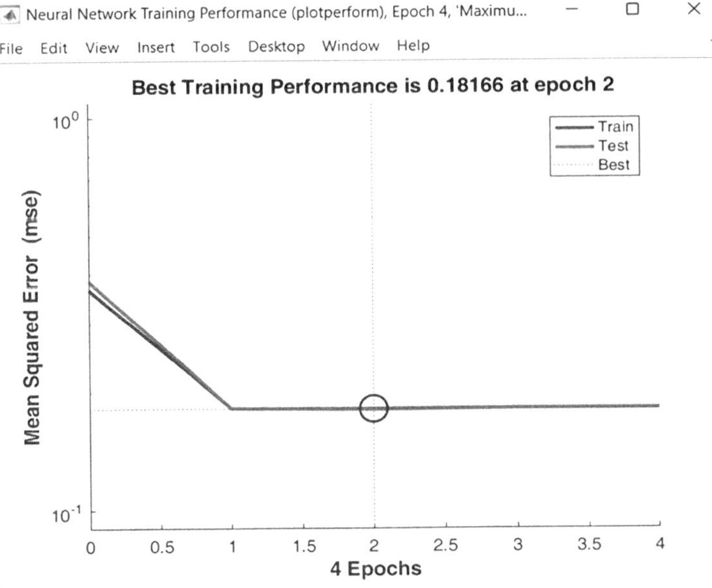


*Figure 13.7* Performance curve of training of six devices' unique identification numbers (two smartphones, tablets, and laptops).

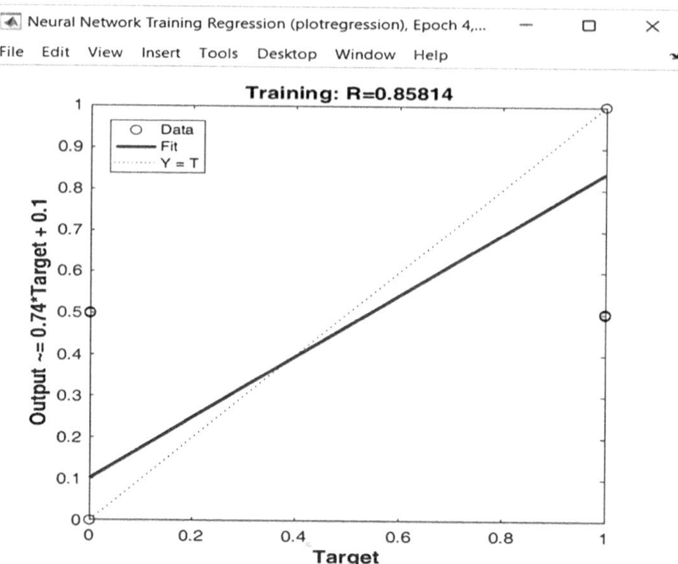


*Figure 13.8* Regression plot of training of unique identification numbers of two smartphones.

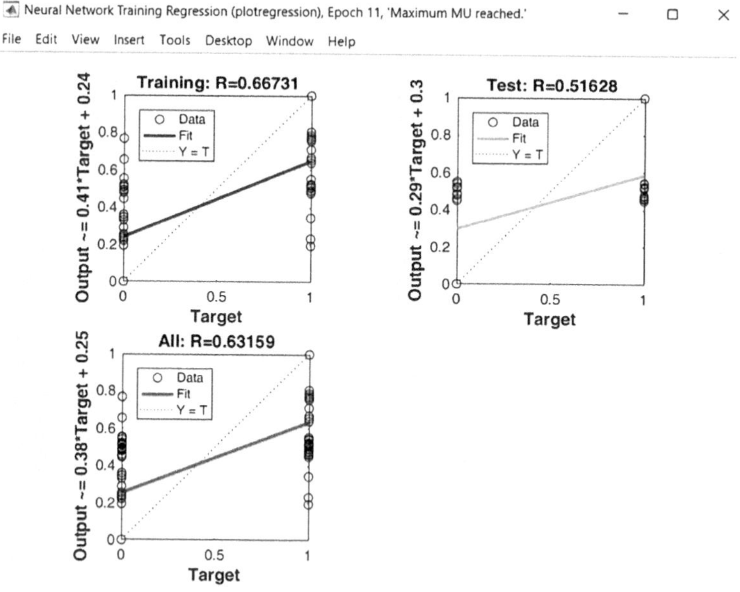


*Figure 13.9* Regression plot of training of four devices' unique identification numbers (two smartphones, tablets).

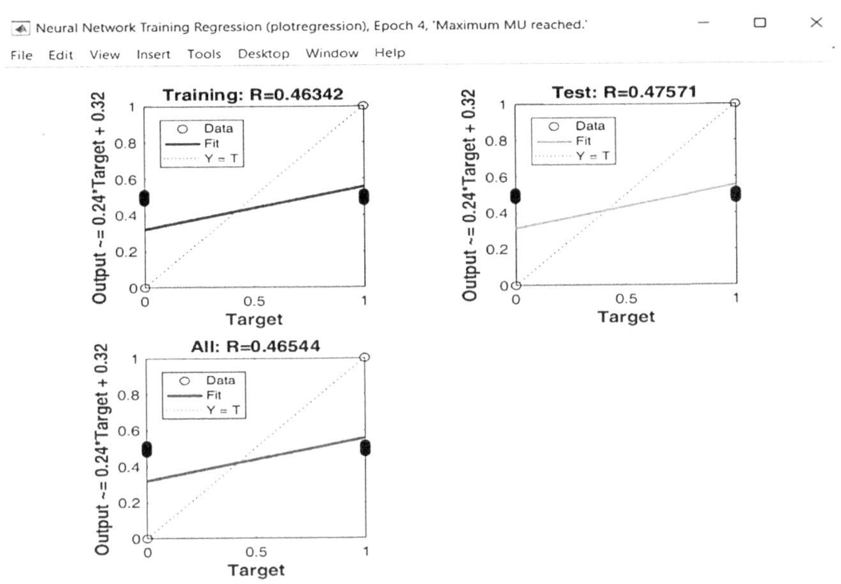


*Figure 13.10* Regression plot of training of six devices' unique identification number (two smartphones, tablets, and laptops).

*Table 13.4* Training results

| *Number of devices* | *Number of epochs* | *Training performance* |
|---|---|---|
| 1 | 2 | 0.06249 |
| 2 | 3 | 0.12802 |
| 3 | 2 | 0.18166 |

to the value of 0.1 throughout 2 out of 4 epochs, and the performance value is 0.18166. Figure 6.4 represents the regression plot of training of unique identification number of two smartphones. During this training, the value of $R$ is 0.85814. A relationship between the output value and the target value is described by the term $R$. Here, 0.0625 is the mean squared error. Figure 13.9 represents the regression plot of training of unique identification numbers (two smartphones and two tablets). Figure 13.10 represents three graphs: training, testing, and validation. Throughout all three graphs, the value of $R$ is about 0.5, which indicates that the machine is not quite accurate in learning the input data from the input data. From all the training curves, regression graphs, and Table 13.4, we have analyzed that a machine can quickly learn the unique identification number of different devices.

## 13.7 CONCLUSIONS

Machine learning enhances the accuracy and speed of identifying malicious activity, reducing the need for manual analysis [40]. Anomalies and vulnerabilities are also detected by this tool, allowing administrators to take immediate action to mitigate risks. A further application of machine learning is the proactive detection and prevention of threats before they cause damage. A few findings have been reached after considering all the factors presented in this paper's introduction part. To keep cloud networks safe, security mechanisms must be implemented for things like multi-factor authentication and intricate algorithmic encryption models. This chapter presents an experimental setup that contributes to the protection of cloud data against DDoS attacks in 6G networks. In our experiment, we have used the FFBP ("feed-forward back-propagation") algorithm of ANN (artificial neural networks) that will help to secure the unique identification numbers of different devices. Through this method, the unique identification number of the device is converted into a weight matrix. It is almost impossible for attackers to determine the dimensions of a weight matrix since it is a multidimensional matrix. In the future, various attacks on 6G networks, storage issues, power supply issues, and other aspects of the cloud environment that require further analysis may be investigated. More research such as fuzzy logic rules can be done in the future to solve the problem of securing 6G networks.

## REFERENCES

[1] K. Chen, "Research on the Future Market Applications of 6G Network Technology," pp. 3576–3584, 2022, doi: 10.2991/978-2-494069-31-2_419.

[2] S. R. Gundu, P. Charanarur, K. K. Chandelkar, D. Samanta, R. C. Poonia, and P. Chakraborty, "Sixth-Generation (6G) Mobile Cloud Security and Privacy Risks for AI System Using High-Performance Computing Implementation," *Wirel. Commun. Mob. Comput.*, vol. 2022, 2022, pp. 1–14. doi: 10.1155/2022/4397610.

[3] R. T. Kamurthi, S. R. Chopra, and R. Sharma, "A Burgeoning 6G Technology and Its Cloud Services," *2022 International Conference on Emerging Smart Computing and Informatics, ESCI 2022*, January 2023, 2022, doi: 10.1109/ESCI53509.2022.9758236.

[4] R. Wang, "Research on Data Security Technology Based on Cloud Storage," *Procedia Eng.*, vol. 174, pp. 1340–1355, 2017, doi: 10.1016/j.proeng.2017.01.286.

[5] Q. Li *et al.*, "6G Cloud-Native System: Vision, Challenges, Architecture Framework and Enabling Technologies," *IEEE Access*, vol. 10, no. August, pp. 96602–96625, 2022, doi 10.1109/ACCESS.2022.3205341.

[6] Q. Liu, S. Sarfraz, and S. Wang, "An Overview of Key Technologies and Challenges of 6G," *Lect. Notes Comput. Sci. (including Subser. Lect. Notes*

*Artif. Intell. Lect. Notes Bioinform.)*, vol. 12487, LNCS, no. November, pp. 315–326, 2020, doi: 10.1007/978-3-030-62460-6_28.
[7] S. Pandi V., A. J. Albert, K. N. K. Thapa, and R. Krishnaprasanna, "A Novel Enhanced Security Architecture for Sixth Generation (6G) Cellular Networks Using Authentication and Acknowledgement (AA) Approach," *Results Eng.*, vol. 21, no. October 2023, p. 101669, 2024, doi: 10.1016/j.rineng.2023.101669.
[8] R. Kumar, A. Anand, N. Singh, D. Singh and P. J. Singh, "An Efficient Machine Learning Approach for Detecting Malwares in IoT Environment," *2023 3rd International Conference on Technological Advancements in Computational Sciences (ICTACS)*, Tashkent, Uzbekistan, pp. 475–481, 2023, doi: 10.1109/ICTACS59847.2023.10390194.
[9] F. J. Abdullayeva, "Distributed Denial of Service Attack Detection in E-Government Cloud via Data Clustering," *Array*, vol. 15, no. December 2021, p. 100229, 2022, doi: 10.1016/j.array.2022.100229.
[10] A. Abusitta, M. Bellaiche, and M. Dagenais, "An SVM-Based Framework for Detecting DoS Attacks in Virtualized Clouds Under Changing Environment," *J. Cloud Comput.*, vol. 7, no. 1, pp. 1–18, 2018, doi: 10.1186/s13677-018-0109-4.
[11] P. Elamparithi, S. Kalaivani, S. Vijayalakshmi, E. Keerthika, S. Koteswari, and R. S. Raaj, "A Machine Learning Approach for Detecting DDOS Attack in IoT Network Using Random Forest Classifier," *Int. J. Intell. Syst. Appl. Eng.*, vol. 12, no. 2s, pp. 495–502, 2024.
[12] A. Bhardwaj, V. Mangat, R. Vig, S. Halder, and M. Conti, "Distributed Denial of Service Attacks in Cloud: State-of-the-Art of Scientific and Commercial Solutions," *Comput. Sci. Rev.*, vol. 39, no. February, pp. 1–32, 2021, doi: 10.1016/j.cosrev.2020.100332.
[13] Y. Shang, "Prevention and Detection of DDOS Attack in Virtual Cloud Computing Environment Using Naive Bayes Algorithm of Machine Learning," *Meas. Sensors*, vol. 31, no. July 2023, p. 100991, 2024, doi: 10.1016/j.measen.2023.100991.
[14] M. Masdari and M. Jalali, "A Survey and Taxonomy of DoS Attacks in Cloud Computing," *Secur. Commun. Netw.*, vol. 9, no. 16, pp. 3724–3751, 2016, doi: 10.1002/sec.1539.
[15] M. Alduailij, Q. W. Khan, M. Tahir, M. Sardaraz, M. Alduailij, and F. Malik, "Machine-Learning-Based DDoS Attack Detection Using Mutual Information and Random Forest Feature Importance Method," *Symmetry (Basel)*, vol. 14, no. 6, pp. 1–15, 2022, doi: 10.3390/sym14061095.
[16] S. Sambangi and L. Gondi, "A Machine Learning Approach for DDoS (Distributed Denial of Service) Attack Detection Using Multiple Linear Regression," *Proceedings*, vol. 63, p. 51, 2020, doi: 10.3390/proceedings2020063051.
[17] M. Zekri, S. El Kafhali, N. Aboutabit, and Y. Saadi, "DDoS attack Detection Using Machine Learning Techniques in Cloud Computing Environments," *Proceedings of the 2017 3rd International Conference of Cloud Computing Technologies and Applications (CloudTech)*, vol. 2018-January, no. October, pp. 1–7, 2017, doi: 10.1109/CloudTech.2017.8284731.

[18] U. A. Butt, R. Amin, H. Aldabbas, S. Mohan, B. Alouffi, and A. Ahmadian, "Cloud-Based Email Phishing Attack Using Machine and Deep Learning Algorithm," *Complex Intell. Syst.*, vol. 9, no. 3, pp. 3043–3070, 2023, doi: 10.1007/s40747-022-00760-3.

[19] S. H. Dhahi, E. H. Dhahi, B. J. Khadhim, and S. T. Ahmed, "Using Support Vector Machine Regression to Reduce Cloud Security Risks in Developing Countries," *Indones. J. Electr. Eng. Comput. Sci.*, vol. 30, no. 2, pp. 1159–1166, 2023, doi: 10.11591/ijeecs.v30.i2.pp1159-1166.

[20] N. Elmrabit, F. Zhou, F. Li, and H. Zhou, "Evaluation of Machine Learning Algorithms for Anomaly Detection," *2020 International Conference on Cyber Security and Protection of Digital Services (Cyber Security)*, pp. 1–8, 2020, doi: 10.1109/CyberSecurity49315.2020.9138871.

[21] A. Jeo San and X. John, "Cloud Security Using Supervised Machine Learning," *Int. J. Adv. Sci. Innov.*, vol. 02, no. 04, pp. 11–14, 2021.

[22] E. K. Subramanian and L. Tamilselvan, "A Focus on Future Cloud: Machine Learning-Based Cloud Security," *Serv. Orient. Comput. Appl.*, vol. 13, no. 3, pp. 237–249, 2019, doi: 10.1007/s11761-019-00270-0.

[23] M. Amitha and M. Srivenkatesh, "DDoS Attack Detection in Cloud Computing Using Deep Learning Algorithms," *Orig. Res. Pap. Int. J. Intell. Syst. Appl. Eng. IJISAE*, vol. 2023, no. 4, pp. 82–90, 2023.

[24] R. V. Deshmukh and K. K. Devadkar, "Understanding DDoS Attack & Its Effect in Cloud Environment," *Procedia Comput. Sci.*, vol. 49, no. 1, pp. 202–210, 2015, doi: 10.1016/j.procs.2015.04.245.

[25] H. F. El-Sofany, S. A. El-Seoud, and I. A. T. F. Taj-Eddin, "A Case Study of the Impact of Denial of Service Attacks in Cloud Applications," *J. Commun.*, vol. 14, no. 2, pp. 153–158, 2019, doi: 10.12720/jcm.14.2.153-158.

[26] K. A. Memon, "Analyzing Distributed Denial of Service Attacks in Cloud Computing Towards the Pakistan Information Technology Industry," *Indian J. Sci. Technol.*, vol. 13, no. 29, pp. 2062–2072, 2020, doi: 10.17485/ijst/v13i29.1040.

[27] O. Pandithurai, C. Venkataiah, S. Tiwari, and N. Ramanjaneyulu, "DDoS Attack Prediction Using a Honey Badger Optimization Algorithm Based Feature Selection and Bi-LSTM in Cloud Environment," *Expert Syst. Appl.*, vol. 241, no. August 2023, p. 122544, 2024, doi: 10.1016/j.eswa.2023.122544.

[28] V. W. Thirumaran, N. Joseph, and U. Srikanth, "Intrusion Detection System in Cloud Computing by Utilizing VTR-HLSTM Based on Deep Learning," vol. 33, no. 3, pp. 1829–1842, 2024, doi: 10.11591/ijeecs.v33.i3.pp1829-1842.

[29] G. S. R. Emil Selvan, R. Ganeshan, I. D. J. Jingle, and J. P. Ananth, "FACVO-DNFN: Deep Learning-Based Feature Fusion and Distributed Denial of Service Attack Detection in Cloud Computing," *Knowledge-Based Syst.*, vol. 261, p. 110132, 2023, doi: 10.1016/j.knosys.2022.110132.

[30] A. Thangasamy, B. Sundan, and L. Govindaraj, "A Novel Framework for DDoS Attacks Detection Using Hybrid LSTM Techniques," *Comput. Syst. Sci. Eng.*, vol. 45, no. 3, pp. 2553–2567, 2023, doi: 10.32604/csse.2023.032078.

[31] S. Farhat, M. Abdelkader, A. Meddeb-Makhlouf, and F. Zarai, "Evaluation of DoS/DDoS Attack Detection with ML Techniques on CIC-IDS2017 Dataset," *Proceedings of the 9th International Conference on Information Systems Security and Privacy*, ICISSP, pp. 287–295, 2023, doi: 10.5220/0011605700003405.

[32] R. S. Aashmi and T. Jaya, "Detecting and Preventing of Attacks in Cloud Computing Using Hybrid Algorithm," *Intell. Autom. Soft Comput.*, vol. 35, no. 1, pp. 79–95, 2023, doi: 10.32604/iasc.2023.024291.

[33] G. Bano and B. Mastoi, "Framework for Monitoring and Detection of DDOS Attacks Using ML Algorithms," Acta Scientific COMPUTER SCIENCES, vol. 5, no. 1, December 2022, 2022 [Online]. Available at: www.researchgate.net/publication/372235478.

[34] S. Azizpour and M. R. Majma, "NADA: New architecture for detecting DoS and DDoS attacks in fog computing," *J. Comput. Virol. Hacking Tech.*, vol. 19, no. 1, pp. 51–64, 2023, doi: 10.1007/s11416-022-00431-4.

[35] D. A. Kelechi, "Machine Learning Algorithms for Cloud Security," *Adv. Image Video Process.*, vol. 11, no. 4, 2023, doi: 10.14738/aivp.114.15210.

[36] A. Kumar, S. Dutta, and P. Pranav, "A Comparative Study of DDoS Attack in Cloud Computing Environment," *SSRG Int. J. Electron. Commun. Eng.*, vol. 10, no. 7, pp. 87–96, 2023, doi: 10.14445/23488549/IJECE-V10I7P109.

[37] C. Sekhar and P. S. Meghana, "A Study on Backpropagation in Artificial Neural Networks," *Asia-Pacific J. Neural Netw. Appl.*, vol. 4, no. 1, pp. 21–28, 2020, doi: 10.21742/ajnnia.2020.4.1.03.

[38] M. U. Sana, Z. Li, F. Javaid, H. Bin Liaqat, and M. U. Ali, "Enhanced Security in Cloud Computing Using Neural Network and Encryption," *IEEE Access*, vol. 9, pp. 145785–145799, 2021, doi: 10.1109/ACCESS.2021.3122938.

[39] B. A. Manjunatha, A. Shastry, G. Nishchala, N. Pavithra, and S. Raheela Banu, "Network Intrusion Detection Using Neural Network Techniques," *Lect. Notes Electr. Eng.*, vol. 928, no. April, pp. 137–147, 2023, doi: 10.1007/978-981-19-5482-5_13.

[40] R. Rana and R. Kumar, "Performance Analysis of AODV in Presence of Malicious Node", *Acta Electronica Malaysia (AEM)*, vol. 3, no. 1, pp. 1–5, 2019

# Index

For Product Safety Concerns and Information please contact our EU representative GPSR@taylorandfrancis.com Taylor & Francis Verlag GmbH, Kaufingerstraße 24, 80331 München, Germany

Batch number: 10397790

Printed by Printforce, the Netherlands